T0360822

Frequency Analysis of Vibration Energy Harvesting Systems

Frequency Analysis of Vibration Energy Harvesting Systems

Xu Wang

School of Engineering
College of Science Engineering and Health
RMIT University

AMSTERDAM • BOSTON • HEIDELBERG • LONDON
NEW YORK • OXFORD • PARIS • SAN DIEGO
SAN FRANCISCO • SINGAPORE • SYDNEY • TOKYO

Academic Press is an imprint of Elsevier

Academic Press is an imprint of Elsevier
125 London Wall, London EC2Y 5AS, United Kingdom
525 B Street, Suite 1800, San Diego, CA 92101-4495, United States
50 Hampshire Street, 5th Floor, Cambridge, MA 02139, United States
The Boulevard, Langford Lane, Kidlington, Oxford OX5 1GB, United Kingdom

Library of Congress Cataloging-in-Publication Data
A catalog record for this book is available from the Library of Congress

British Library Cataloguing-in-Publication Data
A catalogue record for this book is available from the British Library

ISBN: 978-0-12-802321-1

For information on all Academic Press publications
visit our website at https://www.elsevier.com/

Working together
to grow libraries in
developing countries

www.elsevier.com • www.bookaid.org

Publisher: Joe Hayton
Acquisitions Editor: Brian Guerin
Editorial Project Manager: Edward Payne
Production Project Manager: Lisa Jones
Designer: Maria Ines Cruz

Typeset by TNQ Books and Journals

Contents

List of Figures

List of Tables

About the Author

Dr. Xu Wang is an associate professor in School of Engineering, RMIT University. He has worked in international automotive industries such as General Motors Australia Ltd and Continental AG, Germany, for more than 11 years before his academic career in 2005. He has published 2 books; 60 journal papers; 30 peer-reviewed conference papers; and 5 patents on vibration or wave energy harvesting, noise and vibration refinement, engine exhaust waste heat conversion, and vehicle design since then. He is a fellow of Engineers Australia and the Institution of Engineers Australia and Society of Automotive Engineers, Australasia. He is a member of American Society of Mechanical Engineers and International Institute of Acoustics and Vibration. Dr. Xu Wang received his PhD degree in Engineering in Monash University, Australia, in 1995.

Preface

With the global energy crisis and environment concerns, many technologies in energy harvesting, such as solar, wind, geothermal, and hydraulic power plants or farms, have been developed. Vibration or wave exists everywhere, such as the vibration of floor and wall, machines, pumps, vehicle chassis, railway train or tracks, and human motions, and ocean waves, the vibration or wave becomes a good alternative energy source and receives more and more attention in recent years. The global energy potential represented by waves hitting all coasts may be estimated to be of the same order of magnitude as the world's present electricity production, approximately 1 TW. Conversion of vibration or wave energy of structures, machines, vehicles, or ocean waves into electric energy can improve reliability, comfort, energy utilization efficiency, and energy cost, and can potentially reduce the amplitude of vibration and associated damaging effects. The goal of this technology is to provide power sources, recharge storage devices or transmit the converted power to the grid. The concept has ecological ramifications which reduce the green house emissions and usage of batteries for a clean environment.

This book focuses on developing a dimensionless frequency analysis method and aims to provide research scientists and engineers, postgraduate students, and product design and development managers with a new and effective performance prediction and design method for vibration or wave energy harvesters regardless of their sizes and types.

This book starts with an introduction of the fundamental Laplace and Fourier transform and transfer function method in Chapter 1, where a single degree of freedom spring-mass-dashpot oscillator system is analyzed and modeled as an example. The Matlab Simulink transfer function, time domain integration, state space, and Matlab frequency response methods are applied in the analysis and modeling. A piezoelectric material insert is then added into the single degree of freedom spring-mass-dashpot oscillator system in Chapter 2, where both the mechanical system and energy harvesting circuit system are modeled. A hybrid of frequency analysis and time domain integration methods is then developed. Dimensionless harvested power and energy harvesting efficiency are developed from two dimensionless variables: dimensionless resistance and dimensionless force factors. The dimensionless frequency analysis method is then applied to investigate single degree of freedom piezoelectric and electromagnetic vibration energy harvesters connected to a single load resistor and four conventional energy harvesting interface circuits for their energy harvesting performances in Chapters 3 and 4. The similarity and duality of the electromagnetic and piezoelectric energy harvesters are disclosed in Chapter 5 from a comparison of the dimensionless harvested resonant power and energy harvesting efficiencies predicted from the dimensionless resistance and dimensionless force factor. Two, three, and multiple degrees of freedom piezoelectric energy harvesters are investigated for their energy harvesting performances in Chapters 6 and 7 using the dimensionless frequency analysis method.

The vibration energy harvesting performances such as harvesting frequency band-width, harvested power and its density, energy harvesting efficiency of the vibration energy harvesters of different degrees of freedom have been compared. The application of the two degrees of freedom piezoelectric energy harvesters in a vehicle suspension system is illustrated in Chapter 6. The dimensionless frequency analysis method is then validated by experimental testing of one and two degrees of freedom piezoelectric vibration energy harvesters in Chapter 8. In the above chapters, all the analyses are limited to the harmonic excitations. To develop the dimensionless frequency analysis for random ambient environment excitations, coupling loss factor of the electrical and mechanical subsystems is defined and related to the dimensionless force factor or the critical coupling strength in Chapter 9. Frequency ranges of the weak coupling of the electrical and mechanical subsystems are defined for piezoelectric and electromagnetic vibration energy harvesters. The frequency ranges enable statistical energy analysis of piezoelectric or electromagnetic vibration energy harvesting systems. The statistical energy analysis allows for a reliable performance prediction of many piezoelectric or electromagnetic vibration energy harvesters of the same design under random ambient environment excitations. A novel dimensionless frequency response analysis of power variables is proposed in Chapter 10 to enable a direct measurement of mean input and harvested power and energy harvesting efficiency. The novel dimensionless frequency response analysis of power variables is applicable to the vibration energy harvesters under harmonic, narrow band and broad band random excitations. Dimensionless nonlinear frequency response analysis of electromagnetic vibration energy harvesters is studied using the harmonic balancing and perturbation methods in Chapter 11. The applications of the electromagnetic vibration energy harvesters in ocean wave energy conversion are illustrated. The applications include the buoy-type point absorbers or Archimedes Wave Swing ocean wave energy converters. In the end, the dimensionless frequency analysis method is applied to the two, four, and multiple degrees of freedom electromagnetic vibration energy harvesters for evaluation of their performances. Various applications of the multiple degrees of freedom electromagnetic vibration energy harvesters are illustrated in Chapter 12.

This book is a sincere attempt to establish a mathematical foundation for design and development of future vibration or wave-based energy harvesters.

The research area of energy harvesting has captivated both academics and industrialists due to previously mentioned potential and motivations. This has resulted in an explosion of academic research and new products. This book provides a review of the recent research studies on vibration or wave energy harvesting sciences and principles, including dynamic analysis, control methods, and advanced experimental and modeling techniques. The research area of vibration or wave-based energy harvesting encompasses mechanics, materials science, and electrical circuitry. Researchers from all the three of these disciplines contribute heavily to the energy harvesting literature. A comprehensive reference list is provided to direct further studies and to assist readers to develop in-depth knowledge of the vibration or wave energy harvesting, storing, or transmitting processes. The prerequisite of this book is the knowledge of vibration and structural dynamics, fundamental linear and nonlinear differential equations and their solutions, Laplace and Fourier Transform, basic electrics and control theory, Matlab software at the level of bachelor degree in an engineering curriculum.

Acknowledgments

Foremost, I would like to express my sincere gratitude to my PhD students Mr. Han Xiao, Mr. Ran Zhang, and Mr. Zhenwei Liu for their graphic works and research discussions. Also my gratitude is to Professor Len Koss, Professor Reza Jazar, Professor Simon Watkins, and Professor Sabu John for their review of the manuscript and English corrections.

I would like to thank Mr. Peter Tkatchyk, Mr. Julian Bradler, Mr. Patrick Wilkins, and Dr. Greg Oswald for their assistance in the noise, vibration, and harshness lab and the workshop.

Last but not least, I would like to thank my mother for raising me up, my wife, my son, and my daughter for their spiritual support throughout my writing.

Analysis of a Single Degree of Freedom Spring-Mass-Dashpot System Using Transfer Function, Integration, State Space, and Frequency Response Methods

1

CHAPTER OUTLINE

1.1 INTRODUCTION

Vibration analysis based on Laplace or Fourier transform has been well studied [1–4]. Different from a conventional vibration analysis of a spring-mass-dashpot oscillator system, a transducer is added to the oscillator system and used to convert vibration energy into electrical energy. A circuit is added to the oscillator system and used to collect and store the energy converted from vibration energy. The circuit interacts with the oscillator system. The interaction is equivalent to a "shunt" damping added to the oscillator system.

Vibration affects the reliability of a structure or machine, causes damages or failures of the structure or machine. Vibration degrades the human comfort and induces the fatigue and health related problems. Vibration energy is usually dissipated through damping loss and lost in a form of waste heat energy. The motivation of vibration energy harvesting is to convert the vibration energy into a useful electrical energy through the added transducer mechanism and store the converted energy in an electric circuit instead of the vibration energy that is being dissipated and lost. The transducer mechanism may be piezoelectric, electromagnetic, or other mechanism. The electric circuit consists of resistance, coil inductor, capacitor, switches, and rectifiers. "Harvesting" means converting vibration energy, collecting and storing the converted energy in an electric circuit. "Harvesting" converts the wasted vibration energy into useful electric energy. The electric circuit has influences on the natural frequency of the oscillator system. When the effect of the "shunt damping" of the electric circuit or mechanical damping on the natural frequency of the oscillator system is small, the coupling between the oscillator system and electric circuit is weak, which is called "weak coupling." When the effect of the "shunt damping" of the electric circuit or mechanical damping on the natural frequency of the oscillator system is large, the coupling between the oscillator system and electric circuit is strong, which is called "strong coupling." To generalize vibration energy harvesting analysis and make it applicable to vibration energy harvesters of different sizes or dimensions, all function and analysis variables are converted into dimensionless variables. The process of the conversion is called "normalized."

To better understand the analysis in the following chapters, fundamental analysis of the Laplace transform is presented in this chapter; transfer function, state space, and frequency response methods are introduced. Examples are given to illustrate these methods.

1.2 LAPLACE TRANSFORM AND TRANSFER FUNCTION ANALYSIS METHOD

The Laplace transform is often used for calculation of the response of linear systems. The method is used for any type of excitation of a system that includes harmonic, periodic, and nonperiodic excitation time histories. The Laplace transform not only provides an efficient method to solve linear differential equations of constant coefficients, but also allows for the writing of a single algebraic expression relating the excitation input and the response of a system. The Laplace transform method is used to solve for the response of a dynamic system to discontinuous forcing functions and includes the initial conditions.

According to Refs [1,2], the single-sided Laplace transformation of $x(t)$ is defined by the definite integral of

$$X(s) = L[x(t)] = \int_0^\infty e^{-s \cdot t} \cdot x(t) \cdot dt \tag{1.1}$$

where s is a complex variable. The function $e^{-s \cdot t}$ is the kernel of the transformation. The definite integral has t as the variable of the integration, the transformation of $x(t)$ produces a function $X(s)$.

To solve for the response of a linear system such as that shown in Fig. 1.1, the transforms of the derivatives $\frac{dx}{dt}$ and $\frac{d^2x}{dt^2}$ must be evaluated. The transform of the derivative $\frac{dx}{dt}$ leads to

$$L\left[\frac{dx(t)}{dt}\right] = \int_0^\infty e^{-s \cdot t} \cdot \frac{dx(t)}{dt} \cdot dt = e^{-s \cdot t} \cdot x(t)\big|_0^\infty + s \int_0^\infty e^{-s \cdot t} \cdot x(t) \cdot dt$$

$$= s \cdot X(s) - x(0) \tag{1.2}$$

where $x(0)$ is the initial value of the displacement response function $x(t)$ when $t = 0$. Similarly, the transform of the derivative $\frac{d^2x}{dt^2}$ leads to

$$L\left[\frac{d^2x(t)}{dt^2}\right] = \int_0^\infty e^{-s \cdot t} \cdot \frac{d^2x(t)}{dt^2} \cdot dt = s^2 \cdot X(s) - s \cdot x(0) - \dot{x}(0) \tag{1.3}$$

where $\dot{x}(0)$ is the initial velocity. The Laplace transform of the excitation force function $F(t)$ is given by

$$\overline{F}(s) = L[F(t)] = \int_0^\infty e^{-s \cdot t} \cdot F(t) \cdot dt \tag{1.4}$$

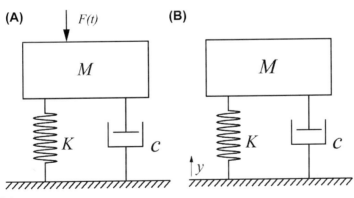

FIGURE 1.1

A single degree of freedom spring-mass-dashpot system driven by (A) direct force and (B) inertial force.

The dynamic equation of motion of the linear single degree of freedom (SDOF) spring-mass-dashpot system as shown in Fig. 1.1(A) can be written as [1].

$$M \cdot \frac{d^2x(t)}{dt^2} + c \cdot \frac{dx(t)}{dt} + K \cdot x(t) = F(t) \tag{1.5}$$

Taking the Laplace transform on the both sides of Eq. (1.5) gives

$$\left(M \cdot s^2 + c \cdot s + K\right) \cdot X(s) = \overline{F}(s) + M \cdot \dot{x}(0) + (M \cdot s + c) \cdot x(0) \tag{1.6}$$

It is assumed that the initial displacement and velocity are equal to zero, Eq. (1.6) becomes

$$\frac{X(s)}{\overline{F}(s)} = \frac{1}{M \cdot s^2 + c \cdot s + K} \tag{1.7}$$

The displacement time response is then calculated by the inverse Laplace transform and is given by:

$$x(t) = L^{-1}[X(s)] = L^{-1}\left[\frac{\overline{F}(s)}{M \cdot s^2 + c \cdot s + K}\right] \tag{1.8}$$

If the excitation is assumed to be a harmonic displacement of the base $y(t)$, the initial displacement and velocity are equal to zero, and assuming $z(t) = x(t) - y(t)$, and the external excitation force $F(t) = 0$ for the SDOF system as shown in Fig. 1.1(B), the equation of the motion of the system becomes

$$M \cdot \frac{d^2z(t)}{dt^2} + c \cdot \frac{dz(t)}{dt} + K \cdot z(t) = -M \cdot \ddot{y}(t) \tag{1.9}$$

The response of $z(t)$ is also harmonic, and is given by

$$\begin{cases} y(t) = Y(s) \cdot e^{s \cdot t} \\ z(t) = Z(s) \cdot e^{s \cdot t} \end{cases} \tag{1.10}$$

where it is assumed $s = i\omega$, ω is the radial frequency in radians per seconds. $Y(s)$ and $Z(s)$ are the Fourier transforms of the excitation displacement and response relative displacement, respectively. Substituting Eq. (1.10) into Eq. (1.9) gives

$$\left(M \cdot s^2 + c \cdot s + K\right) \cdot Z(s) \cdot e^{s \cdot t} = -M \cdot s^2 \cdot Y(s) \cdot e^{s \cdot t}$$

and the transfer function $H_{\ddot{y}z}(s)$

$$H_{\ddot{y}z}(s) = \frac{Z(s)}{Y(s) \cdot s^2} = \frac{-M}{M \cdot s^2 + c \cdot s + K} \tag{1.11}$$

Given the Laplace transform of the excitation acceleration $\ddot{y}(t)(Y(s) \cdot s^2)$, from Eq. (1.11), the Laplace transform of the relative response displacement $z(t)$ can be calculated for both the amplitude and phase using Matlab Simulink transfer function method as shown in Fig. 1.2. A transfer function block is defined according to

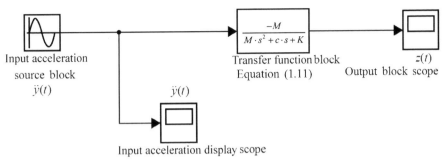

FIGURE 1.2

Prediction of the mass relative displacement response from the base displacement excitation using Matlab Simulink transfer function method.

Eq. (1.11) which has an input source block and an output sink block scope. Given a sine wave input of $\ddot{y}(t)$ (excitation acceleration) with a frequency and amplitude defined in the source block, the output $z(t)$ (response relative displacement) can be predicted using the Matlab Simulink transfer function method and displayed by a sink block scope.

1.3 TIME DOMAIN INTEGRATION METHOD

From Eq. (1.9), the relative acceleration response can be written as

$$\frac{d^2 z(t)}{dt^2} = -\ddot{y}(t) - \frac{c}{M} \cdot \frac{dz(t)}{dt} - \frac{K}{M} \cdot z \tag{1.12}$$

The relative displacement response can be obtained by integrating Eq. (1.12) two times as shown in Fig. 1.3. There is a round sum block in Fig. 1.3. The sum block has three positive inputs and one output. The three terms on the right-hand side of Eq. (1.12) are presented by the three inputs in the sum block in Fig. 1.3. The output of the sum block is $\frac{d^2 z(t)}{dt^2}$. $\frac{d^2 z(t)}{dt^2}$ is integrated once to give the relative velocity $\frac{dz(t)}{dt}$. $\frac{d^2 z(t)}{dt^2}$ is integrated two times to give the relative displacement $z(t)$. $\frac{dz(t)}{dt}$ and $z(t)$ multiplied by $-c/M$ and $-K/M$ contribute to the two positive inputs of the sum block in Fig. 1.3. The other input is the input excitation acceleration multiplied by -1. A source block is used to provide the input excitation signal which is displayed by a scope and a sink block scope is used to display the output relative displacement signal. Given a sine wave input of $\ddot{y}(t)$ (excitation acceleration) with a frequency and amplitude defined in the source block, the output $z(t)$ (relative displacement response) can be predicted using the Matlab Simulink integration method and displayed by the sink block scope.

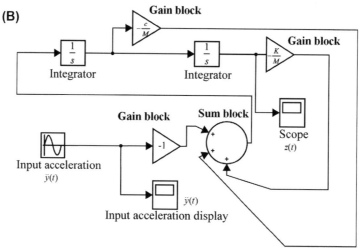

FIGURE 1.3

Prediction of the relative displacement response from the base acceleration excitation input using the Matlab Simulink integration method: (A) Integration schematic of Eq. (1.12); (B) Matlab Simulink schematic following Eq. (1.12).

1.4 STATE SPACE METHOD

There is a third method available to calculate the relative displacement response from the excitation input that is called the state space method [3,4]. The state space method allows for a quick calculation of the output relative displacement response from the input (the excitation displacement). In general, a state-space representation is given by:

$$\dot{\mathbf{x}} = \mathbf{A} \cdot \mathbf{x} + \mathbf{B} \cdot \mathbf{u}$$
$$\mathbf{y} = \mathbf{C} \cdot \mathbf{x} + \mathbf{D} \cdot \mathbf{u} \tag{1.13}$$

where: $\mathbf{x}$ is a column vector of dimension n, called the "state vector"; $\mathbf{u}$ is the "input" to the system; $\mathbf{y}$ is the "output" of the system; $\mathbf{A}$ is the "state matrix"; $\mathbf{B}$ is the "input matrix"; $\mathbf{C}$ is the "output matrix"; $\mathbf{D}$ is the "feed through matrix."

It is assumed that

$$\begin{cases} W_1 = z(t) \\ W_2 = \dfrac{dz(t)}{dt} \end{cases} \tag{1.14}$$

From Eq. (1.9), and using the definition of Eq. (1.14), the following equation is obtained.

$$M\frac{dW_2}{dt} + c \cdot W_2 + K \cdot W_1 = -M \cdot \ddot{y}$$

This equation can be written as the following set matrix equations

$$\left\{ \begin{array}{c} \dfrac{dW_1}{dt} \\ \dfrac{dW_2}{dt} \end{array} \right\} = \left[\begin{array}{cc} 0 & 1 \\ -\dfrac{K}{M} & -\dfrac{c}{M} \end{array} \right] \left\{ \begin{array}{c} W_1 \\ W_2 \end{array} \right\} + \left\{ \begin{array}{c} 0 \\ -1 \end{array} \right\} \{\ddot{y}(t)\} \qquad (1.15)$$

and

$$\left\{ \begin{array}{c} z \\ \dot{z} \end{array} \right\} = \left[\begin{array}{cc} 1 & 0 \\ 0 & 1 \end{array} \right] \left\{ \begin{array}{c} W_1 \\ W_2 \end{array} \right\} + \left\{ \begin{array}{c} 0 \\ 0 \end{array} \right\} \{\ddot{y}(t)\} \qquad (1.16)$$

From Eqs. (1.15) and (1.16), given values of K, M, and c, the following matrix can be determined:

$$\left(\begin{array}{l} \mathbf{A} = \left[\begin{array}{cc} 0 & 1 \\ -\dfrac{K}{M} & -\dfrac{c}{M} \end{array} \right]; \\[4ex] \mathbf{B} = \left[\begin{array}{c} 0 \\ -1 \end{array} \right]; \\[3ex] \mathbf{C} = \left[\begin{array}{cc} 1 & 0 \\ 0 & 1 \end{array} \right]; \\[3ex] \mathbf{D} = \left[\begin{array}{c} 0 \\ 0 \end{array} \right]; \end{array} \right. \qquad (1.17)$$

Given input $y(t)$ (excitation displacement), the output $z(t)$ (response displacement) can be predicted using Matlab Simulink state space method as shown in Fig. 1.4. The matrices of $\mathbf{A}$, $\mathbf{B}$, $\mathbf{C}$, and $\mathbf{D}$ are defined for their values in the Matlab software. The state space block is defined in the Matlab Simulink code which has an input source block and an output sink block scope. Given a sine wave input of $\ddot{y}(t)$ (excitation acceleration) with a frequency and amplitude defined in the source block, the output $z(t)$ (relative displacement response) can be predicted using the Matlab Simulink state space method and is displayed by the sink block scope.

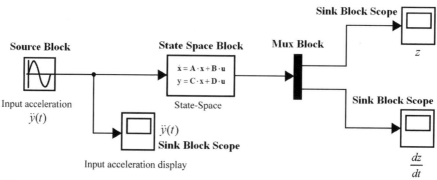

FIGURE 1.4

Schematic of the Matlab Simulink state space method for prediction of the relative displacement response from the base displacement excitation.

1.5 FREQUENCY RESPONSE METHOD

Frequency response function of a linear mechanical system is defined as the Fourier transform of the time domain response divided by the Fourier transform of the time domain input [1]. As mentioned previously, when $s = i\omega$, Eq. (1.11) becomes the frequency response function of the SDOF spring-mass-dashpot system. The frequency response function amplitude versus frequency of the SDOF can be obtained from executing the programming codes as shown in Figs. 1.2–1.4 where a sine wave signal of fixed excitation amplitude is used as input, the excitation frequency of the sine wave signal changes from the low to high to cover the resonant frequency. For each given excitation frequency input of the sine wave signal, the response amplitude is recorded. The response amplitude is then plotted against the excitation frequency which presents the frequency response of the oscillator displacement versus the excitation frequency. The frequency response amplitude frequency curve can also be obtained from plotting the modulus of Eq. (1.11) versus frequency f is in Hz where $s = i2\pi f$.

1.6 AN EXAMPLE OF THE TIME DOMAIN INTEGRATION SIMULATION AND FREQUENCY RESPONSE ANALYSIS METHODS USING MATLAB SIMULINK PROGRAM CODES

To illustrate the time domain integration simulation and frequency response analysis methods, an SDOF spring-mass-dashpot system shown in Fig. 1.1B is studied as an example where the system parameters are listed in Table 1.1.

Table 1.1 The Parameters of the Single Degree of Freedom Spring-Mass-Dashpot System [5]

Parameter	Values	Units
M	8.4×10^{-3}	kg
c	0.154	N·s/m
K	2.5×10^4	N/m
The excitation frequency f	275	Hz
The natural frequency f_n	275	Hz
The excitation displacement amplitude	0.0033	mm

1.6.1 TIME DOMAIN INTEGRATION SIMULATION METHOD USING THE MATLAB SIMULINK PROGRAM CODES

If the above parameters in Table 1.1 are entered into Matlab software, it is assumed that the base excitation acceleration RMS value 9.8 m/s^2 is used with an excitation frequency of 275 Hz. The excitation acceleration is a sine wave signal and is shown in Fig. 1.5. The parameter input template for the excitation sine wave in the Matlab Simulink program code is given in Fig. 1.6. The ODE 23t (mod. stiff/Trapezoidal) algorithm was used in the Matlab Simulink program code. From the Matlab

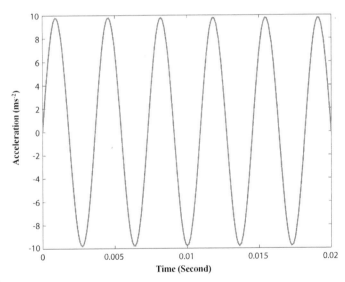

FIGURE 1.5

A sine wave excitation signal input with a frequency of 275 Hz and RMS acceleration of 9.8 m/s^2.

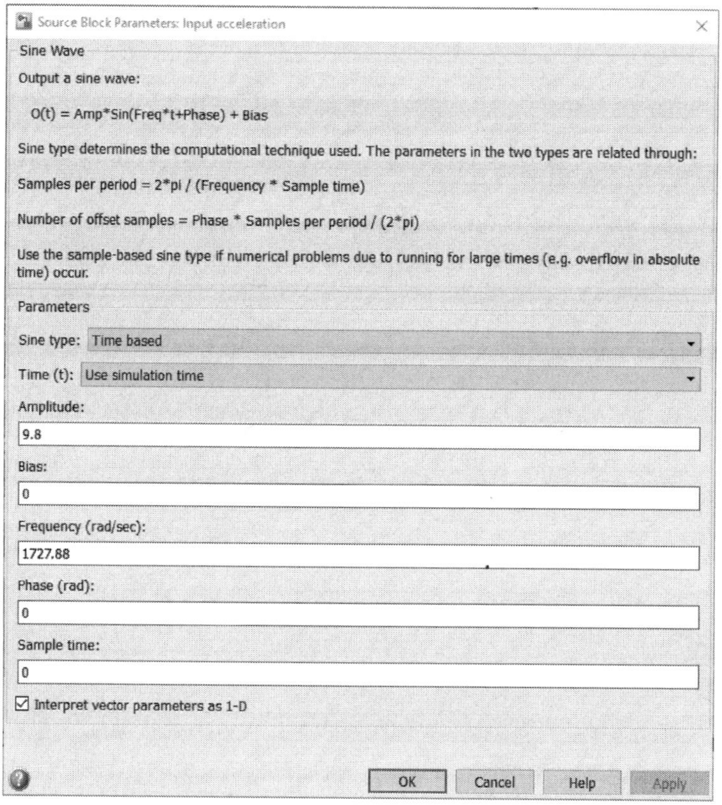

FIGURE 1.6

Parameter inputs of sine wave excitation acceleration for the source block of the Matlab Simulink.

Simulink program code shown in Figs. 1.2−1.4, the response displacement results from the codes are identical and plotted in Fig. 1.7.

1.6.2 THE FREQUENCY RESPONSE ANALYSIS METHOD USING A MATLAB CODE

If the above parameters in Table 1.1 are entered into a Matlab code given in Fig. 1.8, the frequency response function amplitude can be calculated and is plotted in Fig. 1.9. It is seen from Fig. 1.9 that a resonance occurs at around 275 Hz which is same as that listed in Table 1.1. It is noticed that the resonant peak value of the frequency response curve in Fig. 1.9 is about 3.1×10^{-5} s^2. The peak value is multiplied by the input RMS acceleration of 9.8 m/s^2 and divided by 1.414 which gives the RMS value of 2.15×10^{-4} m. This value is same as the peak reading of the sine

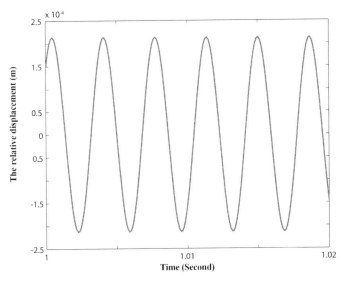

FIGURE 1.7

The time trace of the relative displacement response.

```
*******************************************************************

    CLEAR ALL
    CLC

    K=25000;
    M=0.0084;
    C=0.154;

    F=0:1:500;
    W=2*F*PI;

FOR N=1:LENGTH(W)
    S=1I*W(N);
    X=M*S^2+C*S+K;

    AY=-M;

    Z1(N)=AY/X;

END

    FIGURE (2)
    AXIS AUTO
    PLOT(F,ABS(Z1)')
    XLABEL('FREQUENCY (HZ)')
    YLABEL('DISPLACEMENT RATIO (Z/Y)')

*******************************************************************
```

FIGURE 1.8

A Matlab code for calculation of the relative displacement frequency response function amplitude versus frequency curve.

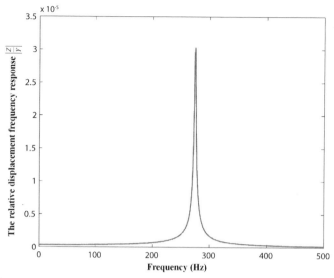

FIGURE 1.9

Relative displacement frequency response function (the relative displacement amplitude over the excitation acceleration amplitude).

wave curve in Fig. 1.7. Therefore the result of the time domain simulation methods has verified that of the frequency response analysis method.

NOMENCLATURE

$L, L[\ldots]$	The Laplace transform
e	2.718281828
f_n	The natural frequency
ω	The excitation frequency
π	3.1415926
$F, F(t)$	The excitation force
$\overline{F}(s)$	The Laplace transform of F or $F(t)$
y	The base excitation displacement
$Y, Y(s)$	The Laplace transform of y or $y(t)$
$\ddot{y}$	The base excitation acceleration
z	The relative displacement of the oscillator mass (M) with respect to the base
$Z, Z(s)$	The Laplace transform of z or $z(t)$
$\dot{z}$	The relative velocity of the oscillator mass (M) with respect to the base
$\ddot{z}$	The relative acceleration of the oscillator mass (M) with respect to the base
$x, x(t)$	The displacement of oscillator (M)
x	A column vector of dimension n, called the "state vector"
y	The "output" of the system

A	The "state matrix"
B	The "input matrix"
C	The "output matrix"
D	The "feed through matrix"
W_i $(i = 1, 2, \ldots)$	The sate variable
$\frac{d}{dt}$	The first differential
$\frac{d^2}{dt^2}$	The second differential
$\int_0^\infty \cdot \cdot \, dt$	The integration from 0 to infinity
$\cdots \big\|_0^\infty$	The function difference at the variables of 0 and infinity
$X, X(s)$	The Laplace transform of x
$L^{-1}, L^{-1}[\ldots]$	The inverse Laplace transform
M	The oscillator mass
K	The short circuit stiffness between the base and the oscillator (M)
c	The short circuit mechanical damping between the base and the oscillator (M)
s	The complex variable
t	The variable of the integration or time variable
i	The square root of -1
∞	Infinity

Superscripts

$\cdot$	The first differential
$\cdot\cdot$	The second differential
-1	Inverse

REFERENCES

[1] Meirovitch L. Elements of vibration analysis. New York: McGraw-Hill, Inc.; 1975, ISBN 0-07-041340-1.

[2] Thomson WT. Theory of vibration with applications. New Jersey: Prentice-Hall, Inc.; 1998, ISBN: 0-13-651068.

[3] Harrison M. Vehicle refinement — controlling noise and vibration in road vehicles. Warrendale (PA, USA): SAE International; 2004, ISBN: 07686 15057.

[4] Wang X. Vehicle noise and vibration refinement. Cambridge (UK): Woodhead Publishing; 2010, ISBN 978-1-84569-744-0.

[5] Guyomar D, et al. Energy harvester of 1.5 cm^3 giving output power of 2.6 mW with only 1 G acceleration. J Intell Mater Syst Struct 2010;22(5):415–20.

Analysis of a Single Degree of Freedom Piezoelectric Vibration Energy Harvester System Using the Transfer Function, Integration, State Space, and Frequency Response Methods

2

CHAPTER OUTLINE

2.1 INTRODUCTION

A vibration harvester converts vibration energy into electrical energy. The harvester consists of a spring-mass oscillating system, a piezoelectric element, and a load resistor as shown in Fig. 2.1. The input energy is due to the vibration of the base $y(t)$ and output energy (power) is generated through the output voltage across the load resistor due to the compression of the lead zirconate titanate $(Pb[Zr(x)Ti(1-x)]O3)$ (PZT) material. In

Frequency Analysis of Vibration Energy Harvesting Systems. http://dx.doi.org/10.1016/B978-0-12-802321-1.00002-9

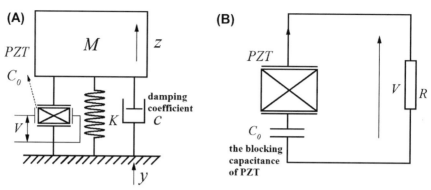

FIGURE 2.1

Schematic of single degree of freedom piezoelectric vibration energy harvester system connected to a single electric load resistor (A) the harvester oscillator electromechanical system (B) energy harvesting circuit.

other words, the charge is generated across the poles of the PZT material, the charge varies with time due to the vibration of the base of the harvester system $y(t)$ which produces the varying current or output voltage across the load resistor.

Because of the ubiquitous existence of vibration of operating machinery and of the environment, energy harvesting from ambient vibration has attracted a lot of attentions in recent years, as indicated by a wealth of literature in which the optimization of the piezoelectric vibration harvester for the maximum harvested resonant power output has been most reported. The piezoelectric vibration energy harvester (PVEH) with open circuit AC output or connected to a single load resistor was also intensively studied. Harvested resonant power was calculated from a reference power multiplied by a dimensionless expression. The dimensionless expression was called dimensionless harvested resonant power. The process which converts the harvested resonant power into the dimensionless harvested resonant power is called "normalization." In other words, the harvested resonant power is normalized in a dimensionless form. The resonant energy-harvesting efficiencies of PVEHs can also be expressed as the functions of dimensionless variables and normalized. The parameters of the vibration energy harvesters were optimized for the maximum energy harvesting efficiency. However, the harvested resonant power and energy harvesting efficiency were rarely optimized and normalized in a dimensionless form simultaneously as a pair.

Researchers have been seeking optimization design methods to maximize the harvested energy. To maximize power generation, the vibration deflection should be as large as possible [1,7−16] as well as the mass should be as large as possible within the available volume of the harvester. The spring should be designed so that the resonant frequency of the harvester matches the excitation frequency of the application. Unwanted damping should be minimized so that it does not affect electrical power generation. However, the effects of the force factor of the piezoelectric material and external electric load resistance on the harvested resonant power and energy harvesting efficiency have not been studied. The force factor of the piezoelectric material is defined as the piezoelectric constant multiplied by the cross-sectional area of the piezoelectric material disk and divided by the thickness of the piezoelectric material disk (see Eq. (2.3)). The force factor reflects how large a force is required to produce the voltage of 1 V. It was believed that with a high electrical power density, the piezoelectric system is particularly well suited to microsystems, in comparison with the electromagnetic system which is recommended for medium scale applications [2,7−16]. It was shown that a highly damped system has the advantage of harvesting energy over a wider frequency bandwidth of excitation, but would harvest marginally less power [3].

In past researches, no research group compared the time domain simulation and frequency response analysis methods for analyzing a harvester system. No research papers have been published that reported work to optimize and normalize the harvested resonant power and energy harvesting efficiency together in a pair using dimensionless resistance and force factor, while both the dimensionless variables are critical in reflection of the power loss of interface circuit, mechanical damping dissipation, and the piezoelectric power transfer. Therefore a comparison of the time domain simulation and frequency analysis methods by simultaneous optimization and normalization of the harvested resonant power and energy harvesting efficiency using the two dimensionless resistance and force factor will be the focus of this chapter.

Aiming to develop an effective tool for analysis and design of a vibration energy harvester, a new approach with a hybrid of the frequency response analysis and time domain simulation methods will be proposed in this chapter. To illustrate the approach, the frequency response analysis and time domain simulation are conducted for a single degree of freedom (SDOF) piezoelectric system, with output voltage and harvested resonant power will be calculated for different system parameters or frequencies. The harvested resonant power is defined as the output voltage square divided by the resistance where the output voltage is harvested when the spring-mass oscillating system is in the resonance. The calculated output voltage and harvested resonant power using the frequency response analysis method will be compared with those using the time domain simulation method that is validated by experimental results. The mechanical−electrical system with built-in piezoelectric material will be connected to a single load resistor and this system is studied using the proposed analysis approach.

The approach allows for a parameter study and optimization of PVEHs. The dimensionless harvested resonant power and energy harvesting efficiency formula developed from this paper are useful for performance evaluation of the vibration energy harvesters ranging from macro- to microscales, even to nanoscales. Potential industrial applications would be in ambient vibration energy-harvesting analysis for machines or structure systems with built-in piezoelectric materials. These systems may include engine torsion vibration absorbers (engine harmonic balancer), vehicle chassis suspension/mounting systems, vehicle power-train suspension/mounting systems, wheels and tires or vehicle body panels. The harvested electric energy is expected to be used to power electronic devices; self-sustaining micro- and nanoscale sensors; or remote sensing systems including electronics, sensors, or wireless sensor nodes.

The focus of this chapter is to develop the normalized formulas for the calculation of the harvested resonant power and energy harvesting efficiency from the two dimensionless variables of the normalized resistance and force factors. These dimensionless variables are independent of the sizes and configurations of PVEHs. A novel analysis approach will be developed from a hybrid of the frequency response analysis method and the time domain simulation method for a SDOF mechanical—electric harvester system.

2.2 ANALYSIS AND SIMULATION OF AN SINGLE DEGREE OF FREEDOM PIEZOELECTRIC VIBRATION ENERGY HARVESTER CONNECTED TO A SINGLE LOAD RESISTANCE

For an SDOF piezoelectric system with a constant excitation magnitude as a function of frequency as shown in Fig. 2.1(A), the mechanical system governing equation is given by:

$$M \cdot \ddot{z} + c \cdot \dot{z} + K \cdot z = -M \cdot \ddot{y} - \alpha \cdot V \tag{2.1}$$

The electrical system governing equation is given by:

$$I = \alpha \cdot \dot{z} - C_0 \cdot \dot{V} \tag{2.2}$$

where y is the base excitation displacement; M is the mass; c is the short circuit mechanical damping coefficient; K is the short circuit stiffness of the SDOF piezo-electric system; $z = x - y$ is the relative displacement of the mass with respect to the base; V is the voltage across the resistance R and I is the current of the energy harvesting circuit as shown in Fig. 2.1(B). According to Guyomar et al. [4], the force factor α and the blocking capacitance of the piezoelectric insert C_0, are respectively defined as:

$$\begin{cases} \alpha = -\dfrac{e_{33} \cdot S_0}{H} \\[2mm] C_0 = \dfrac{\varepsilon_{33}^S \cdot S_0}{H} \end{cases} \tag{2.3}$$

where e_{33} and ε_{33}^S are the piezoelectric constant and permittivity, respectively, and S_0, H are the piezoelectric disk surface area and thickness, respectively. The force factor has a unit of N/V or A s/m. The blocking capacitance of PZT C_0 has a unit of Farad.

2.3 LAPLACE TRANSFORM AND TRANSFER FUNCTION ANALYSIS METHOD

Eqs. (2.1) and (2.2) can be simulated and solved using the transfer function method. From Eq. (2.2), Eq. (2.4) is derived as:

$$C_0 \cdot \dot{V} + \frac{V}{R} = \alpha \cdot \dot{z} \qquad (2.4)$$

where R is the total electrical resistance of the piezoelectric material insert in the unit of Ohm and external load. For the SDOF piezoelectric system connected to a single electrical resistor, if the base excitation is harmonic, $y = Y(s) \cdot e^{s \cdot t}$, and for a linear system, relative oscillator displacement z and output voltage V are assumed to be harmonic and are given by:

$$z = Z(s) \cdot e^{s \cdot t}$$
$$V = \overline{V}(s) \cdot e^{s \cdot t} \qquad (2.5)$$

where $s = i\omega$, i is the square root of -1; ω is circular frequency with a unit of rad/s.

Substituting Eq. (2.5) into Eq. (2.4) gives:

$$\left(C_0 \cdot s + \frac{1}{R}\right) \cdot \overline{V}(s) = \alpha \cdot s \cdot Z(s)$$

This is, in fact, a Laplace transform of Eq. (2.4). The transfer function between the relative oscillator displacement and output voltage is then derived and given by:

$$\frac{\overline{V}(s)}{Z(s)} = \frac{\alpha \cdot s}{C_0 \cdot s + \frac{1}{R}} = \frac{\alpha \cdot s \cdot R}{R \cdot C_0 \cdot s + 1} \qquad (2.6)$$

Substitution of Eqs. (2.5) and (2.6) into Eq. (2.1) gives:

$$\left(K + M \cdot s^2 + c \cdot s + \frac{\alpha^2 \cdot s \cdot R}{R \cdot C_0 \cdot s + 1}\right) \cdot Z(s) = -M \cdot s^2 \cdot Y(s)$$

This is the Laplace transform of Eq. (2.1). The transfer function between the base acceleration excitation and relative oscillator displacement is given by:

$$\frac{Z(s)}{s^2 \cdot Y(s)} = \frac{-M}{K + M \cdot s^2 + c \cdot s + \frac{\alpha^2 \cdot s \cdot R}{R \cdot C_0 \cdot s + 1}} \qquad (2.7)$$

Rearranging Eq. (2.7) gives:

$$\frac{Z(s)}{s^2 \cdot Y(s)} = \frac{-M \cdot R \cdot C_0 \cdot s - M}{(R \cdot C_0 \cdot M \cdot s^3 + (R \cdot C_0 \cdot c + M) \cdot s^2 + (c + R \cdot C_0 \cdot K + \alpha^2 \cdot R) \cdot s + K)}$$

$$(2.8)$$

According to Eq. (2.6), the transfer function between the base excitation acceleration and output voltage is given by:

$$\frac{\overline{V}(s)}{s^2 \cdot Y(s)} = \frac{\overline{V}(s)}{Z(s)} \cdot \frac{Z(s)}{s^2 \cdot Y(s)} = \frac{\alpha \cdot R \cdot s}{R \cdot C_0 \cdot s + 1} \cdot \frac{Z(s)}{s^2 \cdot Y(s)} \tag{2.9}$$

Substitution of Eq. (2.8) into Eq. (2.9) gives:

$$\frac{\overline{V}(s)}{s^2 \cdot Y(s)} = \frac{-\alpha \cdot M \cdot R \cdot s}{(R \cdot C_0 \cdot M \cdot s^3 + (R \cdot C_0 \cdot c + M) \cdot s^2 + (c + R \cdot C_0 \cdot K + \alpha^2 \cdot R) \cdot s + K)} \tag{2.10}$$

To compare the simulation results of Eq. (2.10) with the experimental results in Ref. [5], the SDOF piezoelectric system employed the same parameters as those parameters in Ref. [5], except for a minor correction of the resonant frequency 275 Hz (from 277.4 Hz) and load resistance 30,669.6 Ω (from 30 kΩ) as shown in Table 2.1. A simulation was conducted according to the simulation schedule using the Matlab Simulink transfer function method as shown in Fig. 2.2. Substitution of the

Table 2.1 The Identified Single Degree of Freedom Piezoelectric System Parameters [5]

Parameter	Measurement Type	Values	Units
M	Indirect	8.4×10^{-3}	kg
c	Direct	0.154	N·s/m
K	Indirect	2.5×10^4	N/m
C_0	Direct	1.89×10^{-8}	Farad
α	Indirect	1.52×10^{-3}	N/Volt or Amp·s/m
f_n	Indirect	275	Hz
R	Direct	30,669.6	Ω
Q_i	Direct	95	N/A

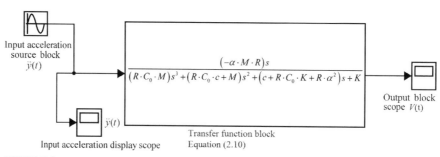

FIGURE 2.2

Simulation schematic for Eq. (2.10) in Matlab Simulink with a sine wave base excitation input and a sinusoidal voltage output at a frequency using the transfer function method.

parameters in Table 2.1 into Eq. (2.10) gives the simulation results. The time domain simulation schedule is arranged to have a 275 Hz sine wave acceleration signal input as defined in Figs. 1.5 and 1.6. It is seen from Fig. 2.5 that Eq. (2.10) is entered into a transfer function block in a Matlab Simulink program code that has one input and one output. The input end of the transfer function block is connected to a source block of a sinusoidal signal of 275 Hz with the RMS acceleration of 9.8 m/s^2. The output end of the transfer function block is connected to a sink block scope which displays and produces a sinusoidal output voltage signal. The simulation results of Eq. (2.10) are displayed in Fig. 2.5.

2.4 TIME DOMAIN INTEGRATION METHOD

Eqs. (2.1) and (2.2) can also be simulated and solved using the integration method and written as

$$
\begin{cases}
\ddot{z} = -\dfrac{c}{M}\cdot\dot{z} - \dfrac{K}{M}\cdot z - \ddot{y} - \dfrac{\alpha}{M}\cdot V & (2.11a) \\[2ex]
\dot{V} = -\dfrac{V}{C_0\cdot R} + \dfrac{\alpha}{C_0}\cdot\dot{z} & (2.11b)
\end{cases}
\qquad (2.11)
$$

Eq. (2.11) can be integrated according to the integration method of the simulation schematic using the Matlab Simulink program code as shown in Fig. 2.3. There are two round sum blocks in Fig. 2.3. The top round sum block has four positive inputs and one output, while the bottom round sum block has two positive inputs and one output. The four terms on the right-hand side (RHS) of Eq. (2.11a) are presented by the four inputs in the top round sum block shown in Fig. 2.3. The output of the top round sum block is $\ddot{z}$. $\ddot{z}$ is integrated once to give the relative velocity $\dot{z}$, $\ddot{z}$ is integrated twice to give the relative displacement z. $\dot{z}$ and z multiplied by $-c/M$ and $-K/M$ contribute to the two positive inputs of the top round sum block in Fig. 2.3. The other two inputs are the input excitation acceleration $-\ddot{y}$ and voltage $V(t)$ multiplied by $-\alpha/M$. The voltage $V(t)$ can be wired from the bottom round sum block. The two terms on the RHS of the second equation of Eq. (2.11) are presented by the two inputs in the bottom round sum block in Fig. 2.3. The output of the bottom round sum block is $\dot{V}$, while $\dot{V}$ is integrated once to give the voltage $V(t)$. $V(t)$ multiplied by $-1/(C_0R)$ and the relative velocity $\dot{z}$ multiplied by α/C_0 are the inputs for the bottom round sum block. The relative velocity $\dot{z}$ can be wired from the top round sum block. A source block is used to provide the input excitation input which is monitored by a scope, and two sink block scopes are used to display the output voltage and relative displacement signals. The excitation acceleration is assumed to be a sine wave signal of excitation frequency of 275 Hz and the excitation acceleration with an RMS value of 9.8 m/s^2 as shown in Fig. 1.5. The parameter input template for the excitation sine wave in the Matlab Simulink program code is given in Fig. 1.6. The Matlab solver type was

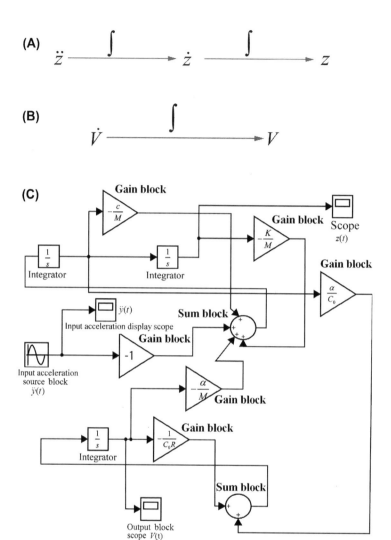

FIGURE 2.3

(A) Relative acceleration integrated to the relative displacement; (B) Derivative of the output voltage integrated to the output voltage; (C) Simulation schematic for Eq. (2.11) with a sine wave base excitation input and a sinusoidal voltage output at a frequency using the integration method.

chosen as the ode23t (mod. stiff/Trapezoidal) algorithm from the Matlab Simulink library. If the above parameters in Table 2.1 are entered into Matlab software, from the Matlab Simulink program code shown in Fig. 2.3, the response output voltage result calculated from the time domain integration method is shown to be same as that calculated from the transfer function method that is plotted in Fig. 2.5.

2.5 **STATE SPACE METHOD**

Eqs. (2.1) and (2.2) can also be simulated and solved using the state space method. Following Eq. (1.13) and the following equations:

$$
\begin{cases}
W_1 = z(t) \\[2mm]
W_2 = \dfrac{dz(t)}{dt} \\[2mm]
W_3 = V(t) \\[2mm]
W_4 = \dfrac{dV(t)}{dt}
\end{cases}
\tag{2.12}
$$

Using the definition of Eq. (2.12), Eq. (2.1) becomes

$$
M \cdot \frac{dW_2}{dt} + c \cdot W_2 + K \cdot W_1 = -M \cdot \ddot{y} - \alpha \cdot W_3
$$

Similarly, Eq. (2.2) becomes

$$
\frac{W_3}{R} = \alpha \cdot W_2 - C_0 \cdot W_4
$$

Then

$$
\frac{dW_4}{dt} = \frac{\alpha}{C_0} \cdot \frac{dW_2}{dt} - \frac{W_4}{C_0 \cdot R}
$$

Consider

$$
\frac{dW_2}{dt} = -\ddot{y} - \frac{\alpha}{M} \cdot W_3 - \frac{c}{M} \cdot W_2 - \frac{K}{M} \cdot W_1
$$

then

$$
\frac{dW_4}{dt} = -\ddot{y} \cdot \frac{\alpha}{C_0} - \frac{\alpha^2}{M \cdot C_0} \cdot W_3 - \frac{c \cdot \alpha}{M \cdot C_0} \cdot W_2 - \frac{K \cdot \alpha}{M \cdot C_0} \cdot W_1 - \frac{W_4}{C_0 \cdot R}
\tag{2.13}
$$

Eqs. (2.12) and (2.13) can be written as

$$
\begin{Bmatrix}
\dfrac{dW_1}{dt} \\[2mm]
\dfrac{dW_2}{dt} \\[2mm]
\dfrac{dW_3}{dt} \\[2mm]
\dfrac{dW_4}{dt}
\end{Bmatrix}
=
\begin{bmatrix}
0 & 1 & 0 & 0 \\[2mm]
-\dfrac{K}{M} & -\dfrac{c}{M} & -\dfrac{\alpha}{M} & 0 \\[2mm]
0 & 0 & 0 & 1 \\[2mm]
-\dfrac{K \cdot \alpha}{M \cdot C_0} & -\dfrac{c \cdot \alpha}{M \cdot C_0} & -\dfrac{\alpha^2}{M \cdot C_0} & -\dfrac{1}{C_0 \cdot R}
\end{bmatrix}
\begin{Bmatrix}
W_1 \\[2mm]
W_2 \\[2mm]
W_3 \\[2mm]
W_4
\end{Bmatrix}
+
\begin{bmatrix}
0 \\[2mm]
-1 \\[2mm]
0 \\[2mm]
\dfrac{\alpha}{C_0}
\end{bmatrix}
\{\ddot{y}\}
$$

$$
\tag{2.14}
$$

and

$$
\left\{ \begin{array}{c} z \\ v \end{array} \right\} = \begin{bmatrix} 1 & 0 & 0 & 0 \\ 0 & 0 & 1 & 0 \end{bmatrix} \left\{ \begin{array}{c} W_1 \\ W_2 \\ W_3 \\ W_4 \end{array} \right\} + \begin{bmatrix} 0 \\ 0 \end{bmatrix} \{ \ddot{y} \}
\tag{2.15}
$$

From Eqs. (2.14) and (2.15), given values of K, M, c, C_0, R, and α, the following matrices can be determined:

$$
\left(\mathbf{A} = \begin{bmatrix} 0 & 1 & 0 & 0 \\ -\dfrac{K}{M} & -\dfrac{c}{M} & -\dfrac{\alpha}{M} & 0 \\ 0 & 0 & 0 & 1 \\ -\dfrac{K \cdot \alpha}{M \cdot C_0} & -\dfrac{c \cdot \alpha}{M \cdot C_0} & -\dfrac{\alpha^2}{M \cdot C_0} & -\dfrac{1}{C_0 \cdot R} \end{bmatrix} ; \mathbf{B} = \begin{bmatrix} 0 \\ -1 \\ 0 \\ -\dfrac{\alpha}{C_0} \end{bmatrix} ; \right.
$$

$$
\left. \mathbf{C} = \begin{bmatrix} 1 & 0 & 0 & 0 \\ 0 & 0 & 1 & 0 \end{bmatrix} ; \mathbf{D} = \begin{bmatrix} 0 \\ 0 \end{bmatrix} ; \right.
$$

The Simulink schematic of the state space method is shown in Fig. 2.4. In the Simulink codes of the transfer function and state space methods, the solver type

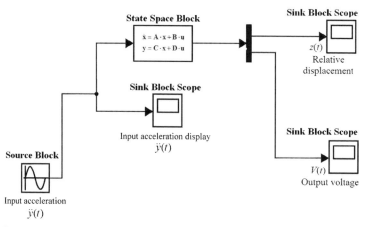

FIGURE 2.4

The Matlab Simulink state space method schematic for the prediction of the response relative displacement and output voltage from the base excitation acceleration.

was chosen as the variable step ode23-Bogacki-Shampine (or ode23t (mod. stiff/Trapezoidal)) from the Matlab Simulink library. The matrices of **A**, **B**, **C**, and **D** are defined for their values in the Matlab software. The state space block is defined in the Matlab Simulink code which has an input source block and an output sink block scope. Given input excitation acceleration as defined in Figs. 1.5 and 1.6, the outputs $z(t)$ and $V(t)$ (the response relative displacement and output voltage) can be predicted using the Matlab Simulink state space method. The output voltage response is presented in Fig. 2.5. It can be seen from Fig. 2.5 that the output voltage signal is shown to be a sine wave signal, this is expected since the input excitation acceleration/displacement signal is a sine wave and the system is linear. It can be seen from Fig. 2.5 that the peak output voltage is 12.57 V, which is equivalent to the RMS voltage value of 8.89 V from which the harvested resonant power (RMS) is calculated to be $V^2/R = 2.576$ mW where the power loss of energy extraction and storage is not considered here and the resonant frequency was at 275 Hz. From Ref. [5], a vibration energy harvester with the same device parameters was able to generate a maximum RMS power of 2.6 mW at 277.4 Hz with an RMS acceleration value of 1 g (9.8 m/s^2). The difference is very small. The reasons for the difference could be (1) the 2.4-Hz shift from the resonant frequency; (2) the energy extraction and storage circuit in Ref. [5] itself consumed energy and caused a power loss.

The time domain simulation methods can be applied to calculate and evaluate the output voltage and harvested resonant power for variations of the mechanical damping, the resistance, and the force factor. When one of the selected parameters was

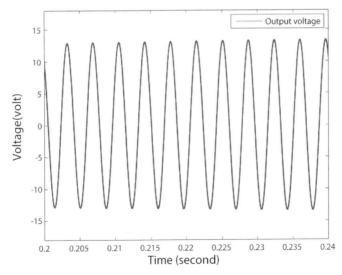

FIGURE 2.5

Output sinusoidal voltage signal (volt) from an input excitation acceleration signal with a root mean square acceleration value of 1 g (9.8 m/s^2) at an excitation frequency of 275 Hz.

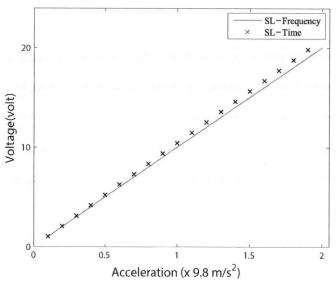

FIGURE 2.6

The output voltage amplitude versus the base excitation acceleration amplitude at an excitation frequency of 275 Hz.

changed, the other parameters in Table 2.1 were kept constant. The amplitudes of the output voltage time traces were read from the sink block scopes. The simulated results of the output voltage and harvested resonant power are plotted in the discrete star marks in Figs. 2.6–2.12. In those figures, the legend of the "SL Time" represents the time domain simulation results for the SDOF system connected to single load resistor. For example, if the base excitation acceleration amplitude changes from 0.1 g (0.98 m/s^2) to 2 g (19.6 m/s^2), other parameters in Table 2.1 are kept unchanged, following the simulation schematics of Figs. 2.2–2.4, the output voltage amplitudes were obtained from which the harvested resonant power amplitudes are calculated. The output voltage and harvested resonant power amplitudes are plotted in discrete star marks in Figs. 2.6 and 2.7.

2.6 FREQUENCY RESPONSE ANALYSIS METHOD

Eqs. (2.1) and (2.2) can also be simulated and solved using the frequency response method. Frequency response function is defined as the Fourier transform of the time domain response divided by the Fourier transform of the time domain excitation input. Eqs. (2.1) and (2.2) can be written in matrix form as:

$$\begin{bmatrix} M & 0 \\ 0 & 0 \end{bmatrix} \cdot \begin{Bmatrix} \ddot{z} \\ \ddot{V} \end{Bmatrix} + \begin{bmatrix} c & 0 \\ -\alpha & C_0 \end{bmatrix} \cdot \begin{Bmatrix} \dot{z} \\ \dot{V} \end{Bmatrix} + \begin{bmatrix} K & \alpha \\ 0 & 1/R \end{bmatrix} \cdot \begin{Bmatrix} z \\ V \end{Bmatrix} = \begin{bmatrix} 0 & -M \\ 0 & 0 \end{bmatrix} \cdot \begin{Bmatrix} \dot{y} \\ \ddot{y} \end{Bmatrix}$$

(2.16)

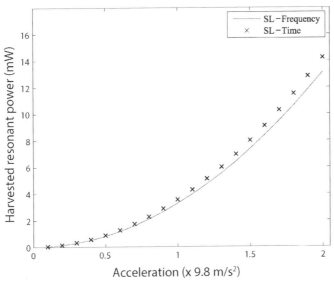

FIGURE 2.7

Harvested resonant power versus the base excitation acceleration amplitude at an excitation frequency of 275 Hz.

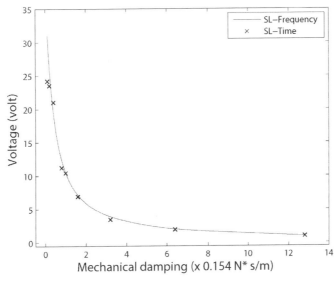

FIGURE 2.8

The output voltage amplitude versus the mechanical damping under an input excitation acceleration signal of an RMS value of 1 g (9.8 m/s^2) at an excitation frequency of 275 Hz.

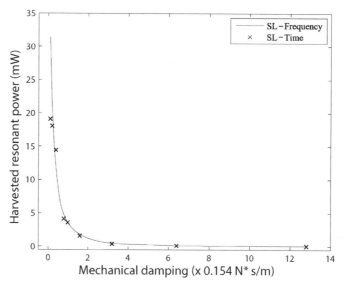

FIGURE 2.9

Harvested resonant power versus the mechanical damping under an input excitation acceleration signal of an RMS value of 1 g (9.8 m/s^2) at an excitation frequency of 275 Hz.

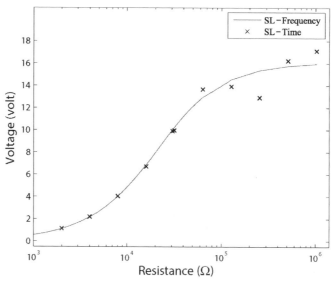

FIGURE 2.10

The output voltage amplitude versus the electrical load resistance under an input excitation acceleration signal of an RMS value of 1 g (9.8 m/s^2) at an excitation frequency of 275 Hz.

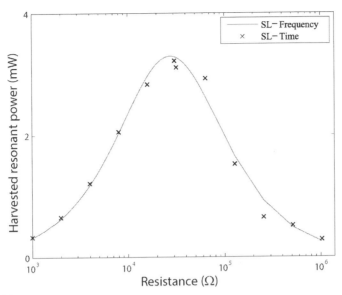

FIGURE 2.11

Harvested resonant power versus the electrical load resistance under an input excitation acceleration signal of an RMS value of 1 g (9.8 m/s^2) at an excitation frequency of 275 Hz.

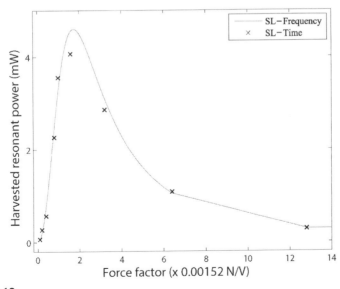

FIGURE 2.12

Harvested resonant power versus the force factor under an input excitation acceleration signal of an RMS value of 1 g (9.8 m/s^2) at an excitation frequency of 275 Hz.

From the definition of Eq. (2.5), Eq. (2.16) becomes

$$\begin{bmatrix} M \cdot s^2 + c \cdot s + K & \alpha \\ -\alpha \cdot s & C_0 \cdot s + 1/R \end{bmatrix} \cdot \left\{ \begin{matrix} Z \\ V \end{matrix} \right\} = \left\{ \begin{matrix} -M \\ 0 \end{matrix} \right\} \cdot s^2 \cdot Y \qquad (2.17)$$

The displacement and voltage frequency response functions are given by:

$$\left\{ \begin{matrix} \dfrac{Z}{s^2 \cdot Y} \\ \dfrac{V}{s^2 \cdot Y} \end{matrix} \right\} = \begin{bmatrix} M \cdot s^2 + c \cdot s + K & \alpha \\ -\alpha \cdot s & C_0 \cdot s + 1/R \end{bmatrix}^{-1} \cdot \left\{ \begin{matrix} -M \\ 0 \end{matrix} \right\} = \mathbf{E}^{-1} \cdot \mathbf{F} \qquad (2.18)$$

where

$$\left\{ \begin{matrix} \mathbf{E} = \begin{bmatrix} M \cdot s^2 + c \cdot s + K & \alpha \\ -\alpha \cdot s & C_0 \cdot s + 1/R \end{bmatrix} \\ \mathbf{F} = \left\{ \begin{matrix} -M \\ 0 \end{matrix} \right\} \end{matrix} \right.$$

Eq. (2.18) can be simulated according to the Matlab code as shown in Fig. 2.13, where $s = i\omega$ and the parameters in Table 2.1 are applied. The amplitudes of the

```
**************************************************************************
CLEAR ALL
CLC

C=0.154;
K=25000;
C0=1.89E-8;
M=0.0084;
R=30669.6;
ALPHA=0.00152;

F=0:0.01:300;
W=2*F*PI;
FOR N=1:LENGTH(W)
  S=1I*W(N);
  E=[M*S^2+C*S+K ALPHA;-ALPHA*S C0*S+1/R];
  F=[-M;0];
  Z=E\F;
    Z1(N)=Z(1); % X1/Y
    Z2(N)=Z(2); % X2/Y
  END

FIGURE (1)
PLOT(F,ABS(Z1))
XLABEL('FREQUENCY (HZ)')
YLABEL('DISPLACEMENT RATIO (X1/Y)')

FIGURE (2)
PLOT(F,ABS(Z2))
XLABEL('FREQUENCY (HZ)')
YLABEL('DISPLACEMENT RATIO (X2/Y)')
**************************************************************************
```

FIGURE 2.13

Matlab codes for simulation of the frequency response functions of the output relative displacement and voltage over the input excitation acceleration.

relative displacement and output voltage over input excitation acceleration are plotted in Figs. 2.14 and 2.15.

It is seen from Figs. 2.14−2.15 that the maximum amplitudes of the relative displacement and output voltage transfer functions versus the input excitation acceleration occur at the resonant frequency of 275 Hz. The transfer function amplitude versus frequency can also be obtained from the time domain simulation programming codes shown in Figs. 2.2−2.4 where the excitation amplitude is fixed to be unity, the excitation frequency changes from low values to high values to cover the resonant frequency. For each given excitation frequency of a sine wave signal input from a source block of the Matlab Simulink library, the response amplitudes are recorded. The response amplitudes are then plotted against the excitation frequency. The calculated amplitude curves of the output displacement and voltage transfer functions versus the excitation acceleration should coincide with those in Figs. 2.14−2.15. From Eqs. (2.9) and (2.10), the following equations are obtained:

$$\left| \frac{V}{\omega^2 \cdot Y} \right| = \frac{\alpha \cdot R \cdot \omega}{\sqrt{1 + R^2 \cdot C_0^2 \cdot \omega^2}} \left| \frac{Z}{\omega^2 \cdot Y} \right|$$

$$= \frac{M \cdot \sqrt{1 + R^2 \cdot C_0^2 \cdot \omega^2}}{\sqrt{(\alpha \cdot R \cdot \omega \cdot C_0)^2 + \left[\frac{c}{R \cdot \alpha} \left(1 + R^2 \cdot C_0^2 \cdot \omega^2 \right) + \alpha \right]^2}} \tag{2.19}$$

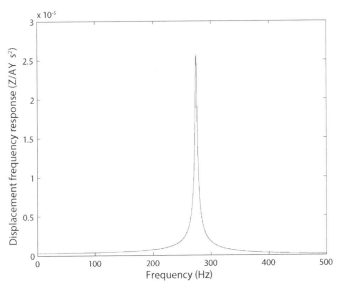

FIGURE 2.14

Amplitude of the relative displacement frequency response function—the output relative displacement divided by the input excitation acceleration.

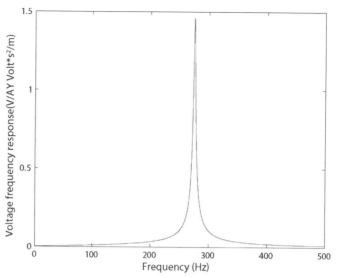

FIGURE 2.15

Amplitude of the voltage frequency response function—the output voltage divided by the input excitation acceleration.

According to Ref. [6], the mean resonant harvested power is given by

$$\frac{P_h}{|\omega^2 \cdot Y|^2} = \frac{1}{2} \cdot \frac{\left|\frac{\overline{V}}{\omega^2 \cdot Y}\right|^2}{R} = \frac{1}{2} \cdot \frac{R \cdot \alpha^2 \cdot \omega^2}{\left(1 + R^2 \cdot C_0^2 \cdot \omega^2\right)} \frac{|z|^2}{|\ddot{y}|^2}$$

$$= \frac{1}{2} \cdot \frac{M^2}{c} \cdot \frac{R \cdot \alpha^2 \cdot c}{\alpha^4 \cdot R^2 + 2 \cdot R \cdot \alpha^2 \cdot c + \left(1 + R^2 \cdot C_0^2 \cdot \omega^2\right) \cdot c^2} \qquad (2.20)$$

where $A = |\omega^2 \cdot Y|$ is the input acceleration excitation amplitude, but not a force. If one of the parameters in Table 2.1 is varied and the other parameters are kept constant, substitution of the constant parameters from Table 2.1 into Eqs. (2.19) and (2.20) gives the amplitude ratio of the output voltage versus the excitation acceleration amplitude and the amplitude ratio of harvested resonant power versus the squared excitation acceleration amplitude for variations of the mechanical damping, the resistance, and the force factor. The results of the parameter study are plotted in the solid curves in Figs. 2.6–2.12. If field vibration acceleration measurement data is used as the excitation acceleration input, the output voltage and harvested resonant power of the SDOF vibration energy harvester can be predicted from the input excitation data using this approach.

As the base excitation acceleration amplitude increases, the output voltage amplitude increases linearly as shown in Fig. 2.6. As the base excitation acceleration amplitude increases, the harvested resonant power can be seen to increase in a parabolic curve as shown in Fig. 2.7. It is seen that the time domain simulation results represented by discrete star marks are very close to the frequency analysis results represented by the solid curves for this case.

If the mechanical damping is changed from 0.1 times to 12.8 times of the original mechanical damping in steps of a factor of 2, and the other parameters in Table 2.1 are kept constant, the output voltage amplitude and harvested resonant power calculated from Eqs. (2.19) and (2.20) can be plotted and shown in Figs. 2.8 and 2.9. Again the time domain simulation results represented by discrete star marks are very close to the frequency analysis results represented by the solid curves. As expected, the output voltage and harvested resonant power amplitudes are shown to decrease as the system mechanical damping increases.

If the resistance increases from 1000 to 1,024,000 Ω in steps of a factor of 2, with the other parameters in Table 2.1 held constant, the output voltage amplitude and harvested resonant power calculated from Eqs. (2.19) and (2.20) are plotted in Figs. 2.10 and 2.11. It can be seen from Fig. 2.11 that the harvested resonant power first increases up to a maximum value, then decreases. This means that for the SDOF system, if only the resistance changes, there exists an optimized electrical load resistance to achieve a peak harvested resonant power. This optimized electrical load resistance is related to the electrical impedance matching of the piezoelectric material insert and external load. The discrete star marks of the time domain simulation results are very close to the solid curves of the frequency analysis results in the low- and high-load resistance ranges. There are differences for the time domain simulation and frequency analysis results in the middle load resistance range. The simulation errors are caused by a coarse step size of the numerical simulation and the solver type using the Runge—Kuta method.

If just the force factor is changed from 0.1 times to 10 times the original value in steps of a factor of 2, the excitation frequency is fixed at 275 Hz which is close to the resonant frequency of the system. The harvested resonant power for variations of the force factors are calculated from Eq. (2.20) and plotted in solid curves in Fig. 2.12. It can be seen that the discrete star marks of the time domain simulation results are very close to the solid curves of the frequency analysis results in the low and high force factor ranges. There are differences for the time domain simulation and frequency analysis results in the middle force factor range where the peak of harvested resonant power is reached. The simulation errors are caused by the relatively coarse step size of the numerical simulation and the solver type of the Runge—Kuta method.

It can be seen from Fig. 2.12 that for the SDOF system, if only the force factor changes, there does exist an optimized force factor or optimized amount of selected piezoelectric material or size which would produce the peak harvested resonant power. This is because the force factor depends on types of materials, section area, and thickness of a selected piezoelectric insert according to Eq. (2.3). In other words,

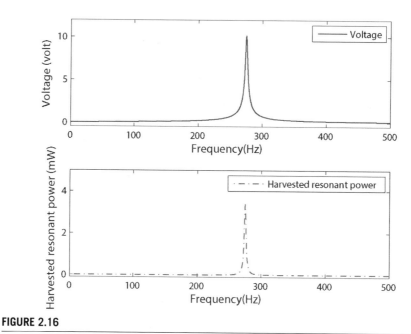

FIGURE 2.16

Harvested resonant power and output voltage versus frequency.

excessive amount of piezoelectric material or size will not help to produce more harvested resonant power at the resonance. This reflects the importance of the frequency analysis and time domain simulation approach, as it can be used to determine the optimized amount of piezoelectric materials or size for harvesting more power. Thus it can provide a tool for design optimization of the vibration energy harvester. It is seen from Figs. 2.6–2.12 that the variation of the base excitation acceleration amplitude or the mechanical damping does not give a peak value of either the voltage or the power, whereas a peak harvested resonant power value can be obtained when the load resistance or force factor is varied.

Substitution of the constant parameter values in Table 2.1 into Eqs. (2.19) and (2.20) gives the output voltage amplitude and harvested resonant power versus the frequency as shown in Fig. 2.16. It is clear that the peaks of the harvested resonant power and output voltage amplitude are only available at the resonance frequency of the system. It is seen from Fig. 2.16 that the harvested resonant power curve reaches the maximum RMS value of 3.38 mW at the natural frequency of 275 Hz for the constant parameters given in Table 2.1 and a base acceleration amplitude value of 9.8 m/s². The maximum harvested RMS power from the time domain simulation is 3.557 mW mentioned before. The slight RMS power difference of 0.18 mW between the frequency analysis and the time domain simulation is caused by a coarse step size of the time domain simulation and the solver type using the Runge–Kuta method.

2.7 DIMENSIONLESS FREQUENCY RESPONSE ANALYSIS AND HARVESTING PERFORMANCE OPTIMIZATION

To conduct a dimensionless frequency response analysis, the following definitions are used:

$$\begin{cases} R_N = R \cdot C_0 \cdot \omega \\ \alpha_N = \sqrt{\dfrac{\alpha^2}{c \cdot C_0 \cdot \omega}} \end{cases} \tag{2.21}$$

where R_N is the dimensionless resistance and α_N is the dimensionless force factor.

Thus, Eq. (2.19) becomes

$$\left| \frac{\overline{V}}{\frac{M}{\alpha} \cdot \omega^2 \cdot Y} \right| = \frac{\sqrt{1 + R_N^2}}{\sqrt{R_N^2 + \left[\frac{1}{R_N \cdot \alpha_N^2} \left(1 + R_N^2 \right) + 1 \right]^2}} \tag{2.22}$$

and Eq. (2.20) becomes

$$\frac{P_h}{\left(\frac{M^2 \cdot A^2}{c} \right)} = \frac{1}{2} \cdot \frac{R_N \cdot \alpha_N^2}{\alpha_N^4 \cdot R_N^2 + 2 \cdot R_N \cdot \alpha_N^2 + \left(1 + R_N^2 \right)} \tag{2.23}$$

The output voltage and harvested power are now normalized. According to Ref. [6], the mean input power at the resonant frequency is given by

$$P_{in} = -\frac{1}{2} \cdot M \cdot \left| \omega^2 \cdot Y \right|^2 \cdot \text{Re} \left(\frac{Z}{-\omega^2 \cdot Y} \right) = \frac{1}{2} \frac{M^2}{c} \cdot A^2 \cdot \frac{c + R \cdot \alpha^2 + R^2 \cdot C_0^2 \cdot c \cdot \omega^2}{R^2 \cdot C_0^2 \cdot c \cdot \omega^2 + c \cdot \left(1 + \frac{R \cdot \alpha^2}{c} \right)^2} \tag{2.24}$$

Substitution of Eq. (2.21) into Eq. (2.24) gives

$$\frac{P_{in}}{\left(\frac{M^2 \cdot A^2}{c} \right)} = \frac{1}{2} \cdot \frac{1 + R_N \cdot \alpha_N^2 + R_N^2}{R_N^2 + \left(1 + R_N \cdot \alpha_N^2 \right)^2} \tag{2.25}$$

From Eqs. (2.23) and (2.25), the resonant energy harvesting efficiency of the SDOF system connected with a single load resistor is normalized and given by

$$\eta = \frac{\frac{P_h}{\left(\frac{M^2 \cdot A^2}{c} \right)}}{\frac{P_{in}}{\left(\frac{M^2 \cdot A^2}{c} \right)}} = \frac{R_N \cdot \alpha_N^2}{1 + R_N \cdot \alpha_N^2 + R_N^2} \tag{2.26}$$

Eqs. (2.23) and (2.26) are very important dimensionless formulas for calculation of the harvested resonant power and energy harvesting efficiency and are applicable to many similar piezoelectric systems ranging from macro, micro, even to nano-scales regardless of configurations and dimensions. For given dimensionless resistance and force factor, the dimensionless harvested resonant power and energy harvesting efficiency of the systems can be predicted from these formulae.

If the dimensionless load resistance and force factor change from 0.0 to 10.0 in a step size of 0.1, the dimensionless harvested resonant power as a function of the dimensionless resistance and force factor can be plotted in Fig. 2.17. It can be seen from Eq. (2.23) that when the dimensionless force factor and the dimensionless resistance tends to be very large or zero, the dimensionless harvested resonant power tends to be zero. In the specific example of the RC oscillation circuit at resonance $R_N = 1$, if $\alpha_N = 1$ and $P_h \left/ \left(\frac{M^2 \cdot A^2}{c} \right) \right. = 0.1$. The dimensionless harvested resonant power is typically about 0.1.

Using Eq. (2.23) with only the dimensionless resistance being varied, to find the peak value of the dimensionless harvested resonant power, the partial differential of the dimensionless harvested resonant power with respect to the dimensionless resistance must be set equal to zero, which gives

$$\frac{\partial \left[\frac{P_h}{\left(\frac{M^2 \cdot A^2}{c} \right)} \right]}{\partial R_N} = 0$$

This leads to

$$\begin{cases} R_N^2 = \dfrac{1}{\left(\alpha_N^4 + 1 \right)} \\ \dfrac{P_{h \, max}}{\left(\dfrac{M^2 \cdot A^2}{c} \right)} = \dfrac{\alpha_N^2}{4 \cdot \sqrt{\alpha_N^4 + 1 + 4 \cdot \alpha_N^2}} \end{cases} \tag{2.27}$$

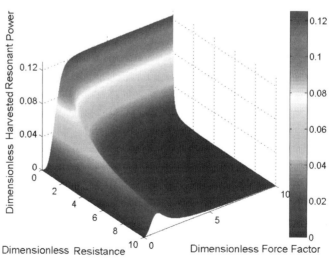

FIGURE 2.17

Dimensionless harvested resonant power versus dimensionless resistance and force factors for the single degree of freedom system connected to a load resistor.

From Eq. (2.27), it is observed that $P_{h\,max}\Big/\left(\frac{M^2 \cdot A^2}{c}\right)$ is a monotonically increasing function of α_N^2. When the dimensionless force factor tends to be very large, the peak dimensionless harvested resonant power tends to be a value of 1/8. In other words, the peak harvested resonant power is limited to $\frac{M^2 \cdot A^2}{8 \cdot c}$.

From Eq. (2.23), if only the dimensionless force factor is changed, then to find the peak value of the dimensionless harvested resonant power, the partial differential of the dimensionless harvested resonant power with respect to the dimensionless force factor must be set equal to zero, which gives

$$\frac{\partial \left[\frac{P_h}{\left(\frac{M^2 \cdot A^2}{c}\right)}\right]}{\partial \alpha_N} = 0$$

This leads to

$$\begin{cases} \alpha_N^4 = \dfrac{\left(1 + R_N^2\right)}{R_N^2} \\[4mm] \dfrac{P_{h\,max}}{\left(\dfrac{M^2 \cdot A^2}{c}\right)} = \dfrac{1}{4 + 4 \cdot \sqrt{1 + R_N^2}} \end{cases} \qquad (2.28)$$

From Eq. (2.28), it is observed that $P_{h\,max}\Big/\left(\frac{M^2 \cdot A^2}{c}\right)$ is a monotonically decreasing function of R_N. When the dimensionless resistance tends to be zero, the peak dimensionless harvested resonant power tends to be 1/8. In other words, the peak harvested resonant power is limited to $\frac{M^2 \cdot A^2}{8 \cdot c}$. Substituting Eqs. (2.27) or (2.28) into Eq. (2.26) gives the corresponding energy harvesting efficiency of 50%.

It is seen from Eqs. (2.27) and (2.28) that the peak harvested resonant power is proportional to the squared magnitude of the applied force or excitation acceleration and inversely proportional to the mechanical damping. It can be seen from Eq. (2.23) that the partial differentials of the dimensionless harvested resonant power with respect to mechanical damping are not equal to zero. There does not exist a mechanical damping value of c, which produces the peak harvested resonant power. This is shown by the results in Fig. 2.9 where the solid curves indicate that the harvested resonant power does not have any peak values.

There is no unique pair of dimensionless resistance and force factor which produces a peak value of the dimensionless harvested resonant power across a full range of the two variables. If variable range limits are specified for the dimensionless resistance and force factor, the dimensionless harvested resonant power could reach its maximum value at the range limits of the dimensionless resistance and force factor.

If the dimensionless resistance and force factor change from 0.0 to 10.0 in a step size of 0.1, the energy harvesting efficiency as a function of the dimensionless resistance and force factor can be plotted in Fig. 2.18. It can be seen from Eq. (2.26) that when the dimensionless force factor tends to be very large, the

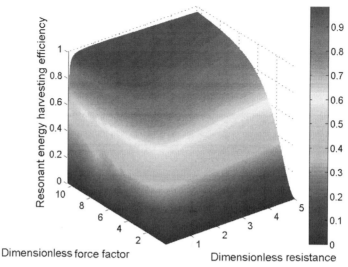

FIGURE 2.18

Resonant energy harvesting efficiency versus dimensionless resistance and force factors for the single degree of freedom system connected to a load resistor.

dimensionless resistance is not zero, the energy harvesting efficiency tends to be 100%, when the dimensionless force factor or the dimensionless resistance tends to be zero, the efficiency tends to be zero. When the dimensionless resistance tends to be very large, the dimensionless force factor is a finite constant, the efficiency tends to be zero. When both the dimensionless resistance and force factor tend to be very large, the efficiency tends to be zero or 100% depending on which is larger for the dimensionless resistance and squared dimensionless force factor. In the specific example of the RC oscillation circuit at the resonance, $R_N = 1$, if $\alpha_N = 1$, $\eta = 33.3\%$. The resonant energy harvesting efficiency is typically about 33.3% in this case.

It is seen from Fig. 2.18 and from Eq. (2.26) that for a full variation range of the dimensionless resistance R_N and force factor α_N, there is no unique pair of R_N and α_N, which produce a peak value of the efficiency.

From Eq. (2.26), if only the dimensionless resistance R_N is changed, then to find a peak value of the vibration energy harvesting efficiency, the partial differential of the efficiency with respect to the dimensionless resistance R_N must be set equal to zero, which gives

$$\frac{\partial \eta}{\partial R_N} = 0 \tag{2.29}$$

which leads to

$$
\begin{cases}
R_N^2 = 1 \\
\eta_{max} = \dfrac{\alpha_N^2}{2 + \alpha_N^2}
\end{cases}
\tag{2.30}
$$

From Eq. (2.30), it is observed that η_{max} is a monotonically increasing function of α_N^2. As α_N^2 tends to be very large, the peak energy harvesting efficiency tends to be 100%. In other words, the peak energy harvesting efficiency is limited to 100%.

From Eq. (2.26), if only the dimensionless force factor is changed, it is observed that the partial differential of the efficiency with respect to the dimensionless force factor is not equal to zero, which means there does not exist a peak value of the resonant energy harvesting efficiency when only the dimensionless force factor is changed.

The harvested resonant power can also be derived from Eqs. (2.23) and (2.26) and is given by

$$
\begin{aligned}
P_h &= \frac{1}{2} \cdot A^2 \cdot \frac{M^2}{c} \cdot \frac{R_N \cdot \alpha_N^2}{\alpha_N^4 \cdot R_N^2 + 2 \cdot R_N \cdot \alpha_N^2 + 1 + R_N^2} \\
&= \frac{1}{2} \cdot A^2 \cdot \frac{M^2}{c} \cdot \frac{1 + R_N \cdot \alpha_N^2 + R_N^2}{\alpha_N^4 \cdot R_N^2 + 2 \cdot R_N \cdot \alpha_N^2 + 1 + R_N^2} \cdot \eta
\end{aligned}
\tag{2.31}
$$

The input power can be derived from Eqs. (2.25) and (2.26) and is given by

$$
P_{in} = \frac{1}{2} \cdot M \cdot |\omega^2 \cdot Y|^2 \cdot \mathrm{Re}\left(\frac{Z}{\omega^2 \cdot Y}\right) = \frac{1}{2} \cdot A^2 \cdot \frac{M^2}{c} \cdot \frac{R_N \cdot \alpha_N^2}{\alpha_N^4 \cdot R_N^2 + 2 \cdot R_N \cdot \alpha_N^2 + 1 + R_N^2} \cdot \frac{1}{\eta}
\tag{2.32}
$$

This means that for the SDOF system, given an excitation force amplitude and a mechanical damping, input mechanical power and harvested resonant power depend on the resonant energy harvesting efficiency, dimensionless resistance and dimensionless force factor. The dimensionless resistance and force factor are related to the resistance, the resonance frequency, blocking capacity, and force factor of the piezoelectric insert.

Using Eq. (2.32), the input power with and without a piezoelectric material insert is considered. An equivalent damping reflecting the effect of the electrical load resistance is given by

$$
C_{eq} = 2 \cdot \frac{R_N^2 + \left(1 + R_N \cdot \alpha_N^2\right)^2}{R_N \cdot \alpha_N^2} \cdot \eta \cdot c
\tag{2.33}
$$

The coefficient in front of mechanical damping c in the RHS of Eq. (2.33) reflects the effect of modified system mechanical damping. When the circuit is short circuited, the coefficient becomes unity, the equivalent damping is equal to the mechanical damping, and input power becomes the one of a mechanical system without connection to the load resistor.

This SDOF vibration energy harvester with the constant excitation magnitude would be mainly applied to a machine or a vehicle at constant speeds to reduce vibration similar to a dynamic absorber. For example, a certain amount of selected piezoelectric material could be designed into a harmonic balancer, thus converting a torsion vibration absorber into a vibration energy harvester. It is well known that for a vibration absorber, reducing mechanical damping will improve vibration absorption efficiency, but decrease effective vibration absorption bandwidth. In the same way, for the SDOF vibration energy harvester, reducing mechanical damping will improve vibration energy harvesting efficiency, but decrease effective vibration energy harvesting frequency bandwidth. How to increase the vibration energy harvesting frequency bandwidth will be illustrated in the later chapters.

NOMENCLATURE

A	The acceleration amplitude
S_0	The piezoelectric disk surface area
H	The piezoelectric disk thickness
c	The short circuit mechanical damping coefficient of the single degree of freedom system
e	2.718281828
e_{33}	The piezoelectric constant
ε_{33}^S	The piezoelectric permittivity
K	The short circuit stiffness of the single degree of freedom system
α	The force factor of the piezoelectric material insert
R	The sum of the external load
x	The displacement of the mass (M)
$z, z(t)$	The relative displacement of the mass (M) with respect to the base
$\dot{z}$	The relative velocity of the mass (M) with respect to the base
$\dot{Z}$	The relative velocity amplitude of the mass (M) with respect to the base
$Z, Z(s)$	The Laplace transform of z or $z(t)$
$y, y(t)$	The base excitation displacement
$\ddot{y}$	The base excitation velocity
$Y, Y(s)$	The Laplace transform of y or $y(t)$
i	The square root of -1
t	The variable of the integration or time variable
s	The Laplace variable
C_0	The blocking capacitance of the piezoelectric insert
V	The output voltage of the single degree of freedom system
$|V|$	The amplitude of the output voltage
$\overline{V}(s)$	The Laplace transform of V or $V(t)$
I	The current in the circuit
M	The oscillator mass of the single degree of freedom system
R_N	The normalized resistance

α_N	The normalized force factor
f_n	The natural frequency
ω	The excitation frequency
P_h	The harvested resonant power
$P_{h\ max}$	The maximum harvested resonant power
P_{in}	The input power
η	The resonant energy harvesting efficiency
η_{max}	The maximum resonant energy harvesting efficiency
π	3.1415928
Q_i	The quality factor
C_{eq}	The equivalent damping coefficient
$\frac{d}{dt}$	The first differential
$\frac{d^2}{dt^2}$	The second differential
A	The "state matrix"
B	The "input matrix"
C	The "output matrix"
D	The "feed through matrix"
E	The system eigen matrix
F	The excitation inertia vector
$W_i(i = 1,2,....)$	The state variable

Subscripts

0	The blocking capacity or the piezoelectric disk surface area of the piezoelectric material insert
33	Piezoelectric working mode having the same direction of loading and electric poles
h	Harvested energy
in	Input
max	The maximum value
N	Normalized
eq	Equivalent

Superscripts

-1	Inverse
—	Time average
$\cdot$	The first differential
$\cdot\cdot$	The second differential

Abbreviations

N/A	Not available
SDOF	Single degree of freedom
SL	Single load
RMS	Root mean square

Fig Legends

SL Freq	Simulated results of the single degree of freedom harvester connected to a single load using frequency analysis
SL time	Integration results of the single degree of freedom harvester connected to a single load using time domain simulation

REFERENCES

[1] Williams CB, Yates RB. Analysis of a micro-electric generator for microsystems. Sensors Actuators A Phys 1996;52(1–3):8–11.

[2] Poulin G, et al. Generation of electrical energy for portable devices. Sensors Actuators A Phys 2004;116(3):461–71.

[3] Stephen NG. On energy harvesting from ambient vibration. J Sound Vib 2006; 293(1–2):409–25.

[4] Guyomar D, et al. Energy harvesting from ambient vibrations and heat. J Intelligent Material Syst Struct 2008;20(5):609–24.

[5] Guyomar D, et al. Energy harvester of 1.5 cm^3 giving output power of 2.6 mW with only 1 G acceleration. J Intelligent Material Syst Struct 2010;22(5):415–20.

[6] Wang X, Koss LL. Frequency response function for power, power transfer ratio and coherence. J Sound Vib 1992;155(1):55–73.

[7] Xiao H, Wang X, John S. A dimensionless analysis of a 2DOF piezoelectric vibration energy harvester. Mech Syst Signal Process 2015;58–59(2015):355–75.

[8] Xiao H, Wang X, John S. A multi degree of freedom piezoelectric vibration energy harvester with piezoelectric elements inserted between two nearby oscillators. Mech Syst Signal Process February 2016;68–69:138–54.

[9] Xiao H, Wang X. A review of piezoelectric vibration energy harvesting techniques. Int Rev Mech Eng 2014;8(3). ISSN: 1970-8742:609–20.

[10] Wang X, Lin LW. Dimensionless optimization of piezoelectric vibration energy harvesters with different interface circuits. Smart Mater Struct 2013;22. ISSN: 0964-1726:1–20.

[11] Wang X, Xiao H. Dimensionless analysis and optimization of piezoelectric vibration energy harvester. Int Rev Mech Eng 2013;7(4). ISSN: 1970-8742:607–24.

[12] Cojocariu B, Hill A, Escudero A, Xiao H, Wang X. Piezoelectric vibration energy harvester − design and prototype. In: Proceedings of ASME 2012 international mechanical engineering congress & exposition, ISBN 978-0-7918-4517-2.

[13] Wang X, Liang XY, Hao ZY, Du HP, Zhang N, Qian M. Comparison of electromagnetic and piezoelectric vibration energy harvesters with different interface circuits. Mech Syst Signal Process 2016;72–73:906–24.

[14] Wang X. Coupling loss factor of linear vibration energy harvesting systems in a framework of statistical energy analysis. J Sound Vib 2016;362(2016):125–41.

[15] Wang X, Liang XY, Shu GQ, Watkins S. Coupling analysis of linear vibration energy harvesting systems. Mech Syst Signal Process March 2016;70–71:428–44.

[16] Wang X, John S, Watkins S, Yu XH, Xiao H, Liang XY, et al. Similarity and duality of electromagnetic and piezoelectric vibration energy harvesters. Mech Syst Signal Process 2015;52–53(2015):672–84.

Analysis of Piezoelectric Vibration Energy Harvester System With Different Interface Circuits

3

CHAPTER OUTLINE

3.1 INTRODUCTION

Harvesting energy from the environment is an attractive alternative to battery-operated systems, especially for long-term, low-power, and self-sustaining electronic systems. In addition to the energy generation apparatuses, interface circuits are indispensable elements in these energy harvesting systems to control and regulate the flows of energy. Previously, investigations of the normalized dimensionless power, energy efficiency have been conducted for different electric energy extraction and storage interface circuits to enhance the power outputs of energy harvesters. These interface circuits include the single load resistor interface circuit, standard interface circuit, synchronous electric charge extraction (SECE) interface circuit, series or parallel "synchronous switch harvesting on inductor" (SSHI) circuit. The interface circuits have been compared for their effects on the performance of the single degree of freedom (SDOF) vibration energy harvester. Most of these research papers have discussed optimizations of harvested power with the standard interface circuits while issues in energy harvesting efficiency and dimensionless analyses have only been addressed in limited works.

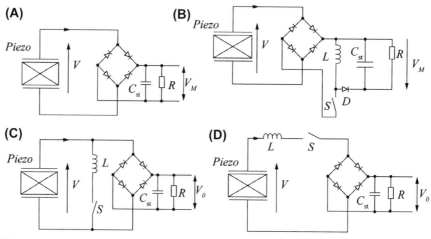

FIGURE 3.1

Extraction and storage interface circuits for vibration energy harvesters, (A) standard; (B) synchronous electric charge extraction (SECE); (C) parallel synchronous switch harvesting on inductor (SSHI); (D) series SSHI circuit [1].

This chapter adopts a dimensionless force factor as one of the performance control variables in the analysis of the electrical and mechanical components which are critically related to the harvested resonant power and efficiency. Since the normalized energy harvesting efficiency provides important design guidelines for vibration-based energy harvesting systems, dimensionless analyses, and optimizations of vibration energy harvesting performance are the focuses of this chapter.

There are four types of energy extraction and storage circuits commonly adopted for energy harvesting devices in the literature: standard energy extraction and storage interface circuit in Fig. 3.1(A), SECE circuit as shown in Fig. 3.1(B), parallel SSHI circuit in Fig. 3.1(C), and series SSHI circuit in Fig. 3.1(D) [2−12]. The standard energy extraction and storage interface circuit with a battery or capacitor is the most frequently used interface circuit in vibration energy harvesting devices.

For the SDOF piezoelectric vibration energy harvester (PVEH) connected with different interface circuits, the displacement, output voltage, the dimensionless harvest resonant power, and resonant energy harvesting efficiency were derived based on Refs. [1,3−13].

3.2 STANDARD INTERFACE CIRCUIT

For the standard energy extraction and storage circuit as shown in Fig. 3.1(A), the voltage V is no more a pure sine wave and has a rectified voltage V_0.

Fig. 3.2 has shown the working principle of a bridge rectification circuit. In Fig. 3.2(A), when the current direction is positive, diode D_1 and D_2 operate in the

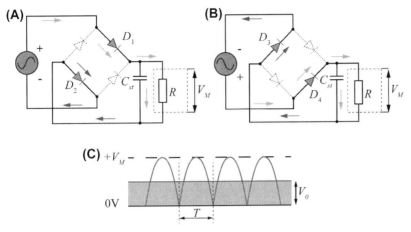

FIGURE 3.2

Working principle of a full cycle of bridge rectification. (A) Positive half-cycle; (B) negative half-cycle; (C) positive output waveform.

circuit, letting the current pass through; in Fig. 3.2(B), for the negative half-cycle, D_3 and D_4 operate, letting the current pass through the circuit. Fig. 3.2(C) has shown the output voltage of the full wave bridge rectification. It is assumed that both the displacement and voltage waves are periodic and change from a trough ($-Z_M$ and $-V_M$) to a crest ($+Z_M$ and $+V_M$) from instant t_0 to $t_0 + T/2$. Integration of Eq. (2.4) with respect to time for a half of the mechanical vibration period ($T/2$) gives

$$\int_{t_0}^{t_0+\frac{T}{2}} C_0 \cdot dV + \int_{t_0}^{t_0+\frac{T}{2}} \frac{V}{R} \cdot dt = \int_{t_0}^{t_0+\frac{T}{2}} \alpha \cdot dz \tag{3.1}$$

Considering $T = 2\pi/\omega$, it gives

$$V_M = \frac{R \cdot \alpha}{R \cdot C_0 \cdot \omega + \frac{\pi}{2}} \cdot \omega \cdot Z_M \tag{3.2}$$

where $Z_M = |Z|$ and $V_M = |\overline{V}|$. The mean harvested power is given by Ref. [14].

$$P_h = \frac{1}{2} \cdot \frac{V_M^2}{R} = \frac{R \cdot \alpha^2 \cdot \omega^2}{2 \cdot \left(R \cdot C_0 \cdot \omega + \frac{\pi}{2}\right)^2} \cdot Z_M^2 \tag{3.3}$$

The system energy equilibrium equation is given by

$$P_{in} - P_C - P_h = 0 \tag{3.4}$$

where P_{in}, P_C, and P_h are the mean input resonant power, mechanical damping dissipation power, and harvested resonant power. From Eqs. (3.3) and (3.4), according to Ref. [14], it gives

$$\frac{1}{2} \cdot \omega \cdot Z_M \cdot M \cdot A = \frac{1}{2} \cdot c \cdot Z_M^2 \cdot \omega^2 + \frac{1}{2} \cdot \frac{R \cdot \alpha^2}{\left(R \cdot C_0 \cdot \omega + \frac{\pi}{2}\right)^2} \cdot \omega^2 \cdot Z_M^2 \qquad (3.5)$$

where $A = \left| \omega^2 \cdot Y \right|$ is the amplitude of the input base excitation acceleration; which gives

$$Z_M = \frac{M \cdot A}{c \cdot \omega + \frac{R \cdot \alpha^2}{\left(R \cdot C_0 \cdot \omega + \frac{\pi}{2}\right)^2} \cdot \omega} \qquad (3.6)$$

According to Ref. [14], the mean input resonant power is given by

$$P_{in} = \frac{1}{2} \cdot \omega \cdot M \cdot |\ddot{y}| \cdot Z_M = \frac{1}{2} \cdot \frac{M^2 \cdot |\ddot{y}|^2}{c + \frac{R \cdot \alpha^2}{\left(R \cdot C_0 \cdot \omega + \frac{\pi}{2}\right)^2}} = \frac{1}{2} \cdot \frac{M^2 \cdot |\ddot{y}|^2}{c} \cdot \frac{1}{1 + \frac{R \cdot \alpha^2}{c}}$$

and

$$\frac{P_{in}}{\left(\frac{M^2 \cdot A^2}{c}\right)} = \frac{\left(R \cdot C_0 \cdot \omega + \frac{\pi}{2}\right)^2}{2 \cdot \left(R \cdot C_0 \cdot \omega + \frac{\pi}{2}\right)^2 + 2 \cdot \frac{R \cdot \alpha^2}{c}} \qquad (3.7)$$

where $\frac{(M \cdot A)^2}{c}$ is the maximum input power which is used as the reference power to normalize the input power. Substituting Eq. (2.21) into Eq. (3.7) gives the dimensionless input resonant power as

$$\frac{P_{in}}{\left(\frac{M^2 \cdot A^2}{c}\right)} = \frac{\left(R_N + \frac{\pi}{2}\right)^2}{2 \cdot \left(R_N + \frac{\pi}{2}\right)^2 + 2 \cdot R_N \cdot \alpha_N^2} \qquad (3.8)$$

Substitution of Eq. (3.6) into Eq. (3.3) gives the harvested power as

$$P_h = \frac{1}{2} \cdot \frac{\frac{R \cdot \alpha^2}{c}}{\left(R \cdot C_0 \cdot \omega + \frac{\pi}{2}\right)^2} \cdot \frac{\frac{M^2 \cdot A^2}{c}}{\left(1 + \frac{\frac{R \cdot \alpha^2}{c}}{\left(R \cdot C_0 \cdot \omega + \frac{\pi}{2}\right)^2}\right)^2}$$

From Eq. (2.21), the dimensionless harvested resonant power becomes

$$\frac{P_h}{\left(\frac{M^2 \cdot A^2}{c}\right)} = \frac{R_N \cdot \alpha_N^2 \cdot \left(R_N + \frac{\pi}{2}\right)^2}{2 \cdot \left[\left(R_N + \frac{\pi}{2}\right)^2 + R_N \cdot \alpha_N^2\right]^2} \qquad (3.9)$$

The efficiency of the standard interface circuit at the resonance frequency is derived from Eqs. (3.8) and (3.9) and is given by

$$\eta = \frac{P_h}{P_{in}} = \frac{R_N \cdot \alpha_N^2}{\left(R_N + \frac{\pi}{2}\right)^2 + R_N \cdot \alpha_N^2} \qquad (3.10)$$

To find the peak efficiency, the partial differential of the efficiency with respect to the dimensionless resistance is set to zero, which gives

$$\begin{cases} R_N = \dfrac{\pi}{2} \\[2ex] \eta_{max} = \dfrac{\alpha_N^2}{2\cdot\pi + \alpha_N^2} \end{cases} \qquad (3.11)$$

When $\alpha_N \to \infty$, $\eta_{max} = 1$, no peak values of the efficiency occur over the ranges of the dimensionless resistance and force factor, there are no peak values of the efficiency. When the dimensionless force factor tends to be very large, while the dimensionless resistance is kept constant, the efficiency will be 100%. When the dimensionless resistance tends to be very large, or when the dimensionless force factor tends be zero, while the dimensionless resistance is kept constant, the efficiency tends to be zero. There is a special case of $R_N \cdot \alpha_N^2 = $ constant. When the dimensionless force factor tends to be very large or when the dimensionless resistance tends to be very small, the efficiency tends to be fixed constant values.

3.3 SYNCHRONOUS ELECTRIC CHARGE EXTRACTION CIRCUIT

For the SECE circuit as shown in Fig. 3.1(B), the charge extraction phase occurs when the electronic switch S is closed: the electrical energy stored on the blocking capacitor C_0 is then transferred into the inductor L. The extraction instants are triggered on the minima and the maxima of the displacement z, synchronously with the mechanical vibration. The inductor L is chosen to get a charge extraction phase duration much shorter than the vibration period. Apart from the extraction phases, the rectifier is blocked and the outgoing current I is null. In this open circuit condition, the mechanical velocity is related to the voltage [2] which is given by

$$\alpha \cdot \dot{z} = C_0 \cdot \dot{V} \qquad (3.12)$$

Integration of Eq. (3.12) with respect to time for the period between t_0 and $t_0 + T/2$ gives,

$$\alpha \cdot \int_{t_0}^{t_0+\frac{T}{2}} \dot{z}\cdot dt = C_0 \cdot \int_{t_0}^{t_0+\frac{T}{2}} \dot{V}\cdot dt \qquad (3.13)$$

This integration gives

$$V_M = \frac{\alpha}{C_0}\cdot Z_M \qquad (3.14)$$

where $Z_M = |Z|$ and $V_M = |\overline{V}|$. The mean harvested power is given according to Ref. [14] as

$$P_h = \frac{1}{2}\cdot\frac{V_M^2}{R} = \frac{1}{2}\cdot\frac{\alpha^2\cdot Z_M^2}{R\cdot C_0^2} \qquad (3.15)$$

From Eqs. (3.4) and (3.15), the system energy equilibrium equation is given by

$$\frac{1}{2} \cdot \omega \cdot M \cdot A \cdot Z_M = \frac{1}{2} \cdot \frac{\alpha^2 \cdot Z_M^2}{R \cdot C_0^2} + \frac{1}{2} \cdot c \cdot Z_M^2 \cdot \omega^2 \qquad (3.16)$$

where $A = |\omega^2 \cdot Y|$ is the input excitation acceleration amplitude.

From Eq. (3.16), the relationship between the input excitation acceleration amplitude and vibration displacement amplitude is derived and given as

$$Z_M = \frac{M \cdot A}{c \cdot \omega + \frac{\alpha^2}{R \cdot \omega \cdot C_0^2}} \qquad (3.17)$$

According to Ref. [14], the mean input resonant power is then given by

$$P_{in} = \frac{1}{2} \cdot \omega \cdot M \cdot A \cdot Z_M = \frac{1}{2} \cdot \frac{(M \cdot A)^2 \cdot \omega}{c \cdot \omega + \frac{\alpha^2}{R \cdot \omega \cdot C_0^2}}$$

and the dimensionless input resonant power is given by

$$\frac{P_{in}}{\frac{(M \cdot A)^2}{c}} = \frac{1}{2} \cdot \frac{1}{1 + \frac{R \cdot \alpha^2}{c \cdot R^2 \cdot \omega^2 \cdot C_0^2}} \qquad (3.18)$$

where $\frac{(M \cdot A)^2}{c}$ is the maximum input power which is used as the reference power. Substituting Eq. (2.21) into Eq. (3.18) gives the normalized dimensionless input resonant power as

$$\frac{P_{in}}{\frac{(M \cdot A)^2}{c}} = \frac{1}{2} \cdot \frac{R_N}{R_N + \alpha_N^2} \qquad (3.19)$$

From Eqs. (3.15) and (3.17), the dimensionless harvested resonant power is then given by

$$\frac{P_h}{\left[\frac{(M \cdot A)^2}{c} \right]} = \frac{1}{2} \cdot \frac{R \cdot \alpha^2}{R^2 \cdot c \cdot C_0^2 \cdot \omega^2} \left(\frac{1}{1 + \frac{R \cdot \alpha^2}{c \cdot R^2 \cdot \omega^2 \cdot C_0^2}} \right)^2 \qquad (3.20)$$

Substituting Eq. (2.21) into Eq. (3.20) gives the normalized dimensionless harvested resonant power as

$$\frac{P_h}{\left[\frac{(M \cdot A)^2}{c} \right]} = \frac{1}{2} \cdot \frac{R_N \cdot \alpha_N^2}{\left(R_N + \alpha_N^2 \right)^2} \qquad (3.21)$$

The efficiency of the SECE circuit at the resonance frequency is derived from Eqs. (3.19) and (3.21), and is given by

$$\eta = \frac{\alpha_N^2}{R_N + \alpha_N^2} \qquad (3.22)$$

No peak values of the efficiency occur over the ranges of the dimensionless resistance and force factor. However, the following condition exists

$$\begin{cases} R_N = 0 \ \text{ or } \ \alpha_N \rightarrow \infty \\ \eta_{max} = 1 \end{cases} \tag{3.23}$$

When the dimensionless resistance is equal to zero or the dimensionless force factor tends to be very large, the efficiency will be 100%. The dimensionless force factor tends to be zero, or the dimensionless resistance tends to be very large, the efficiency tends to be zero.

3.4 PARALLEL SYNCHRONOUS SWITCH HARVESTING ON INDUCTOR CIRCUIT

For a parallel switch harvesting on inductor (parallel SSHI) circuit as shown in Fig. 3.1(C) where the inductor L is in series connected with an electronic switch S, and both the inductor and the electronic switch are connected in parallel with the piezoelectric element electrodes and the diode rectifier bridge. A small part of energy may also be radiated in the mechanical system as the electromagnetic waves due to the resistor, inductor, and capacitor (RLC) electrical circuit oscillation. The radiated energy is quantified by inversion loss. The inversion losses are modeled by the electrical quality factor Q_i of the electrical oscillator. The relation between Q_i and the voltages of the piezoelectric element before and after the inversion process (V_M and V_0, respectively) is given by

$$V_M = V_0 \cdot e^{\frac{-\pi}{2 \cdot Q_i}} \tag{3.24}$$

The electric charge received by the terminal load equivalent resistor R during a half mechanical period $T/2$ is calculated by

$$\int_{t_0}^{t_0+T/2} I \cdot dt + \int_{t_0}^{t_0+T/2} I_S \cdot dt = \frac{V_0}{R} \cdot \frac{T}{2} \tag{3.25}$$

The second integral of the left-hand side of this equation corresponds to the charge stored on the capacitor C_{st} before the voltage inversion plus the charge stored on C_{st} after the inversion, whose expression is given by

$$\int_{t_0}^{t_0+T/2} I_S \cdot dt = C_0 \cdot V_0 \cdot \left(1 + e^{-\frac{\pi}{2 \cdot Q_i}} \right) \tag{3.26}$$

The integration of the piezoelectric outgoing current by

$$\int_{t_0}^{t_0+T/2} I \cdot dt = \int_{t_0}^{t_0+T/2} \alpha \cdot \dot{z} \cdot dt - C_0 \cdot \int_{t_0}^{t_0+T/2} dV \tag{3.27}$$

Substitution of Eqs. (3.26) and (3.27) into Eq. (3.25) gives

$$\int_{t_0}^{t_0+T/2} \alpha \cdot \dot{z} \cdot dt - C_0 \cdot \int_{t_0}^{t_0+T/2} dV + C_0 \cdot V_0 \cdot \left(1 + e^{-\frac{\pi}{2 \cdot Q_i}} \right) = \frac{V_0}{R} \cdot \frac{T}{2} \tag{3.28}$$

Under a harmonic base excitation, it is assumed that both the displacement and voltage waves are periodic and change from a trough ($-Z_M$ and $-V_M$) to a crest ($+Z_M$ and $+V_M$) from instant t_0 to $t_0 + T/2$. Eq. (3.28) becomes

$$2 \cdot \alpha \cdot Z_M = C_0 \cdot V_0 \cdot \left(1 - e^{-\frac{\pi}{2 \cdot Q_i}}\right) + \frac{V_0}{R} \cdot \frac{T}{2}$$

This leads to the expression of the load voltage V_0 as a function of the displacement amplitude Z_M given by

$$V_0 = \frac{2 \cdot \alpha \cdot Z_M \cdot \omega \cdot R}{C_0 \cdot \omega \cdot R \cdot \left(1 - e^{-\frac{\pi}{2 \cdot Q_i}}\right) + \pi} \tag{3.29}$$

where $Z_M = |Z|$ and $V_M = |\overline{V}|$. The mean harvested power is given according to Ref. [14] as

$$P_h = \frac{1}{2} \cdot \frac{V_0^2}{R} = \frac{1}{2} \cdot \frac{4 \cdot \alpha^2 \cdot \omega^2 \cdot R}{\left(C_0 \cdot \omega \cdot R \cdot \left(1 - e^{-\frac{\pi}{2 \cdot Q_i}}\right) + \pi\right)^2} \cdot Z_M^2 \tag{3.30}$$

From Eqs. (3.4) and (3.30), the system energy equilibrium equation is given by

$$\frac{1}{2} \cdot \omega \cdot M \cdot A \cdot Z_M = \frac{1}{2} \cdot \frac{4 \cdot \alpha^2 \cdot \omega^2 \cdot R}{\left(C_0 \cdot \omega \cdot R \cdot \left(1 - e^{-\frac{\pi}{2 \cdot Q_i}}\right) + \pi\right)^2} \cdot Z_M^2 + \frac{1}{2} \cdot c \cdot Z_M^2 \cdot \omega^2 \tag{3.31}$$

where $A = |\omega^2 \cdot Y|$ is the input excitation acceleration amplitude of the base. From Eq. (3.31), the relationship between vibration displacement amplitude and excitation force amplitude is established as

$$Z_M = \frac{M \cdot A}{\frac{4 \cdot \alpha^2 \cdot \omega \cdot R}{\left(C_0 \cdot \omega \cdot R \cdot \left(1 - e^{-\frac{\pi}{2 \cdot Q_i}}\right) + \pi\right)^2} + c \cdot \omega} \tag{3.32}$$

From Eq. (3.32), according to Ref. [14], the mean input resonant power is given by

$$P_{in} = \frac{1}{2} \cdot \omega \cdot M \cdot A \cdot Z_M = \frac{1}{2} \cdot \frac{(M \cdot A)^2}{\frac{4 \cdot \alpha^2 \cdot R}{\left(C_0 \cdot \omega \cdot R \cdot \left(1 - e^{-\frac{\pi}{2 \cdot Q_i}}\right) + \pi\right)^2} + c} \tag{3.33}$$

The dimensionless input resonant power is given by

$$\frac{P_{in}}{\frac{(M \cdot A)^2}{c}} = \frac{1}{2} \cdot \frac{\left(C_0 \cdot \omega \cdot R \cdot \left(1 - e^{-\frac{\pi}{2 \cdot Q_i}}\right) + \pi\right)^2}{4 \cdot \frac{\alpha^2 \cdot R}{c} + \left(C_0 \cdot \omega \cdot R \cdot \left(1 - e^{-\frac{\pi}{2 \cdot Q_i}}\right) + \pi\right)^2} \tag{3.34}$$

where $\frac{(M \cdot A)^2}{c}$ is the reference power. Substituting Eq. (2.21) into Eq. (3.34) gives the normalized dimensionless harvested resonant power as

$$\frac{P_{in}}{\frac{(M \cdot A)^2}{c}} = \frac{1}{2} \cdot \frac{\left(R_N \cdot \left(1 - e^{-\frac{\pi}{2 \cdot Q_i}}\right) + \pi\right)^2}{4 \cdot R_N \cdot \alpha_N^2 + \left(R_N \cdot \left(1 - e^{-\frac{\pi}{2 \cdot Q_i}}\right) + \pi\right)^2} \tag{3.35}$$

Substituting Eq. (3.32) into Eq. (3.30) gives

$$P_h = \frac{1}{2} \frac{V_0^2}{R} = \frac{1}{2} \cdot \frac{4 \cdot \alpha^2 \cdot \omega^2 \cdot R}{\left(C_0 \cdot \omega \cdot R \cdot \left(1 - e^{-\frac{\pi}{2 \cdot Q_i}}\right) + \pi\right)^2} \left(\frac{M \cdot A}{\frac{4 \cdot \alpha^2 \cdot \omega \cdot R}{\left(C_0 \cdot \omega \cdot R \cdot \left(1 - e^{-\frac{\pi}{2 \cdot Q_i}}\right) + \pi\right)^2} + c \cdot \omega}\right)^2 \tag{3.36}$$

The dimensionless harvested resonant power is therefore given by

$$\frac{P_h}{\frac{(M \cdot A)^2}{c}} = \frac{1}{2} \cdot \frac{4 \cdot \frac{\alpha^2 \cdot R}{c}}{\left(C_0 \cdot \omega \cdot R \cdot \left(1 - e^{-\frac{\pi}{2 \cdot Q_i}}\right) + \pi\right)^2} \cdot \left(\frac{1}{\frac{4 \cdot \frac{\alpha^2 \cdot R}{c}}{\left(C_0 \cdot \omega \cdot R \cdot \left(1 - e^{-\frac{\pi}{2 \cdot Q_i}}\right) + \pi\right)^2} + 1}\right)^2 \tag{3.37}$$

Substituting Eq. (2.21) into Eq. (3.37) gives the normalized dimensionless harvested resonant power as

$$\frac{P_h}{\frac{(M \cdot A)^2}{c}} = \frac{1}{2} \cdot \frac{4 \cdot \alpha_N^2 \cdot R_N \cdot \left(R_N \cdot \left(1 - e^{-\frac{\pi}{2 \cdot Q_i}}\right) + \pi\right)^2}{\left[4 \cdot \alpha_N^2 \cdot R_N + \left(R_N \cdot \left(1 - e^{-\frac{\pi}{2 \cdot Q_i}}\right) + \pi\right)^2\right]^2} \tag{3.38}$$

The efficiency of the parallel SSHI circuit at the resonance frequency is derived from Eqs. (3.35) and (3.38), and is given by

$$\eta = \frac{4 \cdot R_N \cdot \alpha_N^2}{\left(R_N \cdot \left(1 - e^{-\frac{\pi}{2 \cdot Q_i}}\right) + \pi\right)^2 + 4 \cdot R_N \cdot \alpha_N^2} \tag{3.39}$$

For variation of the dimensionless resistance, to find the peak efficiency, the partial differential of the efficiency with respect to the dimensionless resistance must be set equal zero, which gives

$$\begin{cases} R_N = \dfrac{\pi}{\left(1 - e^{-\frac{\pi}{2 \cdot Q_i}}\right)} \\[4ex] \eta_{max} = \dfrac{\alpha_N^2}{\left(1 - e^{-\frac{\pi}{2 \cdot Q_i}}\right) \cdot \pi + \alpha_N^2} \end{cases} \tag{3.40}$$

When $\alpha_N \to \infty$, $\eta_{max} = 1$, no peak value of the efficiency occurs over the ranges of the dimensionless resistance and force factor. When the dimensionless force factor tends to be very large, while the dimensionless resistance is kept constant, the efficiency tends to be 100%. When the dimensionless force factor tends to be zero, while the dimensionless resistance is kept constant or when the dimensionless resistance tends to be zero, while the dimensionless force factor is kept constant, the efficiency tends to be zero. When the dimensionless resistance tends to be very large, the efficiency tends to be zero. There is a special case of $R_N \cdot \alpha_N^2 = $ constant, when the dimensionless resistance tends to be zero or when the dimensionless force factor tends to be very large, the efficiency tends to be a constant.

3.5 SERIES SYNCHRONOUS SWITCH HARVESTING ON INDUCTOR CIRCUIT

For a series SSHI circuit as shown in Fig. 3.1(D), in most of the time, the piezoelectric element is in the circuit configuration with the switch opening. Each time the switch is closing, a part of the energy stored in the blocking capacitor C_0 is transferred to the capacitor C_{st} through the rectifier bridge. At those instants, the voltage inversions of V occur. The relation of the piezoelectric voltages V_M and V_m before and after the inversion process, the rectified voltage V_0, and the electrical quality factor Q_i is given by

$$V_m - V_0 = (V_M - V_0) \cdot e^{-\frac{\pi}{2 \cdot Q_i}} \tag{3.41}$$

Under a harmonic base excitation, it is assumed that both the oscillator displacement and output voltage waves are periodic and change from a trough ($-Z_M$ and $-V_M$) to a crest ($+Z_M$ and $+V_M$) from instant t_0 to $t_0 + T/2$. Integration of Eq. (2.4) with respect to time for a half of the mechanical vibration period ($T/2$) gives

$$\int_{t_0}^{t_0+T/2} I \cdot dt = \int_{t_0}^{t_0+T/2} \alpha \cdot \dot{z} \cdot dt - C_0 \cdot \int_{t_0}^{t_0+T/2} dV = 0 \tag{3.42}$$

The open switch circuit evolution of the piezoelectric voltage V between two voltage inversions gives another relation between V_M and V_m as

$$2 \cdot \alpha \cdot Z_M - C_0 \cdot (V_m + V_M) = 0 \quad \text{or} \quad V_M + V_m = \frac{2 \cdot \alpha}{C_0} \cdot Z_M \tag{3.43}$$

Equality of the input energy of the rectification bridge and the energy consumed by the equivalent load resistance R during a semiperiod of vibration $T/2$ leads to

$$V_0 \cdot \int_{t_0}^{t_0+\frac{T}{2}} I \cdot dt = V_0 \cdot C_0 \cdot (V_m + V_M) = \frac{V_0^2}{R} \cdot \frac{\pi}{\omega} \tag{3.44}$$

which leads to

$$C_0 \cdot (V_m + V_M) = \frac{V_0}{R} \cdot \frac{\pi}{\omega} \tag{3.45}$$

Substitution of Eq. (3.43) into Eq. (3.45) gives

$$V_0 = \frac{2 \cdot \alpha \cdot R \cdot \omega}{\pi} \cdot Z_M \tag{3.46}$$

where $Z_M = |Z|$ and $V_M = |\overline{V}|$. The mean harvested power is given according to Ref. [14] as

$$P_h = \frac{1}{2} \cdot \frac{V_0^2}{R} = \frac{1}{2} \cdot \frac{4 \cdot \alpha^2 \cdot R \cdot \omega^2}{\pi^2} \cdot Z_M^2 \tag{3.47}$$

From Eqs. (3.4) and (3.47), according to Ref. [14], the system energy equilibrium equation is given by

$$\frac{1}{2} \cdot \omega \cdot M \cdot A \cdot Z_M = \frac{1}{2} \cdot \frac{4 \cdot \alpha^2 \cdot R \cdot \omega^2}{\pi^2} \cdot Z_M^2 + \frac{1}{2} \cdot c \cdot Z_M^2 \cdot \omega^2 \tag{3.48}$$

where $A = |\omega^2 \cdot Y|$ is the input excitation acceleration amplitude of the base. The relationship between the displacement amplitude and the input excitation acceleration amplitude is then derived from Eq. (3.48) as

$$Z_M = \frac{M \cdot A}{c \cdot \omega + \frac{4 \cdot \alpha^2 \cdot R \cdot \omega}{\pi^2}} \tag{3.49}$$

Substitution of Eq. (3.49) into Eq. (3.47) gives the harvested resonant power amplitude as

$$P_h = \frac{2 \cdot \alpha^2 \cdot R}{\pi^2} \cdot \left(\frac{M \cdot A}{c + \frac{4 \cdot \alpha^2 \cdot R}{\pi^2}} \right)^2 \tag{3.50}$$

From Eq. (2.21), the dimensionless harvested resonant power is then given by

$$\frac{P_h}{\frac{(M \cdot A)^2}{c}} = \frac{2 \cdot \pi^2 \cdot \alpha_N^2 \cdot R_N}{\left(\pi^2 + 4 \cdot \alpha_N^2 \cdot R_N \right)^2} \tag{3.51}$$

where $\frac{(M \cdot A)^2}{c}$ is the reference power. According to Ref. [14], the mean input resonant power is given by

$$P_{in} = \frac{1}{2} \cdot \omega \cdot M \cdot A \cdot Z_M = \frac{1}{2} \cdot \frac{(M \cdot A)^2}{c + \frac{4 \cdot \alpha^2 \cdot R}{\pi^2}} \tag{3.52}$$

Substituting Eq. (2.21) into Eq. (3.52) gives the normalized dimensionless input resonant power as

$$\frac{P_{in}}{\frac{(M \cdot A)^2}{c}} = \frac{1}{2} \cdot \frac{\pi^2}{\pi^2 + 4 \cdot \alpha_N^2 \cdot R_N} \tag{3.53}$$

The efficiency of the series SSHI circuit is derived from Eqs. (3.53) and (3.51) and given by

$$\eta = \frac{4 \cdot R_N \cdot \alpha_N^2}{\pi^2 + 4 \cdot R_N \cdot \alpha_N^2} \tag{3.54}$$

No peak values of the efficiency occur over the ranges of the dimensionless resistance and force factor, although the following condition exists

$$\begin{cases} R_N = \to \infty \ \ \text{or} \ \ \alpha_N \to \infty \\ \eta_{max} = 1 \end{cases} \tag{3.55}$$

When the dimensionless resistance tends to very large, while the dimensionless force factor is kept constant or when the dimensionless force factor tends to be very large, while the dimensionless resistance is kept constant, the efficiency tends to be 100%. When the dimensionless resistance tends to zero, while the dimensionless force factor is kept constant or when the dimensionless force factor tends to be zero, while the dimensionless resistance is kept constant, the efficiency tends to be zero. There is a special case of $R_N \cdot \alpha_N^2 = $ constant, where and when the dimensionless force factor tends to be zero or very large, while the dimensionless resistance tends to be very large or zero, the efficiency tends to be fixed constant values.

3.6 ANALYSIS AND COMPARISON

In a typical condition that the dimensionless resistance and the dimensionless force factor are equal to unity, the resonant energy harvesting efficiencies of the piezoelectric harvesters with the SECE, series SSHI, parallel SSHI, standard interface circuits are 50, 29, 29, and 13%, and the dimensionless harvested resonant power values are 0.125, 0.103, 0.102, and 0.057, respectively. On the other hand, as mentioned in Chapter 2, replacing the electrical interface circuits by a single load resistor, the resonant energy harvesting efficiency is 33% (from Eq. (2.26) and $\alpha_N = R_N = 1$) and the dimensionless harvested resonant power is 0.10. Clearly, in the case, the SECE setup gives the highest efficiency and harvested resonant power and the standard interface setup gives the lowest efficiency and harvested resonant power. It should be noted that the harvested power using the SSHI technique is better than that using the standard interface in the case when α_N^2 is small (for example, $\alpha_N^2 = 1$). The same conclusion can be drawn from Fig. 12 in Ref. [15] where SECE is better than other interfaces only in the case of weak electromechanical coupling. The weak electromechanical coupling is defined as the condition where the effects of mechanical damping and external load resistance on the natural frequency are small, or α_N^2 is small. It is seen from Table 3.1 and Eqs. (2.23) and (2.26) that for a piezoelectric harvester, the resonant energy harvesting efficiency and dimensionless harvested resonant power depend on the system resonant frequency, mechanical damping, load resistance, force factor, and blocking capacitance of the piezoelectric insert. The dimensionless resistance and force factor as defined in Eq. (2.21) are chosen to reflect all these parameters in this work for the system optimization analyses. Table 3.1 lists all

Table 3.1 Formulas for Dimensionless Harvested Resonant Power and the Energy Harvesting Efficiency of a Piezoelectric Harvester for Four Different Interface Circuits

Dimensionless Force Factor $\alpha_N = \sqrt{\frac{\alpha^2}{c\cdot C_0\cdot\omega}}$ **Dimensionless Resistance** $R_N = R\cdot C_0\cdot\omega$	**Dimensionless Displacement Amplitude** $Z_M/|Y|$	**Dimensionless Voltage** $\frac{v_M}{R\cdot\alpha\cdot\omega\cdot Z_M}$	**Dimensionless Harvested Resonant Power** $P_h \Big/ \left[\frac{(M\cdot A)^2}{c}\right]$	**Dimensionless Resonant Energy Harvesting Efficiency** η
Standard interface	$\dfrac{\frac{M\cdot\omega}{c}}{1+\frac{R_N\cdot\alpha_N^2}{\left(R_N+\frac{\pi}{2}\right)^2}}$	$\dfrac{1}{R_N+\frac{\pi}{2}}$	$\dfrac{R_N\cdot\alpha_N^2\cdot\left(R_N+\frac{\pi}{2}\right)^2}{2\cdot\left[\left(R_N+\frac{\pi}{2}\right)^2+R_N\cdot\alpha_N^2\right]^2}$	$\dfrac{R_N\cdot\alpha_N^2}{\left(R_N+\frac{\pi}{2}\right)^2+R_N\cdot\alpha_N^2}$
SECE	$\dfrac{\frac{M\cdot\omega}{c}}{1+\frac{\alpha_N^2}{R_N}}$	$\dfrac{1}{R_N}$	$\dfrac{R_N\cdot\alpha_N^2}{2\cdot\left(R_N+\alpha_N^2\right)^2}$	$\dfrac{\alpha_N^2}{R_N+\alpha_N^2}$
Parallel SSHI	$\dfrac{\frac{M\cdot\omega}{c}}{\dfrac{4\cdot\alpha_N^2\cdot R_N}{\left(R_N\cdot\left(1-e^{-\frac{\pi}{2Q_i}}\right)+\pi\right)^2}+1}$	$R_N\cdot\left(1-e^{-\frac{\pi}{2Q_i}}\right)+\pi$ ÷ 2 $\Rightarrow \dfrac{2}{R_N\cdot\left(1-e^{-\frac{\pi}{2Q_i}}\right)+\pi}$	$\dfrac{2\cdot\alpha_N^2\cdot R_N\cdot\left(R_N\cdot\left(1-e^{-\frac{\pi}{2Q_i}}\right)+\pi\right)^2}{\left[4\cdot\alpha_N^2\cdot R_N+\left(R_N\cdot\left(1-e^{-\frac{\pi}{2Q_i}}\right)+\pi\right)^2\right]^2}$	$\dfrac{4\cdot R_N\cdot\alpha_N^2}{\left(R_N\cdot\left(1-e^{-\frac{\pi}{2Q_i}}\right)+\pi\right)^2+4\cdot R_N\cdot\alpha_N^2}$
Series SSHI	$\dfrac{\frac{M\cdot\omega}{c}}{1+\frac{4\cdot\alpha_N^2\cdot R_N}{\pi^2}}$	$\dfrac{2}{\pi}$	$\dfrac{2\cdot\pi^2\cdot\alpha_N^2\cdot R_N}{\left(\pi^2+4\cdot\alpha_N^2\cdot R_N\right)^2}$	$\dfrac{4\cdot R_N\cdot\alpha_N^2}{\pi^2+4\cdot R_N\cdot\alpha_N^2}$

SECE, synchronous electric charge extraction; SSHI, synchronous switch harvesting on inductor.

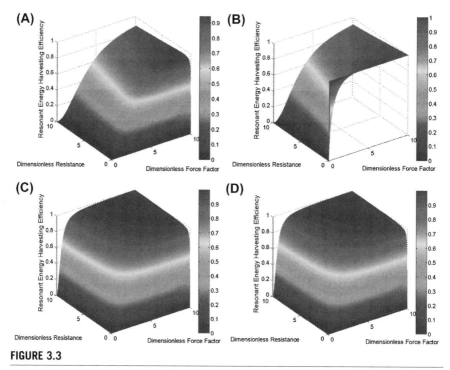

FIGURE 3.3

The resonant energy harvesting efficiency versus dimensionless resistance and force factors for the single degree of freedom piezoelectric harvester connected to the four types of interface circuits. (A) standard; (B) synchronous electric charge extraction; (C) parallel synchronous switch harvesting on inductor (SSHI); (D) series SSHI.

important formulas of dimensionless harvested resonant power and energy harvesting efficiency for the different circuit types.

Shown in Fig. 3.3 are plots of the efficiency versus the dimensionless resistance and dimensionless force factor for the four types of interface circuits. For the cases of standard, series/parallel SSHI interface circuits, it is observed that dimensionless force factor dominates the efficiency as the resonant energy harvesting efficiency gets close to 100 or 0% under large or small dimensionless force factors, respectively. For the SECE circuit, when the dimensionless resistance is kept nonzero constant, dimensionless force factor dominates the efficiency as the resonant energy harvesting efficiency gets close to 100 or 0% under large or small dimensionless force factors, respectively. This is because from the last column of Table 3.1, the energy harvesting efficiency is a monotonically increasing function with respect to α_N^2.

Shown in Fig. 3.4 are plots of the dimensionless harvested resonant power versus the dimensionless resistance and dimensionless force factor for the four types of interface circuits. It is observed from Fig. 3.4 that when the dimensionless force

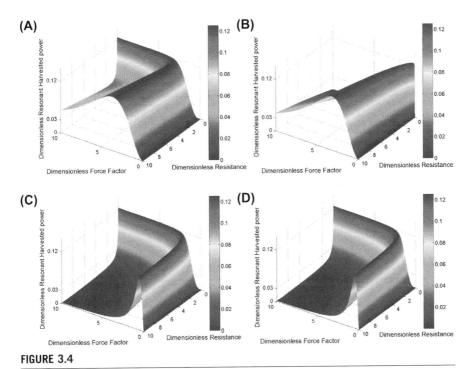

FIGURE 3.4

The dimensionless harvested resonant power versus dimensionless resistance and force factors for the single degree of freedom piezoelectric harvester connected to the four types of interface circuits. (A) standard; (B) synchronous electric charge extraction; (C) Parallel synchronous switch harvesting on inductor (SSHI); (D) series SSHI.

factor becomes either very large or very small, while the dimensionless resistance is kept nonzero constant, the dimensionless harvested power for the four interface circuits goes close to zero. On the other hand, it is observed that when the dimensionless resistance becomes small, while the dimensionless force factor is kept nonzero constant, the dimensionless harvested power goes to close to zero. There is a special case in which $R_N \cdot \alpha_N^2 = $ constant, when the dimensionless resistance becomes very large or very small, while the dimensionless force factor becomes either very small or very large, the dimensionless harvested power tends to be fixed constant values. This has been illustrated in Eq. (49) of Ref. [16].

These can be obtained from the formulas in the last and second last columns of Table 3.1. Physically when the piezoelectric insert is removed, $\alpha_N = 0$, the harvested resonant power is zero. When the dimensionless force factor becomes very large, the dimensionless harvested resonant power would become small and goes to zero. This is because a small mechanical damping results in a large dimensionless force factor according to Eq. (2.21). However, a small mechanical damping would make $\frac{M^2 \cdot A^2}{c}$ large and lead to a small dimensionless harvested resonant power $P_h / \frac{M^2 \cdot A^2}{c}$.

Table 3.2 Formulas for the Peak Dimensionless Harvested Resonant Power and Resonant Energy Harvesting Efficiency of a Piezoelectric Harvester With Four Different Interface Circuits With the Resistance Variation

Dimensionless Force Factor $\alpha_N = \sqrt{\frac{\alpha^2}{c \cdot C_0 \cdot \omega}}$ Dimensionless Resistance $R_N = R \cdot C_0 \cdot \omega$	Optimized Resistance for the Harvested Resonant Power $R_{N_{opt}}$	Peak Dimensionless Harvested Resonant Power $P_{hmax} / \left[\frac{(M \cdot A)^2}{c}\right]$	Optimized Resistance for the Resonant Energy Harvesting Efficiency $R_{N_{opt}}$	Peak Resonant Energy Harvesting Efficiency η_{max}
Standard interface	$\frac{\pi}{2}$	0.125	$\frac{\pi}{2}$	$\frac{\alpha_N^2}{2\cdot\pi + \alpha_N^2}$
SECE	α_N^2	0.125	N/A	N/A
Parallel SSHI	$\dfrac{\pi}{1-e^{-\frac{\pi}{2\cdot Q_i}}}$	0.125	$\dfrac{\pi}{1-e^{-\frac{\pi}{2\cdot Q_i}}}$	$\dfrac{\alpha_N^2}{\left(1-e^{-\frac{\pi}{2\cdot Q_i}}\right)\cdot\pi + \alpha_N^2}$
Series SSHI	$\pi^2/\left(4\cdot\alpha_N^2\right)$	0.125	N/A	N/A

SECE, synchronous electric charge extraction; SSHI, synchronous switch harvesting on inductor.

It is further observed from Figs. 3.3 and 3.4 that it is impossible to obtain a peak harvested resonant power and peak energy harvesting efficiency at a unique pair of the optimal dimensionless resistance and force factors. However, given range limits of the dimensionless resistance and force factor, the dimensionless harvested resonant power or the resonant energy harvesting efficiency may reach its maximum at the range limits of the dimensionless resistance and force factor.

Table 3.2 lists the peak energy harvesting efficiency and peak harvested resonant power with respect to the optimized dimensionless resistance. The peak dimensionless harvesting efficiency and peak harvested resonant power are obtained from $\partial \eta / \partial R_N = 0$ and $\partial\left[P_h/\frac{M^2 \cdot A^2}{c}\right]/\partial R_N = 0$ where η and $P_h/\frac{M^2 \cdot A^2}{c}$ can be calculated from the last and the second last columns in Table 3.1. The peak dimensionless harvested resonant power and peak resonant energy harvesting efficiency are listed in the third and fifth columns of Table 3.2. It is found that there exists no optimized dimensionless resistance for a peak energy harvesting efficiency in the cases of SECE and series SSHI circuits. This is because the energy harvesting efficiency is a monotonically increasing function with respect to the dimensionless resistance for the series SSHI circuit and a monotonically decreasing function with respect to the dimensionless resistance for the SECE circuit. On the other hand, for either the SECE or series SSHI circuits, the peak dimensionless harvested resonant power is calculated as 0.125. For the parallel SSHI and standard interface circuits, the peak dimensionless harvested resonant power is also calculated as 0.125. In other words, for all the four types of extraction circuits, the limit of the peak harvested resonant

Table 3.3 Formulas for the Peak Dimensionless Harvested Resonant Power and Resonant Energy Harvesting Efficiency of a Piezoelectric Harvester Connected With Four Different Interface Circuits With the Force Factor Variation

Dimensionless Force Factor $\alpha_N = \sqrt{\frac{\alpha^2}{c \cdot C_0 \cdot \omega}}$ Dimensionless Resistance $R_N = R \cdot C_0 \cdot \omega$	Optimized Dimensionless Force Factor for the Harvested Resonant Power $\alpha_{N_{opt}}$	Peak Dimensionless Harvested Resonant Power $P_{hmax} / \left[\frac{M^2 \cdot A^2}{c} \right]$	Resonant Energy Harvesting Efficiency at the Peak Power η (%)
Standard interface	$\dfrac{\left(R_N + \frac{\pi}{2}\right)}{\sqrt{R_N}}$	0.125	50
SECE	$\sqrt{R_N}$	0.125	50
Parallel SSHI	$\dfrac{R_N \cdot \left(1 - e^{-\frac{\pi}{2 \cdot Q_i}}\right) + \pi}{2\sqrt{R_N}}$	0.125	50
Series SSHI	$\dfrac{\pi}{2\sqrt{R_N}}$	0.125	50

SECE, *synchronous electric charge extraction*; SSHI, *synchronous switch harvesting on inductor*.

power, P_{hmax} is $0.125\,\frac{M^2 \cdot A^2}{c}$. This conclusion coincides with that in the previous works [1,2,13,15,17]. For the harvester connected to a single load resistor, it is seen from Eqs. (2.27) or (2.28) that the peak harvested resonant power P_{hmax} is also $0.125\frac{M^2 \cdot A^2}{c}$. Therefore it is concluded that the peak harvested resonant power is $0.125\frac{M^2 \cdot A^2}{c}$ in all five external interface circuits analyzed in this work. This implies that the peak harvested resonant power depends on the excitation force magnitude $(M \cdot A)$ and the mechanical damping losses (c) in the structure instead of other parameters.

Table 3.3 lists the peak energy harvesting efficiency and peak harvested resonant power with respect to the optimized dimensionless force factor. The peak dimensionless harvested resonant power and peak resonant energy harvesting efficiency are obtained from $\partial\left[P_h / \frac{M^2 \cdot A^2}{c}\right] / \partial\alpha_N = 0$ and $\partial\eta/\partial\alpha_N = 0$ where η and $P_h / \frac{M^2 \cdot A^2}{c}$ can be calculated from the last and the second last columns in Table 3.1. The peak dimensionless harvested resonant power and its corresponding resonant energy harvesting efficiency are listed in the second last and last columns of Table 3.3. It is seen from Table 3.3 that the peak dimensionless harvested resonant power is 0.125 and the corresponding resonant energy harvesting efficiency is 50% for all four interface circuits under different dimensionless force factors. For the harvester connected to a single load resistor under different dimensionless force factors, it is seen from Eqs. (2.27) or (2.28) that the limit of harvested resonant power P_{hmax} is $0.125\frac{M^2 \cdot A^2}{c}$ and the corresponding energy harvesting efficiency for the single load resistor is 50%. Furthermore, it is observed that optimized dimensionless force factor for the cases of the SECE, series SSHI, and parallel SSHI interface circuits is much less than that for the case of the standard interface circuit. This implies that nonlinear SECE and SSHI techniques could require less piezoelectric material than the standard interface techniques and similar results have been obtained in previous works [1,13].

3.7 CASE STUDY

The developed dimensionless analyses are applied to an experimental case previously published in Ref. [18], where an SDOF piezoelectric harvester was the demonstration example. Table 3.4 lists the key parameters of the energy harvester where slight different resonant frequency of 275 Hz (instead of 277.4 Hz) and load resistance of 30,669.6 Ω (instead of 30 kΩ) are used in this chapter.

Shown in Fig. 3.5(A) and (B) are the output voltages and output power from different interface circuits versus the variable input acceleration, respectively, with all other parameters kept at constant values as listed in Table 3.4. In general, for a particular value of resistance—not the optimal resistance, the output voltage is linearly proportional to the input acceleration in all cases. The discrete square dots in Fig. 3.5 represent the measured harvested power for standard interface circuit [18]. Systems with SECE interface circuit produced highest voltage and power; systems with single load resistor (solid line, SL) had the second largest voltage and power. The next is the series and parallel SSHI circuits as they generated about the same voltage and power while the standard interface circuit (dashed line—stand in) had the smallest voltage and power. Since the simple load resistor interface circuit does not have power losses due to inductance, the system with the simple load resistor interface circuit has higher voltage and power outputs than the system with standard, series SSHI, and parallel SSHI interface circuits.

Fig. 3.6(A) and (B) illustrate the output voltage and harvested resonant power versus the mechanical damping of the piezoelectric harvester with a base excitation

Table 3.4 Parameters of the Single Degree of Freedom Piezoelectric Harvester for the Case Study [18]

Parameter	Measurement Type	Values	Units
M	Indirect	8.4×10^{-3}	kg
c	Direct	0.154	N s/m
K	Indirect	2.5×10^{4}	N/m
C_0	Direct	1.89×10^{-8}	Farads
α	Indirect	1.52×10^{-3}	N/V or Amp s/m
f_n	Indirect	275	Hz
R	Direct	30,669.6	Ω
Q_i	Direct	95	N/A

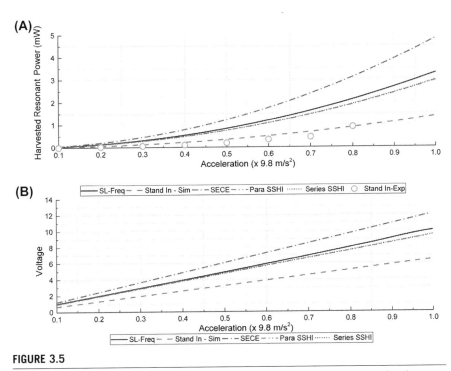

FIGURE 3.5

Output voltage and harvested resonant power versus the input excitation acceleration.
(A) harvested resonant power versus the excitation acceleration; (B) output voltage versus
the excitation acceleration.

acceleration of 9.8 m/s^2 and all other parameters kept at constant values as listed in Table 3.4. As expected, for a particular value of resistance—not the optimal resistance, both outputs decrease as the mechanical damping increases. Harvesters with SECE circuit (dotted line—SECE) has the highest outputs, the single load resistor (solid line, SL) setup has the next highest outputs; both the series SSHI and parallel SSHI circuits are the next; and the standard interface circuit (dashed line—stand in) has the smallest outputs. The high outputs from the SECE interface circuit are the results of the nonlinear amplification circuit. The connection of the electrical circuits to the SDOF piezoelectric harvester equivalently adds resistive-shunt damping to the system which could bring down both the output voltage and harvested power. As such, harvesters with a single load resistor have larger outputs than those with series/parallel SSHI and standard interface circuits as the simple load resistor does not introduce inductance power loss.

Shown in Fig. 3.7(A) and (B) are output voltages and harvested resonant power versus the electrical load resistance under the excitation acceleration of 9.8 m/s^2 and all other parameters kept at constant values as listed in Table 3.4. In the voltage output results in Fig. 3.7(A), output voltages increase as the resistance increases.

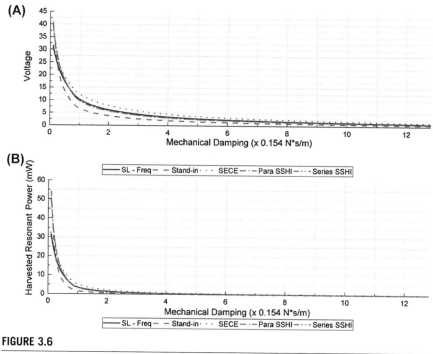

FIGURE 3.6

(A) Output voltage and (B) harvested resonant power versus mechanical damping under the base acceleration of 9.8 m/s^2.

On the other hand, the harvested resonant power can be optimized based on a specific load resistance and the magnitude is related to electrical impedance matching of the piezoelectric material insert and the interface circuit.

Shown in Fig. 3.8 are plots of the harvested resonant power versus the force factor under the excitation acceleration of 9.8 m/s^2 and all other constants fixed as listed in Table 3.4. As the force factor increases, for a particular value of resistance—not the optimal resistance, the harvested power increases initially and reaches an optimal value before decreases. It is observed that a right amount of force factor would produce maximal harvested resonant power. According to Eq. (2.3), the optimal amount of force factor could be achieved by choosing the right piezoelectric material and size. The peak harvested resonant power depends on multiple parameters, including resonant frequency, mechanical damping, load resistance, and the blocking capacity of the piezoelectric material insert. Specifically, peak values of the harvested resonant power for standard, SECE, parallel and Series SSHI interface circuits have are all at 5.555 mW while the value for single load resistor is 4.60 mW. Since the reference power, $\frac{M^2 \cdot A^2}{c}$ is calculated as 44.4 mW, the dimensionless harvested resonant power is then calculated to be 0.125 for the standard, SECE, parallel and series SSHI interface circuits and 0.103 for the single load resistor circuit. These results coincide with those in the previous section and in Table 3.3.

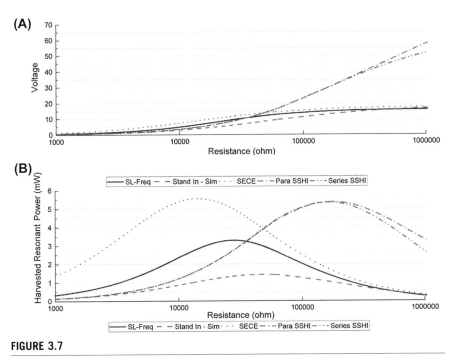

FIGURE 3.7

Output voltage (A) and harvested resonant power (B) versus electric resistance under the base acceleration of 9.8 m/s^2.

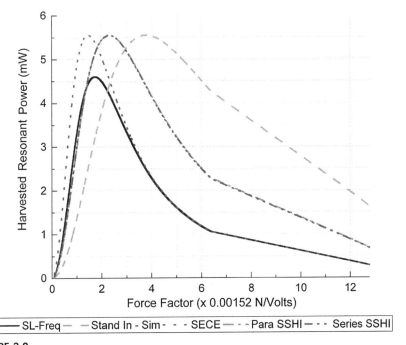

FIGURE 3.8

Harvested resonant power versus force factor under the base acceleration of 9.8 m/s^2.

Table 3.5 Harvested Resonant Power and Energy Harvesting Efficiency for a Single Degree of freedom Piezoelectric Harvester Connected With Different Interface Circuits Under a Base Acceleration of 9.8 m/s^2

Type of Interfaces	SL	Stand In	SECE	Para SSHI	Series SSHI
$R_N = 1$, $\alpha_N = 0.6816$					
Harvesting efficiency	0.19	0.07	0.32	0.16	0.16
Dimensionless harvested power	0.07	0.03	0.11	0.07	0.07
Harvested power (mW)	3.28	1.36	4.81	2.94	2.96

SECE, synchronous electric charge extraction; SL, single load; SSHI, synchronous switch harvesting on inductor.

Substituting parameters given in Table 3.4 into Eqs. (2.23) and (2.26) and into the formulas in the forth and fifth columns of Table 3.1, the dimensionless harvested resonant power and energy harvesting efficiency can be calculated. The dimensionless resistance and force factor are calculated from the parameters listed in Table 3.4 as $R_N = 1$ and $\alpha_N = 0.682$. Table 3.5 summarizes key optimization results under the base acceleration of 9.8 m/s^2. Specifically, the maximum harvested resonant power is 4.81 mW for the case of SECE circuit and 3.28, 2.96, 2.94, and 1.36 mW for the cases of single load resistor, series SSHI, parallel SSHI, and standard interface circuits, respectively.

Furthermore, the SECE circuit provides the best resonant energy harvesting efficiency at 32% and the standard interface provides the lowest efficiency at 7%. The single load resistor circuit provides the resonant energy harvesting efficiency at 19% and both the series and parallel SSHI circuit have the same efficiency at 16%. To further improve the efficiency, the mechanical damping should be reduced.

3.8 SUMMARY

A SDOF PVEH connected to a single load resistor and four types of electrical energy extraction and storage circuits has been studied and investigated based on the dimensionless analyses in case of weak electromechanical coupling. The following conclusions have been reached:

1. By defining a dimensionless resistance and a dimensionless force factor, harvested resonant power and resonant energy harvesting efficiency for the SDOF piezoelectric harvester have been normalized and expressed in a dimensionless form. The dimensionless harvested resonant power and energy harvesting efficiency are found to depend on the harvester resonant frequency, mechanical damping, load resistance, force factor, and blocking capacitance of the piezoelectric insert.

2. There is no unique pair of optimal dimensionless resistance and force factor to achieve a peak of the dimensionless harvested resonant power and the peak energy harvesting efficiency. If the lower and upper limits of the dimensionless resistance and force factor are given, the dimensionless harvested resonant power and resonant energy harvesting efficiency may have their local maximum values.

3. The harvested resonant power may reach a high value of $0.125\frac{M^2 \cdot A^2}{c}$ (one-eighth of the squared applied inertial excitation force magnitude divided by the mechanical damping) with a corresponding resonant energy harvesting efficiency of 50% for all five analyzed interface circuits. This is only valid in the case of weak electromechanical coupling or a small α_N^2. This has been verified by the results in Ref. [16].

4. For the five types of interface circuits under a constant and nonoptimal force factor, there exists an optimized dimensionless resistance for an SDOF piezoelectric harvester to reach the peak dimensionless harvested resonant power. For the cases of the standard and parallel SSHI interface circuits under a constant and nonoptimal force factor, there exists an optimized dimensionless resistance for an SDOF piezoelectric harvester to reach the peak energy harvesting efficiency. For the cases of the SECE and series SSHI interface circuits under a constant and nonoptimal force factor, there is no optimized dimensionless resistance for an SDOF piezoelectric harvester to reach the peak energy harvesting efficiency. For five types of interface circuits with a constant and nonoptimal load resistance, there exists an optimal force factor to reach the peak harvested resonant power. However, there is no optimized force factor to reach the peak energy harvesting efficiency. Excessive or under amount of piezoelectric material insert or size would decrease the harvested resonant power.

5. When the dimensionless resistance and the dimensionless force factor are equal to one, the dimensionless harvested resonant power and energy harvesting efficiency are largest for the harvester connected with SECE circuit and least for that with the standard interface circuit. As such, in this case, it is recommended that the SECE circuit or SSHI circuits should be used for the piezoelectric harvester instead of the standard interface circuit.

NOMENCLATURE

A	The base acceleration amplitude
c	The open circuit mechanical damping of the single degree of freedom system
e	2.718281828
C_0	The blocking capacity of the piezoelectric material insert; the blocking capacitor
C_{st}	Capacitor
f_n	The natural frequency

i	The square root of -1
I	*The current in the circuit*
I_s	The current from the total charge stored on the capacitor C_{st} before and after the voltage inversion
M	The oscillator mass of the single degree of freedom system
P_h	The harvested resonant power (RMS)
P_{in}	The input power (RMS)
P_C	The mechanical damping dissipation power (RMS)
$\frac{P_h}{\frac{(M \cdot A)^2}{c}}$	The dimensionless harvested resonant power (RMS)
$\frac{P_{hmax}}{\frac{(M \cdot A)^2}{c}}$	The maximum dimensionless harvested resonant power (RMS)
Q_i	The quality factor
R	The sum of the external load resistance and piezoelectric material insert
R_N	The dimensionless resistance
$R_{N_{opt}}$	The optimized dimensionless resistance
s	The Laplace variable
t, t_0	Time instances
S	Switch
$Piezo$	Piezoelectric element
$D_i \ (i = 1, 2, 3, 4)$	Diodes
T	The period of the excitation force signal
V	The output RMS voltage of the SDOF system
$\overline{V}, \overline{V}(s)$	The Laplace transform of V or $V(t)$
$V_M, |V|, |\overline{V}|$	The output voltage amplitude of the SDOF system amplitude
V_M	Piezoelectric voltages before the inversion process
V_0	The rectifier voltage amplitude
V_m	The piezoelectric voltage amplitude after the inversion process
y	The base excitation displacement
$\dot{y}$	The base excitation displacement
$\ddot{y}$	The base excitation displacement
Y	The Laplace transform of y or $y(t)$
$Y_M, |Y|$	The modulus or amplitude of the Laplace transform of y or $y(t)$
z	The relative displacement of the mass with respect to the base
$\dot{z}$	The relative velocity of the mass with respect to the base
$\ddot{z}$	The relative acceleration of the mass with respect to the base
$Z_M, |Z|$	The amplitude of the relative displacement, the modulus of the Laplace transform of z or $z(t)$
α	The force factor of the piezoelectric material insert
α_N	The dimensionless force factor
$\alpha_{N_{opt}}$	The optimized dimensionless force factor
η	The resonant energy harvesting efficiency
η_{max}	The maximum resonant energy harvesting efficiency
ω	The excitation radial frequency in radians per second
π	3.1415928
∞	Infinity
$\int_{t_0}^{t_0+\frac{T}{2}}$	The integration from t_0 to $t_0 + \frac{T}{2}$
L	The inductor

Subscripts

0	The rectified or blocking capacity of the piezoelectric material insert
h	Harvested energy
in	Input
M	Amplitude or before the inversion process
max	The maximum value
m	The mode coordinate or after the inversion process
N	Normalized
opt	The optimized value

Superscripts

-1	Inverse
$\overline{}$	Time average
$\cdot$	The first differential
$\cdot\cdot$	The second differential

Special Function

$< >$	Time averaged
$\vert\ \vert$	Modulus or absolute value

Abbreviations

N/A	Not available
Para	Parallel
RMS	Root mean square
SDOF	Single degree of freedom
SECE	Synchronous electric charge extraction
SL	Single load
SSHI	Synchronous switch harvesting on inductor
Standard in	Standard interface circuit

Figure Legends

Para SSHI	Simulated results of the single degree of freedom harvester connected to a parallel synchronous switch harvesting on inductor circuit using frequency domain analysis.
SECE	Simulated results of the single degree of freedom harvester connected to a synchronous electric charge extraction circuit using frequency domain analysis.
Series SSHI	Simulated results of the single degree of freedom harvester connected to a series synchronous switch harvesting on inductor circuit using frequency domain analysis.
SL Freq	Simulated results of the single degree of freedom harvester connected to a single load using frequency domain analysis.
SL time	Simulated results of the single degree of freedom harvester connected to a single load using time domain simulation.
Stand in − Sim	Simulated results of the single degree of freedom harvester connected to a standard interface circuit using frequency domain analysis.
Stand in − Exp	Experimental measurement results of the single degree of freedom harvester connected to a standard interface circuit.

REFERENCES

[1] Lefeuvre E, Badel A, Richard C, Guyomar D. Piezoelectric energy harvesting device optimization by synchronous electric charge extraction. J Intell Mater Syst Struct 2005;16:865—76.

[2] Guyomar D, Sebald GL, Pruvost SB, Lallart M, Khodayari A, Richard C. Energy harvesting from ambient vibrations and heat. J Intell Mater Syst Struct 2009;20:609.

[3] Xiao H, Wang X, John S. A dimensionless analysis of a 2DOF piezoelectric vibration energy harvester. Mech Syst Signal Process 2015;58—59:355—75.

[4] Xiao H, Wang X, John S. A multi degree of freedom piezoelectric vibration energy harvester with piezoelectric elements inserted between two nearby oscillators. Mech Syst Signal Process February 2016;68—69:138—54.

[5] Xiao H, Wang X. A review of piezoelectric vibration energy harvesting techniques. Int Rev Mech Eng 2014;8(3). ISSN: 1970-8742:609—20.

[6] Wang X, Lin LW. Dimensionless optimization of piezoelectric vibration energy harvesters with different interface circuits. Smart Mater Struct 2013;22. ISSN: 0964-1726:1—20.

[7] Wang X, Xiao H. Dimensionless analysis and optimization of piezoelectric vibration energy harvester. Int Rev Mech Eng 2013;7(4). ISSN: 1970-8742:607—24.

[8] Cojocariu B, Hill A, Escudero A, Xiao H, Wang X. Piezoelectric vibration energy harvester — design and prototype. In: Proceedings of ASME 2012. International Mechanical Engineering Congress & Exposition, ISBN 978-0-7918-4517-2.

[9] Wang X, Liang XY, Hao ZY, Du HP, Zhang N, Qian M. Comparison of electromagnetic and piezoelectric vibration energy harvesters with different interface circuits. Mech Syst Signal Process 2016;72—73. [accepted by October 29, 2015].

[10] Wang X. Coupling loss factor of linear vibration energy harvesting systems in a framework of statistical energy analysis. J Sound Vib 2016;362:125—41.

[11] Wang X, Liang XY, Shu GQ, Watkins S. Coupling analysis of linear vibration energy harvesting systems. Mech Syst Signal Process March 2016;70—71:428—44.

[12] Wang X, John S, Watkins S, Yu XH, Xiao H, Liang XY, et al. Similarity and duality of electromagnetic and piezoelectric vibration energy harvesters. Mech Syst Signal Process 2015;52—53:672—84.

[13] Guyomar D, Badel A, Lefeuvre E, Richard C. Toward energy harvesting using active materials and conversion improvement by nonlinear processing. IEEE Trans Ultrason Ferroelectr Freq Control 2005;52(4):584—95.

[14] Wang X, Koss LL. Frequency response function for power, power transfer ratio and coherence. J Sound Vib 1992;155(1):55—73.

[15] Lefeuvre E, Badel A, Richard C, Petit L, Guyomar D. A comparison between several vibration-powered piezoelectric generators for standalone systems. Sens Actuators A 2006;126:405—16.

[16] Shu YC, Lien IC. Analysis of power output for piezoelectric energy harvesting systems. Smart Mater Struct 2006;15:1499—512.

[17] Wu YP, Badel A, Formosa F, Liu WQ, Agbossou AE. Piezoelectric vibration energy harvesting by optimized synchronous electric charge extraction. J Intell Mater Syst Struct 2012;24(3):1—14.

[18] Guyomar D, Sebald G, Kuwano H. Energy harvester of 1.5 cm^3 giving output power of 2.6 mW with only 1 G acceleration. J Intell Mater Syst Struct 2010;22:415.

Analysis of Electromagnetic Vibration Energy Harvesters With Different Interface Circuits

4

CHAPTER OUTLINE

4.1 INTRODUCTION

Electromagnetic vibration energy harvester (EMVEH) with an interface circuit converts vibration energy of a structure into usable electrical energy, which is an attractive renewable energy source. In addition to the energy generation apparatuses, interface circuits are indispensable elements in these energy harvesting systems to control and regulate the flows of energy. Previously, normalized dimensionless harvested power studies, energy efficiency investigations were conducted for piezoelectric vibration energy harvesters (PVEHs) with different electric energy extraction and storage interface circuits to enhance the performance outputs of energy harvesters. The different electric energy extraction and storage interface circuits include single load resistor interface circuit, standard interface circuit, synchronous electric charge extraction (SECE) interface circuit, series or parallel "synchronous switch harvesting on inductor" (SSHI) circuit. EMVEHs were studied, including those connected with synchronized magnetic flux extraction (SMFE) circuit or other RLC circuits.

Most of previous research studies focus on optimized power generation related only to electrical components. This chapter adds the mechanical components in the analyses by using the equivalent force factor as an optimization element as both electrical and mechanical components are critically related to the harvested resonant power and efficiency. Since the normalized energy harvesting efficiency provides important design guidelines for electromagnetic vibration energy harvesting systems, dimensionless analyses and optimizations are the focuses of this chapter. The objectives of this chapter are to establish general dimensionless analytical formulas applicable to different sizes of EMVEHs with different interface circuits. This is significant as EMVEHs have many applications ranging from machines, structures, and vehicles to ocean wave energy conversion devices, self-sustaining microsensors, or remote-sensing systems. The developed dimensionless analysis will be used to evaluate the performance of the EMVEHs and maximize their output power and efficiency.

4.2 DIMENSIONLESS ANALYSIS OF SINGLE DEGREE OF FREEDOM ELECTROMAGNETIC VIBRATION ENERGY HARVESTER CONNECTED WITH A SINGLE LOAD RESISTANCE

For an SDOF electromagnetic vibration harvester with constant vibration magnitude excitation shown in Fig. 4.1, the mechanical system governing equation is given by

$$M \cdot \ddot{z} + c \cdot \dot{z} + K \cdot z = -M \cdot \ddot{y} - B \cdot l \cdot I \tag{4.1}$$

where y is the excitation displacement; M is the mass; c is the short-circuit mechanical damping coefficient; K is the short circuit stiffness of the SDOF electromagnetic–mechanical system; z is the relative displacement of the mass with respect to the base; I is the current in the circuit; B is the magnetic field constant, l is the total length of the wire that constitutes the coil. If physical model of the electromagnetic harvester is assumed to have a voltage source, the electrical system governing equation is given by

$$V = B \cdot l \cdot \frac{\mathrm{d}z}{\mathrm{d}t} - R_e \cdot I - L_e \cdot \frac{\mathrm{d}I}{\mathrm{d}t} \tag{4.2}$$

where V is the voltage; R_e and L_e are respectively the resistance and the inductance of the coil.

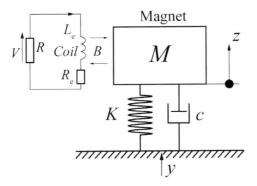

FIGURE 4.1

Schematic of a single degree of freedom electromagnetic vibration energy harvester connected to a single load resistor.

4.3 LAPLACE TRANSFORM AND TRANSFER FUNCTION METHOD

Eqs. (4.1) and (4.2) can be simulated and solved using the transfer function method. From Eq. (4.2), it is derived

$$V + R_e \cdot \frac{V}{R} + \frac{L_e}{R} \cdot \frac{dV}{dt} = B \cdot l \cdot \frac{dz}{dt} \tag{4.3}$$

where R is the electrical resistance of the external load resistor. If the SDOF system connected to a single load resistor is under a sinusoidal displacement excitation, $y = Y(s) \cdot e^{s \cdot t}$, the output voltage and relative oscillator displacement are then assumed to be harmonic and given by:

$$\begin{cases} V = \overline{V}(s) \cdot e^{s \cdot t} \\ y = Y(s) \cdot e^{s \cdot t} \\ z = Z(s) \cdot e^{s \cdot t} \end{cases} \tag{4.4}$$

where $s = i\omega$; ω is the circular frequency; t is time; Y, Z, $\overline{V}$ or $Y(\omega)$, $Z(\omega)$, $\overline{V}(\omega)$ are the Fourier transform of the input displacement, the relative oscillator displacement, and the voltage.

Substituting Eq. (4.4) into Eq. (4.3) gives

$$\overline{V} + \frac{\overline{V}}{R} \cdot (R_e + L_e \cdot s) = B \cdot l \cdot s \cdot Z$$

This is, in fact, the Laplace transform of Eq. (4.3). The transfer function between the relative oscillator displacement and output voltage is then derived and given by

$$\frac{\overline{V}}{Z} = \frac{\alpha \cdot R \cdot s}{1 + \frac{R_e + L_e \cdot s}{R}} \tag{4.5}$$

where $\alpha = Bl/R$ is defined as the equivalent force factor from which Eq. (4.1) can be written as

$$M \cdot \ddot{z} + c \cdot \dot{z} + K \cdot z = -M \cdot \ddot{y} - \alpha \cdot V \tag{4.6}$$

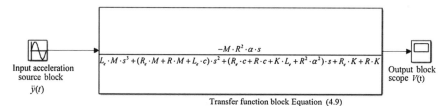

FIGURE 4.2

Simulation schematic for Eq. (4.9) in Matlab Simulink with a sine wave base excitation input and a sinusoidal voltage output at a frequency using the transfer function method.

From Eq. (4.5) it gives the transfer function between the excitation acceleration and output voltage as

$$\frac{\overline{V}}{s^2 \cdot Y} = \frac{\overline{V}}{Z} \cdot \frac{Z}{s^2 \cdot Y} = \frac{\alpha \cdot R \cdot s}{1 + \frac{R_e + L_e \cdot s}{R}} \cdot \frac{Z}{s^2 \cdot Y} \qquad (4.7)$$

Substitution of Eq. (4.4) into Eq. (4.6) gives

$$M \cdot s^2 \cdot (Y + Z) + c \cdot Z \cdot s + K \cdot Z + \alpha \cdot \overline{V} = 0$$

This is a Laplace transfer of Eq. (4.1). The transfer function between the base acceleration and relative oscillator displacement is given by

$$\frac{Z}{s^2 \cdot Y} = \frac{-M \cdot \left[\frac{L_e \cdot s}{R} + 1 + \frac{R_e}{R}\right]}{\left[\frac{R_e + L_e \cdot s + R}{R} \cdot M + \frac{L_e \cdot c}{R}\right] \cdot s^2 + \left[\frac{R_e \cdot c}{R} + c + \frac{K \cdot L_e}{R} + R \cdot \alpha^2\right] \cdot s + \frac{R_e}{R} \cdot K + K} \qquad (4.8)$$

where $s = i\omega$. Substitution of Eq. (4.8) into Eq. (4.7) gives

$$\frac{\overline{V}}{s^2 \cdot Y} = \frac{-M \cdot \alpha \cdot R^2 \cdot s}{L_e \cdot M \cdot s^3 + (R_e \cdot M + R \cdot M + L_e \cdot c) \cdot s^2 + (R_e \cdot c + K \cdot L_e + R \cdot c + R^2 \cdot \alpha^2) \cdot s + R_e \cdot K + R \cdot K}$$
$$(4.9)$$

where the parameters of K, M, c, R, R_e, L_e, B, and l of the system should be measured or identified from experiments. Substitution of the system parameters into Eq. (4.9) allows the simulation using the Matlab Simulink transfer function method as shown in Fig. 4.2. Given input excitation acceleration, the output voltage can be simulated and predicted.

4.4 TIME DOMAIN INTEGRATION METHOD

Eqs. (4.1) and (4.2) can also be simulated and solved using the time domain integration method and written as

$$\begin{cases} \ddot{z} = -\dfrac{c}{M} \cdot \dot{z} - \dfrac{K}{M} \cdot z - \ddot{y} - \dfrac{B \cdot l}{M \cdot R} \cdot V \\[3mm] \dfrac{dV}{dt} = -\dfrac{R + R_e}{L_e} \cdot V + \dfrac{B \cdot l \cdot R}{L_e} \cdot \dfrac{dz}{dt} \end{cases} \qquad (4.10)$$

Eq. (4.10) can be integrated according to the simulation schedule using the Matlab Simulink integration method as shown in Fig. 4.3. Given input excitation acceleration or force and the parameters of K, M, c, R, R_e, L_e, B, and l of the system, the output displacement and voltage can be simulated and predicted.

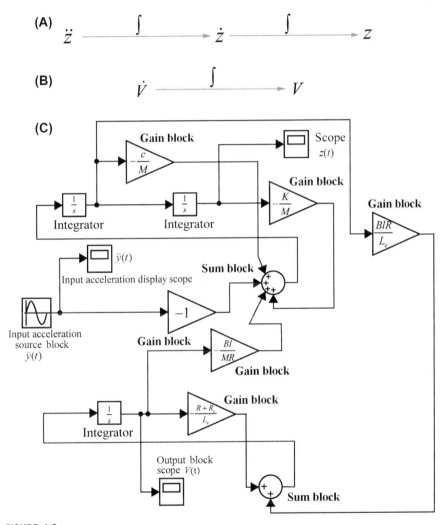

FIGURE 4.3

(A) Relative acceleration integrated to the relative displacement. (B) Derivative of the output voltage integrated to the output voltage. (C) Simulation schematic for Eq. (4.10) with a sine wave base excitation input and a sinusoidal voltage output at a frequency using the time domain integration method.

4.5 STATE SPACE METHOD

Eqs. (4.1) and (4.2) can also be simulated and solved using the state space method. Following Eq. (1.13), it is assumed that

$$
\begin{cases}
W_1 = z(t) \\
W_2 = \dfrac{dz(t)}{dt} \\
W_3 = V(t) \\
W_4 = \dfrac{dV(t)}{dt}
\end{cases}
\tag{4.11}
$$

From Eq. (4.1), it gives

$$
M\frac{dW_2}{dt} + c \cdot W_2 + K \cdot W_1 = -M \cdot \ddot{y} - \frac{B \cdot l}{R} \cdot W_3
$$

From Eq. (4.2), it gives

$$
\frac{R + R_e}{R} \cdot W_3 + \frac{L_e}{R} \cdot W_4 = B \cdot l \cdot W_2
$$

which can be written as

$$
\begin{Bmatrix}
\dfrac{dW_1}{dt} \\
\dfrac{dW_2}{dt} \\
\dfrac{dW_3}{dt} \\
\dfrac{dW_4}{dt}
\end{Bmatrix}
=
\begin{bmatrix}
0 & 1 & 0 & 0 \\
-\dfrac{K}{M} & -\dfrac{c}{M} & \dfrac{B \cdot l}{R \cdot M} & 0 \\
0 & 0 & 0 & 1 \\
0 & B \cdot l & -\dfrac{R + R_e}{R} & -\dfrac{L_e}{R}
\end{bmatrix}
\cdot
\begin{Bmatrix}
W_1 \\
W_2 \\
W_3 \\
W_4
\end{Bmatrix}
+
\begin{Bmatrix}
0 \\
-1 \\
0 \\
0
\end{Bmatrix}
\cdot \{\ddot{y}\}
\tag{4.12}
$$

and

$$
\begin{Bmatrix}
x \\
V
\end{Bmatrix}
=
\begin{bmatrix}
1 & 0 & 0 & 0 \\
0 & 0 & 1 & 0
\end{bmatrix}
\cdot
\begin{Bmatrix}
W_1 \\
W_2 \\
W_3 \\
W_4
\end{Bmatrix}
+
\begin{Bmatrix}
0 \\
0
\end{Bmatrix}
\cdot \{\ddot{y}\}
\tag{4.13}
$$

From Eqs. (4.12) and (4.13), given values of K, M, c, R, B, l, the following matrix can be determined:

$$
\begin{aligned}
\mathbf{A} &= \begin{bmatrix} 0 & 1 & 0 & 0 \\[4pt] -\dfrac{K}{M} & -\dfrac{c}{M} & \dfrac{B \cdot l}{R \cdot M} & 0 \\[8pt] 0 & 0 & 0 & 1 \\[4pt] 0 & B \cdot l & -\dfrac{R + R_e}{R} & -\dfrac{L_e}{R} \end{bmatrix}; \\[12pt]
\mathbf{B} &= \begin{bmatrix} 0 \\ -1 \\ 0 \\ 0 \end{bmatrix}; \\[12pt]
\mathbf{C} &= \begin{bmatrix} 1 & 0 & 0 & 0 \\ 0 & 0 & 1 & 0 \end{bmatrix}; \\[12pt]
\mathbf{D} &= \begin{bmatrix} 0 \\ 0 \end{bmatrix};
\end{aligned}
\tag{4.14}
$$

Given input $\ddot{y}(t)$ (excitation acceleration) and the parameters of K, M, c, R, R_e, L_e, B, and l of the system, the output $z(t)$ and $V(t)$ (response displacement and output voltage) can be simulated and predicted using the Matlab Simulink state space method. The simulation schedule diagram is shown in Fig. 2.4. The system parameters of K, M, c, R, R_e, L_e, B, and l should be measured or identified from experiments.

4.6 FREQUENCY RESPONSE METHOD

Eqs. (4.1) and (4.2) can also be simulated and solved using the frequency response method. Frequency response function is defined as the Fourier transform of the time domain response divided by the Fourier transform of the time domain input. Eqs. (4.1) and (4.2) can be written in a form of matrix:

$$\begin{bmatrix} M & 0 \\ 0 & 0 \end{bmatrix} \cdot \begin{Bmatrix} \ddot{z} \\ \ddot{V} \end{Bmatrix} + \begin{bmatrix} c & 0 \\ -B \cdot l & L_e/R \end{bmatrix} \cdot \begin{Bmatrix} \dot{z} \\ \dot{V} \end{Bmatrix} + \begin{bmatrix} K & B \cdot l/R \\ 0 & (R+R_e)/R \end{bmatrix} \cdot \begin{Bmatrix} z \\ V \end{Bmatrix}$$
$$= \begin{bmatrix} 0 & -M \\ 0 & 0 \end{bmatrix} \cdot \begin{Bmatrix} \dot{y} \\ \ddot{y} \end{Bmatrix} \tag{4.15}$$

From Eq. (4.4), Eq. (4.15) becomes

$$\begin{bmatrix} M \cdot s^2 + c \cdot s + K & B \cdot l/R \\ -B \cdot l \cdot s & (L_e \cdot s + R + R_e)/R \end{bmatrix} \cdot \begin{Bmatrix} Z \\ V \end{Bmatrix} = \begin{Bmatrix} -M \\ 0 \end{Bmatrix} \cdot s^2 \cdot Y \tag{4.16}$$

The displacement and voltage frequency response functions are given by:

$$\begin{Bmatrix} \dfrac{Z}{s^2 \cdot Y} \\ \dfrac{V}{s^2 \cdot Y} \end{Bmatrix} = \begin{bmatrix} M \cdot s^2 + c \cdot s + K & B \cdot l/R \\ -B \cdot l \cdot s & (L_e \cdot s + R + R_e)/R \end{bmatrix}^{-1} \cdot \begin{Bmatrix} -M \\ 0 \end{Bmatrix} = \mathbf{E}^{-1} \cdot \mathbf{F}$$
$$\tag{4.17}$$

where

$$\begin{cases} \mathbf{E} = \begin{bmatrix} M \cdot s^2 + c \cdot s + K & B \cdot l/R \\ -B \cdot l \cdot s & (L_e \cdot s + R + R_e)/R \end{bmatrix} \\ \mathbf{F} = \begin{Bmatrix} -M \\ 0 \end{Bmatrix} \end{cases}$$

Eq. (4.17) can be simulated according to the Matlab code as shown in Fig. 4.4 where $s = i\omega$ and the parameters of K, M, c, R, R_e, L_e, B, and l of the system should be identified and measured from experiments. The amplitudes of the displacement and output voltage over those of input excitation acceleration can be simulated, predicted, and plotted.

The frequency response function amplitude versus frequency of the SDOF can also be obtained from the programming codes shown in Figs. 4.2–4.3 and 2.4, where the excitation amplitude is fixed to be unity, the excitation frequency changes from the low to high to cover the resonant frequency. For each given excitation frequency input of the sine wave signal, the response amplitudes are recorded. The response amplitudes are then plotted against the excitation frequency. The calculated frequency response curves of the output displacement and voltage versus the excitation acceleration should be same as those calculated from Fig. 4.4.

```
**********************************************************************
  CLEAR ALL
  CLC

  C=?;
  K=?;
  L=?;
  M=?;
  R=?;
  RE=?;
  B=?;

  F=0:0.01:300;
  W=2*F*PI;
  FOR N=1:LENGTH(W)
  S=1I*W(N);
  E=[M*S^2+C*S+K B*L/R;-B*L*S (LE*S+R+RE)/R];
  F=[-M;0];
  Z=E\F;
    Z1(N)=Z(1); % X1/Y
    Z2(N)=Z(2); % X2/Y
  END

  FIGURE (1)
  PLOT(F,ABS(Z1))
  XLABEL('FREQUENCY (HZ)')
  YLABEL('DISPLACEMENT RATIO (X1/Y)')

  FIGURE (2)
  PLOT(F,ABS(Z2))
  XLABEL('FREQUENCY (HZ)')
  YLABEL('DISPLACEMENT RATIO (X2/Y)')
**********************************************************************
```

FIGURE 4.4

Matlab codes for simulation of the frequency response functions of the output relative displacement and voltage over the base input excitation acceleration.

From Eq. (4.8), the modulus of the transfer function between the base acceleration and relative oscillator displacement is then given by

$$\left| \frac{Z}{\omega^2 \cdot Y} \right| = \frac{M \cdot \sqrt{\left(\frac{L_e \cdot \omega}{R} \right)^2 + \left(1 + \frac{R_e}{R} \right)^2}}{\sqrt{\left[\frac{R_e}{R} \cdot K + K - \frac{(R_e + R) \cdot M + L_e \cdot c}{R} \cdot \omega^2 \right]^2 + \left[\frac{R_e \cdot c}{R} + c + \frac{K \cdot L_e}{R} + R \cdot \alpha^2 - \frac{L_e \cdot \omega^2 \cdot M}{R} \right]^2 \cdot \omega^2}}$$

(4.18)

If only a weak damping coupling is considered, to simplify Eq. (4.18) and the derivation, the short circuit resonant frequency is approximated by $\omega \approx (K/M)^{0.5}$.

There are two resonant frequencies for the system since the electromagnetic device exhibits both open and short circuit stiffness. If the circuit is switched on or open, the stiffness and damping would become slightly larger which would lead to the open circuit resonance at a slightly larger frequency. The electrical response near the open circuit resonance for the electromagnetic case can be similarly derived from Refs. [1–3,13–18]. Eq. (4.18) is simplified as

$$\left|\frac{Z}{\omega^2 \cdot Y}\right| = \frac{M \cdot \sqrt{\left(\frac{L_e \cdot \omega}{R}\right)^2 + \left(1 + \frac{R_e}{R}\right)^2}}{\sqrt{\left[\frac{L_e \cdot c}{R} \cdot \omega^2\right]^2 + \left[\frac{R_e \cdot c}{R} + c + R \cdot \alpha^2\right]^2 \cdot \omega^2}} \tag{4.19}$$

From Eq. (4.19), the modulus of the transfer function between the base acceleration and output voltage is given by

$$\left|\frac{\overline{V}}{\omega^2 \cdot Y}\right| = \left|\frac{\overline{V}}{Z}\right| \cdot \left|\frac{Z}{\omega^2 \cdot Y}\right| = \left|\frac{B \cdot l \cdot i \cdot \omega}{1 + \frac{R_e + L_e \cdot i \cdot \omega}{R}}\right| \cdot \left|\frac{Z}{\omega^2 \cdot Y}\right| \tag{4.20}$$

Substitution of Eq. (4.19) into Eq. (4.20) gives the modulus of the transfer function between the base acceleration and output voltage as:

$$\left|\frac{\overline{V}}{\omega^2 \cdot Y}\right| = \frac{\frac{M \cdot B \cdot l}{c}}{\sqrt{\left(\frac{L_e \cdot \omega}{R}\right)^2 + \left[\frac{R_e}{R} + 1 + \frac{R \cdot \alpha^2}{c}\right]^2}} \tag{4.21}$$

The ratio of the root mean squared (RMS) harvested resonant power over the squared base acceleration magnitude is given according to Ref. [4] as

$$\frac{P_h}{|\omega^2 \cdot Y|^2} = \frac{1}{2} \cdot \frac{\left|\frac{\overline{V}}{\omega^2 \cdot Y}\right|^2}{R} = \frac{1}{2} \cdot \frac{M^2}{c} \cdot \frac{\frac{R \cdot \alpha^2}{c}}{\left(\frac{L_e \cdot \omega}{R}\right)^2 + \left[\frac{R_e}{R} + 1 + \frac{R \cdot \alpha^2}{c}\right]^2} \tag{4.22}$$

Given the input excitation acceleration and the parameters of K, M, c, R, R_e, L_e, B, and l of the system, Eqs. (4.21) and (4.22) gives the frequency response amplitudes of the output voltage and harvested resonant power over the input acceleration amplitudes. The system parameters should be measured and identified from experiments. If the input acceleration is field vibration acceleration measurement data, the output voltage and harvested resonant power of the SDOF vibration energy harvester can be predicted from the input data using this approach.

4.7 DIMENSIONLESS FREQUENCY RESPONSE ANALYSIS AND HARVESTING PERFORMANCE OPTIMIZATION

To conduct dimensionless frequency response analysis, it is assumed,

$$\begin{cases} R_N = \dfrac{R}{L_e \cdot \omega} \\[2mm] \alpha_N = \sqrt{\dfrac{\alpha^2 \cdot L_e \cdot \omega}{c}} = \dfrac{B \cdot l}{R} \cdot \sqrt{\dfrac{L_e \cdot \omega}{c}} \end{cases} \tag{4.23}$$

where $\alpha = Bl/R$; the squared dimensionless force factor $\alpha_N^2 = \frac{\alpha^2}{\left(\frac{c}{L_e \cdot \omega}\right)}$ is equivalent to one half of the electromagnetic–mechanical coupling factor $k_e^2 = \frac{\Theta^2}{K \cdot \frac{1}{L_e \cdot \omega^2}}$ divided by the mechanical damping ratio $\alpha_N^2 = \frac{k_e^2}{2\cdot\zeta}$. This is similar to Eq. (36) in Ref. [5], where the blocking capacitance of the piezoelectric insert C_0 in PVEH corresponds to $\frac{1}{L_e\cdot\omega^2}$ in EMVEH. α_N reflects the effect of the electromechanical coupling and corresponds to the coupling coefficient k in Ref. [6]. R_N is the dimensionless resistance reflecting the effect of external load which is equal to a double of the load coefficient ξ_e defined in Ref. [6].

From Eq. (4.23), Eqs. (4.21) and (4.22) are normalized as

$$
\begin{cases}
\dfrac{\frac{|\overline{V}|}{M\cdot A}}{\alpha} = \dfrac{R_N\cdot\alpha_N^2}{\sqrt{\dfrac{1}{R_N^2} + \left[\dfrac{R_e}{R} + 1 + R_N\cdot\alpha_N^2\right]^2}} \\[4ex]
\dfrac{P_h}{\left(\dfrac{M^2\cdot A^2}{c}\right)} = \dfrac{1}{2}\cdot\dfrac{R_N\cdot\alpha_N^2}{\left(\dfrac{1}{R_N}\right)^2 + \left[\dfrac{R_e}{R} + 1 + R_N\cdot\alpha_N^2\right]^2}
\end{cases}
\tag{4.24}
$$

where $A = |\omega^2\cdot Y|$ is the input excitation acceleration amplitude of the base. $M\cdot A/\alpha$ is the reference voltage amplitude; $M^2\cdot A^2/c$ is the maximum input power and is used as the reference power. It is assumed that the resistance of the coil wire is very small, comparing with the external load resistance, that is, $R_e \ll R$ or $\frac{R_e}{R}\approx 0$. For example, in the experimental device in Ref. [7], the coil resistance R_c was $0.3\,\Omega$, coil inductance L_c was $2.4\,$mH, load resistance R_L was $10{,}000\,\Omega$. It is obvious that the coil resistance is much less than that of load resistance. Eq. (4.24) therefore becomes

$$
\begin{cases}
\dfrac{\frac{|\overline{V}|}{M\cdot A}}{\alpha} \approx \dfrac{R_N\cdot\alpha_N^2}{\sqrt{\dfrac{1}{R_N^2} + \left[1 + R_N\cdot\alpha_N^2\right]^2}} \\[4ex]
\dfrac{P_h}{\left(\dfrac{M^2\cdot A^2}{c}\right)} \approx \dfrac{1}{2}\cdot\dfrac{R_N\cdot\alpha_N^2}{\left(\dfrac{1}{R_N}\right)^2 + \left[1 + R_N\cdot\alpha_N^2\right]^2}
\end{cases}
\tag{4.25}
$$

For the SDOF electromagnetic–mechanical system shown in Fig. 4.1, the mean input power (RMS) at the resonance frequency is given according to Ref. [4] as

$$
P_{in} = -\frac{1}{2}\cdot M\cdot A^2\cdot\mathrm{Re}\left(\frac{Z}{i\omega\cdot Y}\right)
\tag{4.26}
$$

To simplify the optimization process, it is assumed that at the resonance, $\omega \approx (K/M)^{0.5}$. Considering $s = i\omega$ and substituting Eqs. (4.4) and (4.8) into Eq. (4.26), the dimensionless input power at the resonance can be derived as

$$
\frac{P_{in}}{A^2} = \frac{1}{2}\cdot\frac{M^2}{c}\cdot\frac{\frac{L_e}{R}\cdot\frac{L_e\cdot\omega^2}{R} + \left[\frac{R_e}{R} + 1 + \frac{R\cdot\alpha^2}{c}\right]\left[1 + \frac{R_e}{R}\right]}{\left[\frac{L_e\cdot\omega}{R}\right]^2 + \left[\frac{R_e}{R} + 1 + \frac{R\cdot\alpha^2}{c}\right]^2}
\tag{4.27}
$$

Substituting Eq. (4.23) into Eq. (4.27) gives the dimensionless input resonant power as

$$
\frac{P_{\text{in}}}{\left(\frac{M^2 \cdot A^2}{c}\right)} = \frac{1}{2} \cdot \frac{\frac{1}{R_N^2} + \left[\frac{R_e}{R} + 1 + R_N \cdot \alpha_N^2\right]\left[1 + \frac{R_e}{R}\right]}{\frac{1}{R_N^2} + \left[\frac{R_e}{R} + 1 + R_N \cdot \alpha_N^2\right]^2}
\tag{4.28}
$$

Again, if $R_e \ll R$ or $\frac{R_e}{R} \approx 0$, Eq. (4.28) becomes

$$
\frac{P_{\text{in}}}{\left(\frac{M^2 \cdot A^2}{c}\right)} \approx \frac{1}{2} \cdot \frac{\frac{1}{R_N^2} + \left[1 + R_N \cdot \alpha_N^2\right]}{\frac{1}{R_N^2} + \left[1 + R_N \cdot \alpha_N^2\right]^2}
\tag{4.29}
$$

The electric power generation efficiency is calculated from Eqs. (4.25) and (4.29) and therefore given by

$$
\eta = \frac{\frac{P_h}{\left(\frac{M^2 \cdot A^2}{c}\right)}}{\frac{P_{\text{in}}}{\left(\frac{M^2 \cdot A^2}{c}\right)}} = \frac{R_N \cdot \alpha_N^2}{\frac{1}{R_N^2} + \left[1 + R_N \cdot \alpha_N^2\right]}
\tag{4.30}
$$

It is seen from Eqs. (4.25) and (4.30) that for given dimensionless electrical resistance and force factor, the dimensionless harvested resonant power and energy harvesting efficiency of the system can be predicted. For the EMVEH, the dimensionless harvested resonant power and energy harvesting efficiency only depends on the system resonance frequency, mechanical damping, load resistance, magnet field constant, coil wire length, coil self-inductance, and coil resistance. Eqs. (4.25) and (4.30) are very important dimensionless formulas for calculation of the dimensionless harvested resonant power and energy harvesting efficiency from only two characteristic parameters of the dimensionless resistance and equivalent force factor. Eqs. (4.25) and (4.30) are applicable to many electromagnetic vibration energy harvesting systems of different sizes.

The above analysis is based on the approximation assumption of the weak damping coupling or the short circuit resonance. To consider the case of the strong damping coupling, a similar analysis can be conducted under the assumption of the open circuit resonance following Section 4 in Ref. [5].

The dimensionless energy harvesting efficiency is plotted in Fig. 4.5 for the dimensionless electric load resistance and dimensionless equivalent force factor changing from 0 to 10. It is seen from Fig. 4.5 that there is no local efficiency peak with respect to the dimensionless resistance, R_N, or equivalent force factor, α_N. It can be seen that when the dimensionless equivalent force factor becomes very large and the dimensionless resistance is kept nonzero constant, the efficiency goes higher and gets close to 100%. Physically, the dimensionless equivalent force factor as defined in Eq. (4.23) is affected by various parameters. For example, a small damping coefficient will result in a large dimensionless equivalent force factor and high energy efficiency. On the other hand, when the dimensionless electric load resistance becomes large, while the dimensionless equivalent force factor is kept

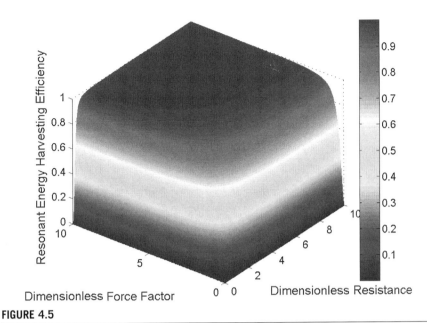

FIGURE 4.5

Resonant energy harvesting efficiency versus dimensionless resistance and equivalent force factors for the single degree of freedom electromagnetic harvester connected to a load resistor.

finite nonzero constant, the efficiency also tends to be 100%. This is different from the trend in Fig. 2 in Ref. [8]. This is because when the dimensionless electric load resistance becomes very large, while the dimensionless equivalent force factor is kept constant, the dimensionless harvested power as shown in Eq. (4.25) tends to be very small while the dimensionless input power as shown in Eq. (4.29) also tends to be very small. When either the dimensionless resistance becomes very small while the dimensionless force factor is kept constant, or the dimensionless force factor becomes very small while the dimensionless resistance is kept constant, the efficiency goes to zero eventually. There is a special case of $R_N \cdot \alpha_N^2 = $ constant where and when the dimensionless resistance tends to be zero and the dimensionless force factor tends to be a very large value, the efficiency tends to be a zero value. When the dimensionless force factor tends to zero and the dimensionless resistance tends to be a very large value, the efficiency tends to be a fixed constant value. When

$R_N = \alpha_N = 1$, $\eta = 33.3\%$ and $P_h \Big/ \left(\frac{M^2 \cdot A^2}{c} \right) = 0.1$. In other words, the energy effi-

ciency of the EMVEH is 33.3% under unity of both the dimensionless resistance and the force factor, and the dimensionless harvested resonant power is 0.1.

The dimensionless harvested resonant power of the energy harvester can be calculated using Eq. (4.25) when both the dimensionless electric load resistance

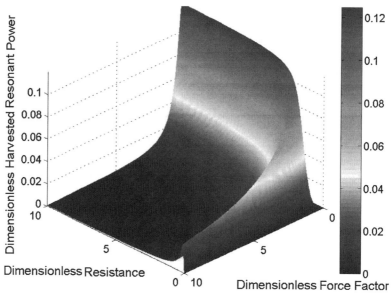

FIGURE 4.6

Dimensionless harvested resonant power versus dimensionless resistance and equivalent force factors for the single degree of freedom electromagnetic harvester connected to a load resistor.

and the dimensionless force factor are changed from 0 to 10, as plotted in Fig. 4.6. It is seen from Eq. (4.25) and Fig. 4.6 that when either the dimensionless resistance becomes very large while the dimensionless force factor is kept constant or/and when the dimensionless force factor becomes very large while the dimensionless resistance is kept constant, the dimensionless harvested resonant power goes to zero. When either the dimensionless resistance becomes very small while the dimensionless force factor is kept constant or/and the dimensionless force factor becomes very small while the dimensionless resistance is kept constant, the dimensionless harvested resonant power goes to zero.

There is a special case of $R_N \cdot \alpha_N^2 = $ constant, where and when the dimensionless resistance tends to be zero and the dimensionless force tends to be a very large value, the dimensionless harvested resonant power tends to a zero value. When the dimensionless resistance tends to be a very large value and the dimensionless force factor tends to be a zero value, the dimensionless harvested resonant power tends to a fixed constant.

To find the peak value of the dimensionless harvested resonant power, the partial differential of the dimensionless harvested resonant power with respect to the dimensionless resistance must be equal to zero, which leads to

$$
\begin{cases}
R_N^2 = \dfrac{1 + \sqrt{1 + 12 \cdot \alpha_N^4}}{2 \cdot \alpha_N^4} \\[3em]
\dfrac{P_{h\,max}}{\left(\dfrac{M^2 \cdot A^2}{c}\right)} = \dfrac{1}{2} \cdot \dfrac{\sqrt{\dfrac{1 + \sqrt{1 + 12 \cdot \alpha_N^4}}{2}}}{\left(\dfrac{2 \cdot \alpha_N^4}{1 + \sqrt{1 + 12 \cdot \alpha_N^4}}\right) + \left[1 + \sqrt{\dfrac{1 + \sqrt{1 + 12 \cdot \alpha_N^4}}{2}}\right]^2}
\end{cases}
\tag{4.31}
$$

From Eq. (4.31), it is observed that $P_{h\,max}\big/\frac{M^2 \cdot A^2}{c}$ is a monotonically decreasing function of α_N^2. When the dimensionless force factor tends to be zero, the limit of the peak dimensionless harvested resonant power tends to be 1/8. In other words, the limit of the peak harvested resonant power $P_{h\,max}$ is $\frac{M^2 \cdot A^2}{8 \cdot c}$ which is one-eighth of the reference power regardless of the dimensionless resistance and force factor where the reference power amplitude is defined as $\frac{M^2 \cdot A^2}{c}$.

From Eq. (4.25), if only the dimensionless force factor is changed, then to find the peak value of the dimensionless harvested resonant power, the partial differential of the dimensionless harvested resonant power with respect to the dimensionless force factor must be equal to zero, which gives

$$
\begin{cases}
\alpha_N^4 = \dfrac{R_N^2 + 1}{R_N^4} \\[2em]
\dfrac{P_{h\,max}}{\left(\dfrac{M^2 \cdot A^2}{c}\right)} = \dfrac{1}{2} \cdot \dfrac{R_N \sqrt{R_N^2 + 1}}{1 + \left[R_N + \sqrt{R_N^2 + 1}\right]^2}
\end{cases}
\tag{4.32}
$$

From Eq. (4.32), it is observed that $P_{h\,max}\big/\frac{M^2 \cdot A^2}{c}$ is a monotonically increasing function of R_N. When the dimensionless resistance tends to be very large, the limit of the peak dimensionless harvested resonant power tends to be 1/8. In other words, the limit of the peak harvested resonant power $P_{h\,max}$ is $\frac{M^2 \cdot A^2}{8 \cdot c}$ under very large α_N, which is one-eighth of the reference power regardless of the dimensionless force factor and resistance where the reference power is the squared magnitude of the applied force divided by the mechanical damping.

The peaks of the dimensionless harvested resonant power can be found when either only the dimensionless resistance or only the dimensionless force factor changes as shown in Eqs. (4.31) and (4.32). If the dimensionless resistance and force factor are simultaneously changed, the partial differential of the dimensionless harvested resonant power with respect to the dimensionless resistance and force factor must be equal to zero, which gives

$$\begin{cases} R_{\mathrm{N}} = \sqrt{\dfrac{1 + \sqrt{1 + 12 \cdot \alpha_{\mathrm{N}}^{4}}}{2 \cdot \alpha_{\mathrm{N}}^{4}}} \\[2em] \alpha_{\mathrm{N}}^{4} = \dfrac{R_{\mathrm{N}}^{2} + 1}{R_{\mathrm{N}}^{4}} \end{cases} \tag{4.33}$$

Eq. (4.33) does not have a solution. This means there is no unique pair of the dimensionless resistance and force factor which produces the peak value of the dimensionless harvested resonant power across a full range of the two variables. If variable range limits are specified for the dimensionless resistance and force factor, the dimensionless harvested resonant power could reach its maximum value at the range limits.

Using Eq. (4.30) with only the dimensionless resistance being varied, the partial differential of the efficiency with respect to the dimensionless resistance is found not equal to zero,

$$\frac{\partial \eta}{\partial R_{\mathrm{N}}} \neq 0$$

which means there is no peak efficiency in this case. Using Eq. (4.30) with only the dimensionless force factor being varied, the partial differential of the efficiency with respect to the dimensionless force factor is found not equal to zero,

$$\frac{\partial \eta}{\partial \alpha_{\mathrm{N}}} \neq 0$$

which means there is no peak efficiency in this case. There is no unique pair of the dimensionless resistance and force factor which produces the peak value of the harvesting efficiency across a full range of the two dimensionless variables.

4.8 DIMENSIONLESS ANALYSIS OF ELECTROMAGNETIC VIBRATION ENERGY HARVESTERS CONNECTED WITH ENERGY EXTRACTION AND STORAGE CIRCUITS

Four types of energy extraction and storage circuits are commonly adopted for energy harvesting devices in the literature: the standard energy extraction and storage interface circuit in Fig. 3.1(A), the SECE circuit in Fig. 3.1(B), the parallel synchronous switch harvesting on inductor (parallel SSHI) circuit in Fig. 3.1(C), and the series synchronous switch harvesting on inductor (series SSHI) circuit in Fig. 3.1(D) [9].

For the SDOF electromagnetic harvester as shown in Fig. 4.1, the displacement, output voltage, dimensionless harvested resonant power, and resonant energy harvesting efficiency for the four interface circuits are derived based on Refs. [8,10,13–18].

4.8.1 STANDARD INTERFACE CIRCUIT

For the standard energy extraction and storage circuit as shown in Fig. 3.1(A), the voltage V is no longer a pure sine wave and has a rectified voltage V_0. It is assumed that both the displacement and voltage waves are periodic and change from a trough ($-Z_M$ and $-V_M$) to a crest ($+Z_M$ and $+V_M$) from instant t_0 to $t_0 + T/2$. Integration of Eq. (4.3) with respect to time for a half of the mechanical vibration period ($T/2$) gives

$$\int_{t_0}^{t_0+\frac{T}{2}} V \cdot dt + \int_{t_0}^{t_0+\frac{T}{2}} R_e \cdot \frac{V}{R} dt + \int_{t_0}^{t_0+\frac{T}{2}} \frac{L_e}{R} dV = \int_{t_0}^{t_0+\frac{T}{2}} B \cdot l dz \tag{4.34}$$

Considering $T = 2\pi/\omega$, this gives

$$V_M = \frac{B \cdot l \cdot \omega}{\frac{\pi}{2} + \frac{\pi}{2} \cdot \frac{R_e}{R} + \frac{L_e \cdot \omega}{R}} \cdot Z_M \tag{4.35}$$

where $Z_M = |Z|$ and $V_M = |\overline{V}|$. The mean harvested power is given according to Ref. [4] as

$$P_h = \frac{1}{2} \cdot \frac{V_M^2}{R} = \frac{B^2 \cdot l^2 \cdot \omega^2}{2 \cdot \left(\frac{\pi}{2} + \frac{\pi}{2} \cdot \frac{R_e}{R} + \frac{L_e \cdot \omega}{R}\right)^2 \cdot R} \cdot Z_M^2 \tag{4.36}$$

The system energy equilibrium equation is given by

$$P_{in} - P_D - P_h = 0 \tag{4.37}$$

where P_{in}, P_D, and P_h are the mean input resonant power, mechanical damping dissipation power, and harvested resonant power. From Eqs. (4.36) and (4.37), according to Ref. [4], it gives

$$\frac{1}{2} \cdot \omega \cdot Z_M \cdot M \cdot A^2 = \frac{1}{2} \cdot c \cdot Z_M^2 \cdot \omega^2 + \frac{1}{2} \cdot \frac{B^2 \cdot l^2}{\left(\frac{\pi}{2} + \frac{\pi}{2} \cdot \frac{R_e}{R} + \frac{L_e \cdot \omega}{R}\right)^2 \cdot R} \cdot \omega^2 \cdot Z_M^2 \tag{4.38}$$

where $A = |\omega^2 \cdot Y|$ is the input excitation acceleration amplitude of the base. From Eq. (4.38), the following equation is derived as

$$Z_M = \frac{M \cdot A^2}{c \cdot \omega + \frac{B^2 \cdot l^2}{\left(\frac{\pi}{2} + \frac{\pi}{2} \cdot \frac{R_e}{R} + \frac{L_e \cdot \omega}{R}\right)^2 \cdot R} \cdot \omega} \tag{4.39}$$

According to Ref. [4], the mean input resonant power is given by

$$P_{in} = \frac{1}{2} \cdot \omega \cdot M \cdot A \cdot Z_M = \frac{1}{2} \cdot \frac{M^2 \cdot A^2}{c + \frac{B^2 \cdot l^2}{\left(\frac{\pi}{2} + \frac{\pi}{2} \cdot \frac{R_e}{R} + \frac{L_e \cdot \omega}{R}\right)^2 \cdot R}} = \frac{1}{2} \cdot \frac{M^2 \cdot A^2}{c} \cdot \frac{1}{1 + \frac{B^2 \cdot l^2}{\left(\frac{\pi}{2} + \frac{\pi}{2} \cdot \frac{R_e}{R} + \frac{L_e \cdot \omega}{R}\right)^2 \cdot c \cdot R}}$$

and

$$\frac{P_{\text{in}}}{\left(\frac{M^2 \cdot A^2}{c}\right)} = \frac{1}{2} \cdot \frac{\left(\frac{\pi}{2} + \frac{\pi}{2} \cdot \frac{R_e}{R} + \frac{L_e \cdot \omega}{R}\right)^2 c \cdot R}{\left(\frac{\pi}{2} + \frac{\pi}{2} \cdot \frac{R_e}{R} + \frac{L_e \cdot \omega}{R}\right)^2 \cdot c \cdot R + B^2 \cdot l^2} \tag{4.40}$$

where $\frac{M^2 \cdot A^2}{c}$ is the maximum input power and is used as reference power. If the parameters are normalized using Eq. (4.23), substituting Eq. (4.23) into Eq. (4.40) gives the dimensionless input resonant power as

$$\frac{P_{\text{in}}}{\left(\frac{M^2 \cdot A^2}{c}\right)} = \frac{1}{2} \cdot \frac{\left(\frac{\pi}{2} + \frac{\pi}{2} \cdot \frac{R_e}{R} + \frac{1}{R_N}\right)^2}{\left(\frac{\pi}{2} + \frac{\pi}{2} \cdot \frac{R_e}{R} + \frac{1}{R_N}\right)^2 + \alpha_N^2 \cdot R_N} \tag{4.41}$$

Substitution of Eq. (4.39) into Eq. (4.36) gives

$$P_h = \frac{1}{2} \cdot \frac{V_M^2}{R} = \frac{B^2 \cdot l^2}{2 \cdot \left(\frac{\pi}{2} + \frac{\pi}{2} \cdot \frac{R_e}{R} + \frac{L_e \cdot \omega}{R}\right)^2 \cdot c \cdot R} \cdot \frac{\frac{M^2 \cdot A^2}{c}}{\left(1 + \frac{B^2 \cdot l^2}{\left(\frac{\pi}{2} + \frac{\pi}{2} \cdot \frac{R_e}{R} + \frac{L_e \cdot \omega}{R}\right)^2 \cdot c \cdot R}\right)^2} \tag{4.42}$$

Substitution of Eq. (4.23) into Eq. (4.42) gives the dimensionless harvested resonant power as

$$\frac{P_h}{\left(\frac{M^2 \cdot A^2}{c}\right)} = \frac{\alpha_N^2 \cdot R_N \cdot \left(\frac{\pi}{2} + \frac{\pi}{2} \cdot \frac{R_e}{R} + \frac{1}{R_N}\right)^2}{2 \cdot \left[\left(\frac{\pi}{2} + \frac{\pi}{2} \cdot \frac{R_e}{R} + \frac{1}{R_N}\right)^2 + \alpha_N^2 \cdot R_N\right]^2} \tag{4.43}$$

It is assumed that the resistance of the coil wire is very small, comparing with the external load resistance, that is, $R_e \ll R$ or $\frac{R_e}{R} \approx 0$, Eq. (4.43) becomes

$$\frac{P_h}{\left(\frac{M^2 \cdot A^2}{c}\right)} = \frac{\alpha_N^2 \cdot R_N \cdot \left(\frac{\pi}{2} + \frac{1}{R_N}\right)^2}{2 \cdot \left[\left(\frac{\pi}{2} + \frac{1}{R_N}\right)^2 + \alpha_N^2 \cdot R_N\right]^2} \tag{4.44}$$

The efficiency of the standard interface circuit at the resonant frequency is derived from Eqs. (4.41) and (4.44) and is given by

$$\eta = \frac{P_h}{P_{\text{in}}} = \frac{\alpha_N^2 \cdot R_N}{\left(\frac{\pi}{2} + \frac{\pi}{2} \cdot \frac{R_e}{R} + \frac{1}{R_N}\right)^2 + \alpha_N^2 \cdot R_N} \tag{4.45}$$

It is assumed that the resistance of the coil wire is very small, comparing with the external load resistance, that is, $R_e \ll R$ or $\frac{R_e}{R} \approx 0$. The efficiency can be given

$$\eta = \frac{\alpha_N^2 \cdot R_N}{\left(\frac{\pi}{2} + \frac{1}{R_N}\right)^2 + \alpha_N^2 \cdot R_N} \tag{4.46}$$

To find the maximum value of the dimensionless harvested power, the partial differential of the dimensionless harvested power with respect to R_N or α_N has to be zero, which gives

$$\frac{\partial\left(\frac{P_h}{\left(\frac{M^2 \cdot A^2}{c}\right)}\right)}{\partial R_N} = 0$$

or

$$\frac{\partial\left(\frac{P_h}{\left(\frac{M^2 \cdot A^2}{c}\right)}\right)}{\partial \alpha_N} = 0$$

which leads to

$$\left(\frac{\pi}{2} + \frac{1}{R_N}\right)^2 = \alpha_N^2 \cdot R_N$$

$$\left(\frac{P_h}{\left(\frac{M^2 \cdot A^2}{c}\right)}\right)_{max} = \frac{1}{8}$$

When $R_N \to \infty$ and α_N is a nonzero finite number, or when $\alpha_N \to \infty$ and R_N is a nonzero finite number, the dimensionless harvested power tends to be zero. When the dimensionless resistance tends to zero while the dimensionless force factor is kept constant, or when the dimensionless force factor tends to zero while the dimensionless resistance is kept constant, the dimensionless harvested power tends to zero. When $R_N \to \infty$ and $\alpha_N \to \infty$, or when $R_N \to 0$ and $\alpha_N \to 0$, the dimensionless harvested power tends to be zero. There is the special case of $\alpha_N^2 \cdot R_N = $ constant where and when the dimensionless force factor tends to zero while the resistance tends to a very large value, or when the dimensionless force factor tends to a very large value while the resistance tends to zero. If $\alpha_N^2 \cdot R_N$ is constant, when the dimensionless force factor tends to zero while the resistance tends to a very large value, the dimensionless harvested power tends to be constant. When the dimensionless force factor tends to a very large value while the resistance tends to zero, the dimensionless harvested power tends to be zero. However, the dimensionless harvested power is always less than 0.125.

From Eq. (4.46), when the dimensionless harvested power reaches its maximum value, the efficiency is 50%. To find the maximum value of the efficiency, the partial differential of the efficiency with respect to R_N or α_N has to be zero, which gives

$$\frac{\partial \eta}{\partial R_N} = \frac{\alpha_N^2 \cdot \left(\frac{\pi}{2} + \frac{1}{R_N}\right) \cdot \left(\frac{\pi}{2} + \frac{3}{R_N}\right)}{\left(\left(\frac{\pi}{2} + \frac{1}{R_N}\right)^2 + \alpha_N^2 \cdot R_N\right)^2} > 0$$

or

$$\frac{\partial \eta}{\partial \alpha_N} = \frac{2 \cdot \alpha_N \cdot R_N \cdot \left(\frac{\pi}{2} + \frac{1}{R_N}\right)^2}{\left(\left(\frac{\pi}{2} + \frac{1}{R_N}\right)^2 + \alpha_N^2 \cdot R_N\right)^2} > 0$$

From Eq. (4.46), it is seen that both the partial differentials of the efficiency with respect to the dimensionless resistance and with respect to the dimensionless force factor are larger than zero. The efficiency is a monotonically increasing function with respect to α_N^2 or R_N. If the variation ranges of the dimensionless resistance and force factor are not limited, there does not exist a point peak of the efficiency for a pair of the dimensionless resistance and force factor. When the dimensionless resistance is kept constant and the dimensionless force factor tends to be very large, the efficiency tends to be 100%. When the dimensionless resistance tends to a very large value, while the dimensionless force factor is kept constant, the efficiency tends to be 100%. There is a special case of $\alpha_N^2 \cdot R_N = $ constant, when the dimensionless force factor tends to very large value, while the dimensionless resistance tends to a very small value, the efficiency tends to zero.

When the dimensionless force factor tends to very small value, while the dimensionless resistance tends to a very large value, the efficiency tends to a fixed constant value.

4.8.2 SYNCHRONOUS ELECTRIC CHARGE EXTRACTION CIRCUIT

For the SECE circuit as shown in Fig. 3.1(B), the charge extraction phase occurs when the electronic switch S is on; the electrical energy stored on the induction coil L_e is then transferred into the inductor L. The extraction instants are triggered on the minima and maxima of the displacement z, synchronously with the mechanical vibration. The inductor L is chosen to get a charge extraction phase duration much shorter than the vibration period. Apart from the extraction phases, the rectifier is blocked and the outgoing current I is null. In this switch off circuit condition, the mechanical velocity is related to the voltage by

$$L_e \cdot \dot{I} + R_e \cdot I = B \cdot l \cdot \dot{z} \tag{4.47}$$

Integration of Eq. (4.47) with respect to time for the period between t_0 and $t_0 + T/2$ becomes

$$\int_{t_0}^{t_0+\frac{T}{2}} L_e \cdot \dot{I} \cdot dt + \int_{t_0}^{t_0+\frac{T}{2}} R_e \cdot I \cdot dt = \int_{t_0}^{t_0+\frac{T}{2}} B \cdot l \cdot \dot{z} \cdot dt$$

Considering $T = 2\pi/\omega$, this gives

$$\frac{V_M}{R} \cdot \left(L_e + R_e \cdot \frac{\pi}{2 \cdot \omega}\right) = B \cdot l \cdot Z_M$$

where $Z_M = |Z|$ and $V_M = |\overline{V}|$.

From Eq. (4.23), it gives

$$\frac{V_M}{R} \cdot L_e \cdot \left(1 + \frac{R_e}{R} \cdot R_N \cdot \frac{\pi}{2}\right) = B \cdot l \cdot Z_M$$

It is assumed that the resistance of the coil wire is very small, comparing with the external load resistance, that is, $R_e \ll R$ or $\frac{R_e}{R} \approx 0$, this gives

$$\frac{V_M}{R} \cdot L_e = B \cdot l \cdot Z_M$$

The mean harvested power is given according to Ref. [4] as

$$P_h = \frac{1}{2} \cdot \frac{V_M^2}{R} = \frac{1}{2} \cdot \frac{B^2 \cdot l^2 \cdot Z_M^2 \cdot R}{L_e^2} \tag{4.48}$$

From Eqs. (4.37) and (4.48), the system energy equilibrium equation is given by

$$\frac{1}{2} \cdot \omega \cdot Z_M \cdot M \cdot A = \frac{1}{2} \cdot c \cdot Z_M^2 \cdot \omega^2 + \frac{1}{2} \cdot \frac{B^2 \cdot l^2 \cdot Z_M^2 \cdot R}{L_e^2} \tag{4.49}$$

where $A = |\omega^2 \cdot Y|$ is the input excitation acceleration amplitude of the base. From Eq. (4.49), the relationship between the input excitation acceleration amplitude and the vibration displacement amplitude is derived and given as

$$Z_M = \frac{M \cdot A}{\left(c \cdot \omega + \frac{B^2 \cdot l^2 \cdot R}{\omega \cdot L_e^2} \right)} \tag{4.50}$$

According to Ref. [4], the mean input resonant power is then given by

$$P_{in} = \frac{1}{2} \cdot \omega \cdot M \cdot A^2 \cdot Z_M = \frac{1}{2} \cdot \frac{M^2 \cdot A^2}{c + \frac{B^2 \cdot l^2 \cdot R}{\omega^2 \cdot L_e^2}} \tag{4.51}$$

From Eq. (4.51), the dimensionless input resonant power is given by

$$\frac{P_{in}}{\frac{M^2 \cdot A^2}{c}} = \frac{1}{2} \cdot \frac{1}{1 + \frac{B^2 \cdot l^2}{R^2} \cdot \frac{L_e \cdot \omega}{c} \cdot \frac{R^3}{L_e^3 \cdot \omega^3}} = \frac{1}{2} \cdot \frac{1}{1 + \alpha_N^2 \cdot R_N^3} \tag{4.52}$$

Substitution of Eq. (4.50) into Eq. (4.48) gives

$$P_h = \frac{1}{2} \cdot \frac{B^2 \cdot l^2 \cdot R}{L_e^2} \cdot Z_M^2 = \frac{1}{2} \cdot \frac{B^2 \cdot l^2 \cdot R \cdot c}{L_e^2} \cdot \frac{\frac{M^2 \cdot A^2}{c}}{\left(c \cdot \omega + \frac{B^2 \cdot l^2 \cdot R}{\omega \cdot L_e^2} \right)^2} \tag{4.53}$$

Substitution of Eq. (4.23) into Eq. (4.53) gives

$$\frac{P_h}{\frac{M^2 \cdot A^2}{c}} = \frac{1}{2} \cdot \frac{\frac{B^2 \cdot l^2 \cdot R}{L_e^2 \cdot c \cdot \omega^2}}{\left(1 + \frac{B^2 \cdot l^2 \cdot R}{c \cdot \omega^2 \cdot L_e^2} \right)^2} = \frac{1}{2} \cdot \frac{\alpha_N^2 \cdot R_N^3}{\left(1 + \alpha_N^2 \cdot R_N^3 \right)^2} \tag{4.54}$$

The efficiency of the SECE circuit at the resonant frequency is derived from Eqs. (4.52) and (4.54), and is given by

$$\eta = \frac{\alpha_N^2 \cdot R_N^3}{1 + \alpha_N^2 \cdot R_N^3} \tag{4.55}$$

For Eq. (4.54), to find the maximum value of the dimensionless harvested power, the partial differential of the dimensionless harvested power with respect to R_N or α_N has to be zero, which gives

$$\frac{\partial \left(\frac{P_h}{M^2 \cdot A^2}{c} \right)}{\partial R_N} = 0$$

or

$$\frac{\partial \left(\frac{P_h}{M^2 \cdot A^2}{c} \right)}{\partial \alpha_N} = 0$$

$$\alpha_N^2 \cdot R_N^3 = 1$$

$$\left(\frac{P_h}{\frac{M^2 \cdot A^2}{c}} \right)_{max} = \frac{1}{8}$$

When $R_N \to \infty$ and α_N is a nonzero finite number, or when $\alpha_N \to \infty$ and R_N is a nonzero finite number, the dimensionless harvested power tends to be zero. When the dimensionless resistance tends to zero while the dimensionless force factor is kept constant, or when the dimensionless force factor tends to zero while the dimensionless resistance is kept constant, the dimensionless harvested power tends to zero. When $R_N \to \infty$ and $\alpha_N \to \infty$, or when $R_N \to 0$ and $\alpha_N \to 0$, the dimensionless harvested power tends to be zero. There is the special case of $\alpha_N^2 \cdot R_N^3 =$ constant where and when the dimensionless force factor tends to zero while the resistance tends to a very large value, or when the dimensionless force factor tends to a very large value while the resistance tends to zero. If $\alpha_N^2 \cdot R_N^3$ is constant, the dimensionless harvested power will always be a constant and less than 0.125.

From Eq. (4.55), when the dimensionless harvested power reaches its maximum value, the efficiency is 50%. To find the maximum value of the efficiency, the partial differential of the efficiency with respect to R_N or α_N has to be zero, which gives

$$\frac{\partial \eta}{\partial R_N} = \frac{3 \cdot \alpha_N^2 \cdot R_N^2}{\left(1 + \alpha_N^2 \cdot R_N^3 \right)^2} > 0$$

and

$$\frac{\partial \eta}{\partial \alpha_N} = \frac{2 \cdot \alpha_N \cdot R_N^3}{\left(1 + \alpha_N^2 \cdot R_N^3 \right)^2} > 0$$

No peak values of the efficiency occur over the ranges of the dimensionless resistance and force factors. The efficiency is a monotonically increasing function with respect to α_N^2 or R_N. When $R_N \to \infty$ and α_N is a nonzero finite number, or when $\alpha_N \to \infty$ and R_N is a nonzero finite number, the efficiency tends to be 100%.

When the dimensionless resistance tends to zero while the dimensionless force factor is kept constant, or when the dimensionless force factor tends to zero while the dimensionless resistance is kept constant, the efficiency tends to zero. When $R_N \rightarrow \infty$ and $\alpha_N \rightarrow \infty$, the efficiency tends to be 100%. The maximum efficiency is 100%. When $R_N \rightarrow 0$ and $\alpha_N \rightarrow 0$, the efficiency tends to be zero. There is special case of $\alpha_N^2 \cdot R_N = $ constant where and when the dimensionless force factor tends to zero while the resistance tends to a very large value. When $R_N \rightarrow \infty$ and $\alpha_N \rightarrow 0$, the efficiency tends to be 100% or when $\alpha_N \rightarrow \infty$ and $R_N \rightarrow 0$, the efficiency tends to be zero; if $\alpha_N^2 \cdot R_N^3$ is constant, the efficiency is constant.

4.8.3 PARALLEL SYNCHRONOUS SWITCH HARVESTING ON INDUCTOR CIRCUIT

For a parallel switch harvesting on inductor (parallel SSHI) circuit as shown in Fig. 3.1(C), the inductor L is in series connected with an electronic switch S, and both the inductor and the electronic switch are connected in parallel with the piezoelectric element electrodes and the diode rectifier bridge. A small part of the energy may also be radiated in the mechanical system. The inversion losses are modeled by the electrical quality factor Q_i of the electrical oscillator. The relation between Q_i and the voltages of the electromagnetic element before and after the inversion process (V_M and V_0, respectively) is given by

$$V_M = V_0 \cdot e^{\frac{-\pi}{2 \cdot Q_i}} \tag{4.56}$$

The electric charge received by the terminal load equivalent resistor R during a half mechanical period $T/2$ is calculated by

$$\int_{t_0}^{t_0 + \frac{T}{2}} I \cdot dt + \int_{t_0}^{t_0 + \frac{T}{2}} I_S \cdot dt = \frac{V_0}{R + R_e} \cdot \frac{T}{2} \tag{4.57}$$

The second integral on the left-hand side of this equation corresponds to the charge stored on the coil L_e before the voltage inversion plus the charge stored on L_e after the inversion, whose expression is given by

$$\int_{t_0}^{t_0 + \frac{T}{2}} I_S \cdot dt = \frac{L_e \cdot V_0}{(R + R_e)^2} \left(1 + e^{-\frac{\pi}{2 \cdot Q_i}} \right) \tag{4.58}$$

The electromagnetic coil outgoing current is integrated by

$$\int_{t_0}^{t_0 + \frac{T}{2}} (R_e + R) \cdot I \cdot dt + \int_{t_0}^{t_0 + \frac{T}{2}} \frac{L_e}{R_e + R} \cdot dV = \int_{t_0}^{t_0 + \frac{T}{2}} B \cdot l \cdot dz \tag{4.59}$$

Substitution of Eqs. (4.57) and (4.58) into Eq. (4.59) gives

$$(R_e + R) \cdot \left[\frac{V_0}{R + R_e} \cdot \frac{T}{2} - \frac{V_0 \cdot L_e}{(R + R_e)^2} \cdot \left(1 + e^{-\frac{\pi}{2 \cdot Q_i}} \right) \right] + \int_{t_0}^{t_0 + \frac{T}{2}} \frac{L_e}{R_e + R} \cdot dV$$

$$= \int_{t_0}^{t_0 + \frac{T}{2}} B \cdot l \cdot dz \tag{4.60}$$

Under a harmonic base excitation, it is assumed that both the displacement and voltage waves are periodic and change from a trough $(-Z_M$ and $-V_M)$ to a crest $(+Z_M$ and $+V_M)$ from instant t_0 to $t_0 + T/2$. Eq. (4.60) becomes

$$V_0 \cdot \frac{T}{2} - \frac{L_e}{R_e + R} \cdot V_0 \cdot \left(1 + e^{-\frac{\pi}{2 \cdot Q_i}}\right) \cdot \frac{T}{2} + 2 \cdot \frac{L_e}{R_e + R} \cdot V_0 = 2 \cdot B \cdot l \cdot Z_M \qquad (4.61)$$

where $Z_M = |Z|$ and $V_M = |\overline{V}|$. This leads to the expression of the load voltage V_0 as a function of the displacement amplitude Z_M given by

$$V_0 = \frac{2 \cdot B \cdot l \cdot \omega}{\pi + \frac{L_e \cdot \omega}{R_e + R} \cdot \left(1 - e^{-\frac{\pi}{2 \cdot Q_i}}\right)} \cdot Z_M \qquad (4.62)$$

According to Ref. [4], the mean harvested resonant power is then given by

$$P_h = \frac{1}{2} \cdot \frac{V_0^2}{R} = \frac{1}{2} \cdot \frac{4 \cdot B^2 \cdot l^2 \cdot \omega^2}{\left[\pi + \frac{L_e \cdot \omega}{R_e + R} \cdot \left(1 - e^{-\frac{\pi}{2 \cdot Q_i}}\right)\right]^2 \cdot R} \cdot Z_M^2 \qquad (4.63)$$

From Eqs. (4.37) and (4.63), the system energy equilibrium equation is given by

$$\frac{1}{2} \cdot \omega \cdot Z_M \cdot M \cdot A = \frac{1}{2} \cdot c \cdot Z_M^2 \cdot \omega^2 + \frac{1}{2} \cdot \frac{4 \cdot B^2 \cdot l^2 \cdot \omega^2}{\left[\pi + \frac{L_e \cdot \omega}{R_e + R} \left(1 - e^{-\frac{\pi}{2 \cdot Q_i}}\right)\right]^2 \cdot R} \cdot Z_M^2 \qquad (4.64)$$

where $A = |\omega^2 \cdot Y|$ is the input excitation acceleration amplitude of the base. From Eq. (4.64), the relationship between the vibration displacement amplitude and the input excitation acceleration amplitude is established as

$$Z_M = \frac{M \cdot A}{c \cdot \omega + \frac{4 \cdot B^2 \cdot l^2 \cdot \omega}{\left[\pi + \frac{L_e \cdot \omega}{R_e + R} \left(1 - e^{-\frac{\pi}{2 \cdot Q_i}}\right)\right]^2 \cdot R}} \qquad (4.65)$$

From Eq. (4.65), according to Ref. [4], the mean input resonant power is given by

$$P_{in} = \frac{1}{2} \cdot \omega \cdot M \cdot A \cdot Z_M = \frac{1}{2} \cdot \frac{M^2 \cdot A^2}{c + \frac{4 \cdot B^2 \cdot l^2}{\left[\pi + \frac{L_e \cdot \omega}{R_e + R} \left(1 - e^{-\frac{\pi}{2 \cdot Q_i}}\right)\right]^2 \cdot R}} \qquad (4.66)$$

The dimensionless input resonant power is given by

$$\frac{P_{in}}{\left(\frac{M^2 \cdot A^2}{c}\right)} = \frac{1}{2} \cdot \frac{\left[\pi + \frac{L_e \cdot \omega}{R_e + R} \left(1 - e^{-\frac{\pi}{2 \cdot Q_i}}\right)\right]^2}{\left[\pi + \frac{L_e \cdot \omega}{R_e + R} \left(1 - e^{-\frac{\pi}{2 \cdot Q_i}}\right)\right]^2 + \frac{4 \cdot B^2 \cdot l^2 \cdot R}{c}}$$

It is assumed that the resistance of the coil wire is very small, comparing with the external load resistance, that is, $R_e \ll R$ or $\frac{R_e}{R} \approx 0$.

$$\frac{P_{in}}{\left(\frac{M^2 \cdot A^2}{c}\right)} = \frac{1}{2} \cdot \frac{\left[\pi + \frac{L_e \cdot \omega}{R}\left(1 - e^{-\frac{\pi}{2 \cdot Q_i}}\right)\right]^2}{\left[\pi + \frac{L_e \cdot \omega}{R}\left(1 - e^{-\frac{\pi}{2 \cdot Q_i}}\right)\right]^2 + \frac{4 \cdot B^2 \cdot l^2}{c \cdot R}} \tag{4.67}$$

where $\frac{M^2 \cdot A^2}{c}$ is the maximum input power and used as reference power. Substitution of Eq. (4.23) into Eq. (4.67) gives the normalized dimensionless harvested resonant power as

$$\frac{P_{in}}{\left(\frac{M^2 \cdot A^2}{c}\right)} = \frac{1}{2} \cdot \frac{\left[\pi + \frac{1}{R_N} \cdot \left(1 - e^{-\frac{\pi}{2 \cdot Q_i}}\right)\right]^2}{\left[\pi + \frac{1}{R_N} \cdot \left(1 - e^{-\frac{\pi}{2 \cdot Q_i}}\right)\right]^2 + 4 \cdot R_N \cdot \alpha_N^2} \tag{4.68}$$

Substituting Eq. (4.65) into Eq. (4.63) gives

$$P_h = \frac{1}{2} \cdot \frac{V_0^2}{R} = \frac{1}{2} \cdot \frac{4 \cdot B^2 \cdot l^2 \cdot \omega^2 \cdot R}{\left[\pi + \frac{L_e \cdot \omega}{R_e + R} \cdot \left(1 - e^{-\frac{\pi}{2 \cdot Q_i}}\right)\right]^2} \cdot \frac{M^2 \cdot A^2}{\left\{ c \cdot \omega + \frac{B^2 \cdot l^2 \cdot \omega \cdot R}{\left[\pi + \frac{L_e \cdot \omega}{R_e + R} \cdot \left(1 - e^{-\frac{\pi}{2 \cdot Q_i}}\right)\right]^2} \right\}^2}$$

$$\frac{P_h}{\left(\frac{M^2 \cdot A^2}{c}\right)} = \frac{1}{2} \cdot \frac{\frac{4 \cdot B^2 \cdot l^2 \cdot R}{c} \cdot \left[\pi + \frac{L_e \cdot \omega}{R_e + R} \cdot \left(1 - e^{-\frac{\pi}{2 \cdot Q_i}}\right)\right]^2}{\left\{\left[\pi + \frac{L_e \cdot \omega}{R_e + R} \cdot \left(1 - e^{-\frac{\pi}{2 \cdot Q_i}}\right)\right]^2 + \frac{4 \cdot B^2 \cdot l^2 \cdot R}{c}\right\}^2} \tag{4.69}$$

It is assumed that the resistance of the coil wire is very small, comparing with the external load resistance, that is, $R_e \ll R$ or $\frac{R_e}{R} \approx 0$. Eq. (4.69) then becomes

$$\frac{P_h}{\left(\frac{M^2 \cdot A^2}{c}\right)} = \frac{1}{2} \cdot \frac{\frac{4 \cdot B^2 \cdot l^2}{c \cdot R} \cdot \left[\pi + \frac{L_e \cdot \omega}{R} \cdot \left(1 - e^{-\frac{\pi}{2 \cdot Q_i}}\right)\right]^2}{\left\{\left[\pi + \frac{L_e \cdot \omega}{R} \cdot \left(1 - e^{-\frac{\pi}{2 \cdot Q_i}}\right)\right]^2 + \frac{4 \cdot B^2 \cdot l^2}{c \cdot R}\right\}^2} \tag{4.70}$$

Substituting Eq. (4.23) into Eq. (4.70) gives the normalized dimensionless harvested resonant power as

$$\frac{P_h}{\left(\frac{M^2 \cdot A^2}{c}\right)} = \frac{1}{2} \cdot \frac{4 \cdot \alpha_N^2 \cdot R_N \cdot \left[\pi + \frac{1}{R_N} \cdot \left(1 - e^{-\frac{\pi}{2 \cdot Q_i}}\right)\right]^2}{\left\{\left[\pi + \frac{1}{R_N} \cdot \left(1 - e^{-\frac{\pi}{2 \cdot Q_i}}\right)\right]^2 + 4 \cdot \alpha_N^2 \cdot R_N\right\}^2} \tag{4.71}$$

The efficiency of the parallel SSHI circuit at the resonance frequency is derived from Eqs. (4.68) and (4.71), and is given by

$$\eta = \frac{4 \cdot \alpha_N^2 \cdot R_N}{\left[\pi + \frac{1}{R_N} \cdot \left(1 - e^{-\frac{\pi}{2 \cdot Q_i}}\right)\right]^2 + 4 \cdot \alpha_N^2 \cdot R_N} \tag{4.72}$$

For Eq. (4.71), to find the maximum value of the dimensionless harvested power, the partial differential of the dimensionless harvested power with respect to R_N or α_N has to be zero, which gives

$$\frac{\partial\left(\frac{P_h}{M^2 \cdot A^2}{c}\right)}{\partial R_N} = 0$$

or

$$\frac{\partial\left(\frac{P_h}{M^2 \cdot A^2}{c}\right)}{\partial \alpha_N} = 0$$

which leads to

$$\left[\pi + \frac{1}{R_N} \cdot \left(1 - e^{-\frac{\pi}{2 \cdot Q_i}}\right)\right]^2 = \alpha_N^2 \cdot R_N$$

and

$$\left(\frac{P_h}{\frac{M^2 \cdot A^2}{c}}\right)_{max} = \frac{1}{8}$$

When $R_N \to \infty$ and α_N is a nonzero finite number, or when $\alpha_N \to \infty$ and R_N is a nonzero finite number, the dimensionless harvested power tends to be zero. When the dimensionless resistance tends to zero while the dimensionless force factor is kept constant, or when the dimensionless force factor tends to zero while the dimensionless resistance is kept constant, the dimensionless harvested power tends to zero. When $R_N \to \infty$ and $\alpha_N \to \infty$, or when $R_N \to 0$ and $\alpha_N \to 0$, the dimensionless harvested power tends to be zero. There is the special case of $\alpha_N^2 \cdot R_N = $ constant where and when the dimensionless force factor tends to zero while the resistance tends to a very large value, or when the dimensionless force factor tends to a very large value while the resistance tends to zero. If $\alpha_N^2 \cdot R_N$ is constant, when the dimensionless force factor tends to zero while the dimensionless resistance tends to a very large value, the dimensionless harvested power is constant. When the dimensionless force factor tends to a very large value while the dimensionless resistance tends to zero, the dimensionless harvested power is zero. However, the dimensionless harvested power will always be less than 0.125.

From Eq. (4.72), when the dimensionless harvested power reaches its maximum value, the efficiency is 50%. To find the maximum value of the efficiency, the partial differential of the efficiency with respect to R_N or α_N has to be zero, which gives

$$\frac{\partial \eta}{\partial R_N} = \frac{4 \cdot \alpha_N^2 \cdot \left[\pi + \frac{1}{R_N} \cdot \left(1 - e^{-\frac{\pi}{2 \cdot Q_i}}\right)\right] \cdot \left[\pi + \frac{1}{R_N}\left(3 - e^{-\frac{\pi}{2 \cdot Q_i}}\right)\right]}{\left(\left[\pi + \frac{1}{R_N} \cdot \left(1 - e^{-\frac{\pi}{2 \cdot Q_i}}\right)\right]^2 + 4 \cdot \alpha_N^2 \cdot R_N\right)^2} > 0$$

$$\frac{\partial \eta}{\partial \alpha_N} = \frac{8 \cdot \alpha_N \cdot R_N \cdot \left[\pi + \frac{1}{R_N} \cdot \left(1 - e^{-\frac{\pi}{2 \cdot Q_i}}\right)\right]^2}{\left(\left[\pi + \frac{1}{R_N}\left(1 - e^{-\frac{\pi}{2 \cdot Q_i}}\right)\right]^2 + 4 \cdot \alpha_N^2 \cdot R_N\right)^2} > 0$$

No peak values of the efficiency occur over the ranges of the dimensionless resistance and force factors. The efficiency is a monotonically increasing function with respect to α_N^2 or R_N. When $R_N \rightarrow \infty$ and α_N is a nonzero finite number, or when $\alpha_N \rightarrow \infty$ and R_N is a nonzero finite number, the efficiency tends to be 100%. When the dimensionless resistance tends to zero while the dimensionless force factor is kept constant, or when the dimensionless force factor tends to zero while the dimensionless resistance is kept constant, the efficiency tends to zero. When $R_N \rightarrow \infty$ and $\alpha_N \rightarrow \infty$, the efficiency tends to be 100%. The maximum efficiency is 100%. When $R_N \rightarrow 0$ and $\alpha_N \rightarrow 0$, the efficiency tends to be zero. There is a special case of $\alpha_N^2 \cdot R_N = $ constant, where and when the dimensionless force factor tends to zero while the dimensionless resistance tends to a very large value. When $R_N \rightarrow \infty$ and $\alpha_N \rightarrow 0$, the efficiency tends to be a fixed constant or when $\alpha_N \rightarrow \infty$ and $R_N \rightarrow 0$, the efficiency tends to be zero for $\alpha_N^2 \cdot R_N = $ constant.

4.8.4 SERIES SYNCHRONOUS SWITCH HARVESTING ON INDUCTOR CIRCUIT

For a series SSHI circuit as shown in Fig. 4.7(D), in most of the time, the electromagnetic element is in switch off circuit configuration. Each time the switch is on, a part of the energy stored in the coil L_e is transferred to the capacitor C_{st} through the rectifier bridge. At those instants, the voltage inversions of V occur. The relation of the voltages V_M and V_m before and after the inversion process, the rectified voltage V_0, and the electrical quality factor Q_i is given by

$$V_m - V_0 = (V_M - V_0) \cdot e^{-\frac{\pi}{2 \cdot Q_i}} \tag{4.73}$$

Under a harmonic base excitation, it is assumed that both the oscillator displacement and output voltage waves are periodic and change from a trough ($-Z_M$ and $-V_M$) to a crest ($+Z_M$ and $+V_M$) from instant t_0 to $t_0 + T/2$. Integration of Eq. (4.3) with respect to time for a half of the mechanical vibration period ($T/2$) gives

$$\int_{t_0}^{t_0+\frac{T}{2}} V \cdot dt + \int_{t_0}^{t_0+\frac{T}{2}} R_e \cdot I \cdot dt + \int_{t_0}^{t_0+\frac{T}{2}} \frac{L_e}{R} \cdot dV = \int_{t_0}^{t_0+\frac{T}{2}} B \cdot l \cdot dz \tag{4.74}$$

The switch off circuit evolution of the voltage V between two voltage inversions gives another relation between V_M and V_m,

$$2 \cdot B \cdot l \cdot Z_M - \frac{L_e}{R} \cdot (V_m + V_M) - \frac{R_e}{R} \cdot (V_m + V_M) \cdot \frac{T}{2} = 0 \quad \text{or}$$

$$V_M + V_m = \frac{2 \cdot B \cdot l \cdot R}{L_e + R_e \cdot \frac{T}{2}} \cdot Z_M$$

where $Z_M = |Z|$ and $V_M = |\overline{V}|$.

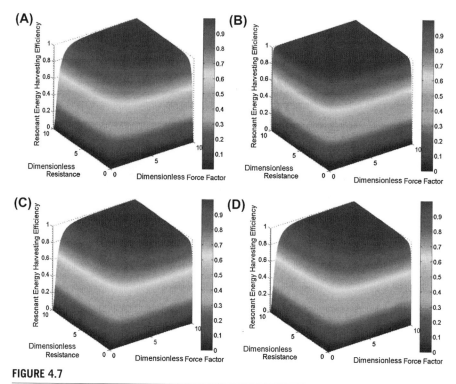

FIGURE 4.7

The resonant energy harvesting efficiency versus dimensionless resistance and force factors for the single degree of freedom piezoelectric harvester connected to the four types of interface circuits. (A) Standard interface; (B) synchronous electric charge extraction; (C) parallel synchronous switch harvesting on inductor (SSHI); (D) series SSHI.

Considering $T = 2 \cdot \pi/\omega$ and Eq. (4.23), this gives

$$V_M + V_m = \frac{2 \cdot B \cdot l \cdot Z_M \cdot R}{L_e \cdot \left(1 + \frac{R_e}{R} \cdot R_N \cdot \pi\right)}$$

It is assumed that the resistance of the coil wire is very small, comparing with the external load resistance, that is, $R_e \ll R$ or $\frac{R_e}{R} \approx 0$, this gives

$$V_M + V_m = \frac{2 \cdot B \cdot l \cdot Z_M \cdot R}{L_e} \tag{4.75}$$

Equality of the input energy of the rectification bridge and the energy consumed by the equivalent load resistance R during a semi-period of vibration $T/2$ leads to

$$V_0 \cdot \int_{t_0}^{t_0 + \frac{T}{2}} I \cdot dt = V_0 \cdot \frac{L_e}{R^2} \cdot (V_m + V_M) = \frac{V_0^2}{R} \cdot \frac{\pi}{\omega} \tag{4.76}$$

which leads to

$$\frac{L_e}{R^2} \cdot (V_m + V_M) = \frac{V_0}{R} \cdot \frac{\pi}{\omega} \tag{4.77}$$

Substitution of Eq. (4.75) into Eq. (4.77) gives

$$2 \cdot B \cdot l \cdot Z_M = V_0 \cdot \frac{\pi}{\omega} \tag{4.78}$$

According to Ref. [4], the mean harvested resonant power is given by

$$P_h = \frac{1}{2} \cdot \frac{V_0^2}{R} = \frac{1}{2} \cdot \frac{4 \cdot B^2 \cdot l^2 \cdot \omega^2}{\pi^2 \cdot R} \cdot Z_M^2 \tag{4.79}$$

From Eq. (4.37) and Eq. (4.79), according to Ref. [4], the system energy equilibrium equation is given by

$$\frac{1}{2} \cdot \omega \cdot M \cdot A^2 \cdot Z_M = \frac{1}{2} \cdot \frac{4 \cdot B^2 \cdot l^2 \cdot \omega^2}{\pi^2 \cdot R} \cdot Z_M^2 + \frac{1}{2} \cdot c \cdot Z_M^2 \cdot \omega^2 \tag{4.80}$$

where $A = |\omega^2 \cdot Y|$ is the input excitation acceleration amplitude of the base. The relationship between the displacement amplitude and input excitation acceleration amplitude is then derived from Eq. (4.80) as

$$Z_M = \frac{M \cdot A}{c \cdot \omega + \frac{4 \cdot B^2 \cdot l^2 \cdot \omega}{\pi^2 \cdot R}} \tag{4.81}$$

Substitution of Eq. (4.81) into Eq. (4.79) gives the harvested resonant power amplitude as

$$P_h = \frac{2 \cdot B^2 \cdot l^2}{\pi^2 \cdot R} \left(\frac{M \cdot A}{c + \frac{4 \cdot B^2 \cdot l^2}{\pi^2 \cdot R}} \right)^2 \tag{4.82}$$

From Eq. (4.23), the normalized dimensionless harvested resonant power is then given by

$$\frac{P_h}{\frac{(M \cdot A)^2}{c}} = \frac{2 \cdot \pi^2 \cdot \alpha_N^2 \cdot R_N}{\left(\pi^2 + 4 \cdot \alpha_N^2 \cdot R_N \right)^2} \tag{4.83}$$

where $\frac{(M \cdot A)^2}{c}$ is the reference power.

From Eq. (4.81), according to Ref. [4], the mean input resonant power is given by

$$P_{in} = \frac{1}{2} \cdot \omega \cdot M \cdot A \cdot Z_M = \frac{1}{2} \cdot \frac{M^2 \cdot A^2}{c + \frac{4 \cdot B^2 \cdot l^2}{R \cdot \pi^2}} \tag{4.84}$$

Substituting Eq. (4.23) into Eq. (4.84) gives the normalized dimensionless input resonant power as

$$\frac{P_{in}}{\frac{(M \cdot A)^2}{c}} = \frac{1}{2} \cdot \frac{\pi^2}{\pi^2 + 4 \cdot \alpha_N^2 \cdot R_N} \tag{4.85}$$

The efficiency of the series SSHI circuit is derived from Eqs. (4.83) and (4.85) and given by

$$\eta = \frac{4 \cdot R_N \cdot \alpha_N^2}{\pi^2 + 4 \cdot R_N \cdot \alpha_N^2} \tag{4.86}$$

For Eq. (4.83), to find the maximum value of the dimensionless harvested power, the partial differential of the dimensionless harvested power with respect to R_N or α_N has to be zero, which gives

$$\frac{\partial \left(\frac{P_h}{\frac{M^2 \cdot A^2}{c}} \right)}{\partial R_N} = 0$$

or

$$\frac{\partial \left(\frac{P_h}{\frac{M^2 \cdot A^2}{c}} \right)}{\partial \alpha_N} = 0$$

which leads to

$$4 \cdot \alpha_N^2 \cdot R_N = \pi^2$$

and

$$\left(\frac{P_h}{\frac{M^2 \cdot A^2}{c}} \right)_{max} = \frac{1}{8}$$

When $R_N \to \infty$ and α_N is a nonzero finite number, or when $\alpha_N \to \infty$ and R_N is a nonzero finite number, the dimensionless harvested power tends to be zero. When the dimensionless resistance tends to zero while the dimensionless force factor is kept constant, or when the dimensionless force factor tends to zero while the dimensionless resistance is kept constant, the dimensionless harvested power tends to zero. When $R_N \to \infty$ and $\alpha_N \to \infty$, or when $R_N \to 0$ and $\alpha_N \to 0$, the dimensionless harvested power tends to be zero. There is the special case of $\alpha_N^2 \cdot R_N = $ constant, where and when the dimensionless force factor tends to zero while the resistance tends to a very large value, or when the dimensionless force factor tends to a very large value while the resistance tends to zero. If $\alpha_N^2 \cdot R_N$ is constant, the dimensionless harvested power is a constant. However, the dimensionless harvested power will always be less than 0.125.

From Eq. (4.86), when the dimensionless harvested power reaches its maximum value, the efficiency is 50%. To find the maximum value of the efficiency, the partial differential of the efficiency with respect to R_N or α_N has to be zero, which gives

$$\frac{\partial \eta}{\partial R_N} = \frac{4 \cdot \alpha_N^2 \cdot \pi^2}{\left(\pi^2 + 4 \cdot R_N \cdot \alpha_N^2 \right)^2} > 0$$

or

$$\frac{\partial \eta}{\partial \alpha_N} = \frac{8 \cdot R_N \cdot \alpha_N \cdot \pi^2}{\left(\pi^2 + 4 \cdot R_N \cdot \alpha_N^2\right)^2} > 0$$

No peak values of the efficiency occur over the ranges of the dimensionless resistance and force factors. The efficiency is a monotonically increasing function with respect to α_N^2 or R_N. When $R_N \rightarrow \infty$ and α_N is a nonzero finite number, or when $\alpha_N \rightarrow \infty$ and R_N is a nonzero finite number, the efficiency tends to be 100%. When the dimensionless resistance tends to zero while the dimensionless force factor is kept constant, or when the dimensionless force factor tends to zero while the dimensionless resistance is kept constant, the efficiency tends to zero. When $R_N \rightarrow \infty$ and $\alpha_N \rightarrow \infty$, the efficiency tends to be 100%. The maximum efficiency is 100%. When $R_N \rightarrow 0$ and $\alpha_N \rightarrow 0$, the efficiency tends to be zero. There is the special case of $\alpha_N^2 \cdot R_N = $ constant, where and when the dimensionless force factor tends to zero while the dimensionless resistance tends to a very large value, or when the dimensionless force factor tends to a very large value while the dimensionless resistance tends to zero. If $\alpha_N^2 \cdot R_N$ is constant, the efficiency is a constant.

4.9 ANALYSIS AND COMPARISON

In a typical condition that the dimensionless resistance and the dimensionless force factor are equal to unity, the resonant energy harvesting efficiencies of the EMVEHs with the SECE, series SSHI, parallel SSHI, standard interface circuits are 50, 29, 29, and 13%, and the dimensionless harvested resonant power values are 0.125, 0.103, 0.102, and 0.057, respectively. On the other hand, replacing the electrical interface circuits by a single load resistor, the resonant energy harvesting efficiency is 33% (from Eq. (2.26) and $\alpha_N = R_N = 1$) and the dimensionless harvested resonant power is 0.10. Clearly, in the case, the SECE setup gives the highest efficiency and harvested resonant power, and the standard interface setup gives the lowest efficiency and harvested resonant power. It should be noted that the harvested resonant power using the SSHI technique is better than that using the standard interface in the case when α_N^2 is small (for example, $\alpha_N^2 = 1$). The same conclusions can be drawn from Chapters 2 and 3 and Fig. 12 in Ref. [11], where SECE is better than other interfaces only in the case of a weak electromechanical coupling, or a small α_N^2.

Table 4.1 lists all important formulas of dimensionless harvested resonant power and energy harvesting efficiency. It is seen from Table 4.1 that for an EMVEH connected with a single load resistance or the four types of interface circuits, the resonant energy harvesting efficiency and dimensionless harvested resonant power depend on the system resonant frequency, mechanical damping, load resistance, magnetic field intensity, and coil length. The dimensionless resistance and force factor as defined in Eq. (4.23) are chosen to reflect all these parameters in this chapter for the system optimization analyses. It is seen from Table 4.1 that for the EMVEH

Table 4.1 Comparison of Dimensionless Harvested Power and Efficiency for Electromagnetic Vibration Energy Harvesters Connected With Four Different Interface Circuits

Vibration Energy Harvester Type	Single Load Resistor	Standard Circuits	SECE Circuits	Parallel SSHI Circuits	Series SSHI Circuits
Normalized dimensionless harvested resonant power	$\dfrac{P_h}{\left(\frac{M^2 \cdot A^2}{c}\right)} = \dfrac{1}{2} \cdot \dfrac{R_N \cdot \alpha_N^2}{\left(\frac{1}{R_N}\right)^2 + \left[1 + R_N \cdot \alpha_N^2\right]^2}$	$\dfrac{P_n}{\left(\frac{M^2 \cdot A^2}{c}\right)} = \dfrac{\alpha_N^2 \cdot R_N \cdot \left(\frac{\pi}{2} + \frac{1}{R_N}\right)^2}{2 \cdot \left[\left(\left(\frac{\pi}{2} + \frac{1}{R_N}\right)^2 + \alpha_N^2 \cdot R_N\right)\right]^2}$	$\dfrac{P_h}{\frac{M^2 \cdot A^2}{c}} = \dfrac{1}{2} \cdot \dfrac{\alpha_N^2 \cdot R_N^3}{\left(1 + \alpha_N^2 \cdot R_N^3\right)^2}$	$\dfrac{P_h}{\left(\frac{M^2 \cdot A^2}{c}\right)} = \dfrac{1}{2} \cdot \dfrac{4 \cdot \alpha_N^2 \cdot R_N \cdot \left[\pi + \frac{1}{R_N} \cdot \left(1 - e^{-\frac{\pi}{2 \cdot Q_i}}\right)\right]^2}{\left\{\left[\pi + \frac{1}{R_N}\left(1 - e^{-\frac{\pi}{2 \cdot Q_i}}\right)\right]^2 + 4 \cdot \alpha_N^2 \cdot R_N\right\}}$	$\dfrac{P_h}{\frac{|M \cdot A|^2}{c}} = \dfrac{2 \cdot \pi^2 \cdot \alpha_N^2 \cdot R_N}{\left(\pi^2 + 4 \cdot \alpha_N^2 \cdot R_N\right)^2}$
Normalized efficiency	$\eta = \dfrac{R_N \cdot \alpha_N^2}{\frac{1}{R_N^2} + \left[1 + R_N \cdot \alpha_N^2\right]}$	$\eta = \dfrac{\alpha_N^2 \cdot R_N}{\left(\frac{\pi}{2} + \frac{1}{R_N}\right)^2 + \alpha_N^2 \cdot R_N}$	$\eta = \dfrac{\alpha_N^2 \cdot R_N^3}{1 + \alpha_N^2 \cdot R_N^3}$	$\eta = \dfrac{4 \cdot \alpha_N^2 \cdot R_N}{\left[\pi + \frac{1}{R_N}\left(1 - e^{-\frac{\pi}{2 \cdot Q_i}}\right)\right]^2 + 4 \cdot \alpha_N^2 \cdot R_N}$	$\eta = \dfrac{4 \cdot R_N \cdot \alpha_N^2}{\pi^2 + 4 \cdot R_N \cdot \alpha_N^2}$
Harvested resonant power limit	$\dfrac{|M \cdot Y|^2 \cdot \omega^2}{8 \cdot c}$	$\dfrac{|M \cdot Y|^2 \cdot \omega^2}{8 \cdot c}$	$\dfrac{|M \cdot Y|^2 \cdot \omega^2}{8 \cdot c}$	$\dfrac{|M \cdot Y|^2 \cdot \omega^2}{8 \cdot c}$	$\dfrac{|M \cdot Y|^2 \cdot \omega^2}{8 \cdot c}$
The efficiency at the peak harvested resonant power	0.5	0.5	0.5	0.5	0.5

SECE, synchronous electric charge extraction; SSHI, synchronous switch harvesting on inductor.

connected to a single resistor or the four types of interface circuits, the maximum harvested power will be the one-eighth of the reference power which is the maximum input power or the squared input force magnitude divided by mechanical damping. The energy harvesting efficiency at the maximum harvested power is 50%. These results are same as those of PVEH as illustrated in Chapters 2 and 3.

Fig. 4.7 plots the efficiency versus the dimensionless resistance and dimensionless force factor for the four types of interface circuits. For the cases of standard, SECE, series/parallel SHHI interface circuits, it is observed that dimensionless force factor dominates the efficiency as the resonant energy harvesting efficiency gets close to 100 or 0% under large or small dimensionless force factors, respectively. This is because from the third row of Table 4.1, the energy harvesting efficiency is a monotonically increasing function with respect to α_N^2. When the dimensionless force factor is kept nonzero constant, dimensionless resistance dominates the efficiency as the resonant energy harvesting efficiency gets close to 100 or 0% under large or small dimensionless resistance, respectively. This is because from the third row of Table 4.1, the energy harvesting efficiency is a monotonically increasing function with respect to R_N.

Fig. 4.8 plots the dimensionless harvested resonant power versus the dimensionless resistance and dimensionless force factor for the four types of interface circuits.

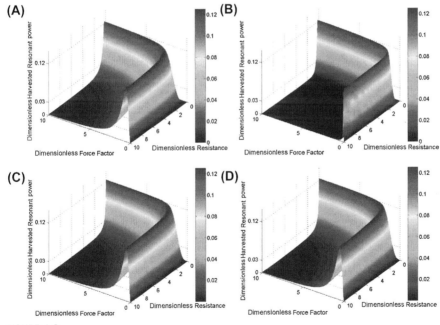

FIGURE 4.8

The dimensionless harvested resonant power versus dimensionless resistance and force factors for the SDOF electromagnetic harvester connected to the four types of interface circuits. (A) Standard interface; (B) synchronous electric charge extraction; (C) parallel synchronous switch harvesting on inductor (SSHI); (D) series SSHI.

It is observed from Fig. 4.8 that when the dimensionless force factor becomes either very large or very small, while the dimensionless resistance is kept nonzero constant, the dimensionless harvested power for the four interface circuits goes close to zero. On the other hand, it is observed that when the dimensionless resistance becomes large, while the dimensionless force factor is kept nonzero constant, the dimensionless harvested resonant power goes to close to zero. The attribute of EMVEH has the opposite trend to that of the PVEH. There is a special case in which $R_N \cdot \alpha_N^2 = $ constant when the dimensionless resistance becomes very large or very small and when the dimensionless force factor becomes either very small or very large, the dimensionless harvested power tends to be fixed constant values. This has been illustrated in Eq. (49) of Ref. [12]. Physically when the electromagnetic device is removed, $\alpha_N = 0$, the harvested resonant power is zero. When the dimensionless force factor becomes very large, the dimensionless harvested resonant power would become small and goes to zero. This is because a small mechanical damping results in a large dimensionless force factor according to Eq. (4.23). However, a small mechanical damping would make $\frac{M^2 \cdot A^2}{c}$ large and leads to a small dimensionless harvested resonant power, $P_h \Big/ \frac{M^2 \cdot A^2}{c}$.

4.10 SUMMARY

In this chapter, EMVEHs connected to a single load resistor, standard interface circuit, SECE interface circuit, the series and parallel SSHI interface circuits have been studied. Through frequency response analysis, the normalized dimensionless harvested resonant power and efficiency can be predicted from only two nondimensional parameters of R_N and α_N, dimensionless resistance and equivalent force factor. EMVEHs are able to be evaluated and compared based on the formulas of the dimensionless harvested power and efficiency regardless of its size.

For the EMVEH connected to a single resistor or one of the four types of interface circuits, the resonant energy harvesting efficiency and dimensionless harvested resonant power depend on the system resonant frequency, mechanical damping, load resistance, magnetic field intensity constant, and coil length. The maximum harvested power will be the one-eighth of the reference power which is the maximum input power or the squared input force magnitude divided by mechanical damping. The energy harvesting efficiency at the maximum harvested power is 50%.

Typically, the sequence of the interface circuits for the resonant energy harvesting efficiency from the large to small is the SECE, single load resistance, series SSHI, parallel SSHI, and standard interface circuits. The sequence of the interface circuits for the dimensionless harvested resonant power sequence from the large to small is the SECE, series SSHI, parallel SSHI, single load resistance, and standard interface circuits.

NOMENCLATURE

B	The magnetic field constant
l	The length of the coil in the electromagnetic generator
c	The short circuit mechanical damping of the single degree of freedom system
e	2.718281828
f_n	The natural frequency
i	The square root of -1
I	The current in the circuit
K	The short circuit stiffness of the single degree of freedom (SDOF) system
L_e	The inductance of the coil
L_c	The coil inductance
M	The oscillator mass of the single degree of freedom system
P_h	The harvested resonant power
P_{in}	The input resonant power
P_D	The mechanical damping dissipation power
$\frac{P_h}{\frac{M^2 \cdot A^2}{c}}$	The dimensionless harvested resonant power
$\frac{P_{h\,max}}{\frac{M^2 \cdot A^2}{c}}$	The maximum dimensionless harvested resonant power
s	Laplace variable
T	The period of the excitation force signal
V	The output voltage of the SDOF system
V_M	The output voltage amplitude of the SDOF system
V_m	The voltage amplitude after the inversion process
V_0	The rectifier voltage amplitude
$|V|$	The modulus or amplitude of the output voltage
$\overline{V}, \overline{V}(s)$	The Laplace transform of V or $V(t)$
R	The sum of the external load resistance and piezoelectric material insert
R_e	The resistance of the coil
R_c	The coil resistance
R_L	The load resistance
y	The base excitation displacement
$\dot{y}$	The base excitation velocity
$\ddot{y}$	The base excitation acceleration
A	The base excitation displacement amplitude
Y_M	The base excitation displacement amplitude
$Y, Y(s)$	The Laplace transform of y or $y(t)$
z	The relative displacement of the mass with respect to the base
$\dot{z}$	The relative velocity of the mass with respect to the base
$\ddot{z}$	The relative acceleration of the mass with respect to the base
$|z|$	The amplitude of the relative displacement
Z_M	The relative displacement amplitude of the mass with respect to the base
$Z, Z(s)$	The Laplace transform of z or $z(t)$

$Z, Z(\omega)$	The Fourier transform of z or $z(t)$
α_N	The dimensionless force factor
α	The equivalent force factor $\alpha = Bl/R$
Θ	The equivalent force factor or effective piezoelectric coefficient
R_N	The dimensionless resistance
η	The resonant energy harvesting efficiency
η_{max}	The maximum resonant energy harvesting efficiency
ω	The excitation radial frequency in radians per second
π	3.1415928
$\frac{d}{dt}$	The first differential
$\frac{\partial P_h}{\partial R_N}$ or $\frac{\partial \eta}{\partial R_N}$	The partial differential of the harvested power or efficiency with respect to the dimensionless resistance
$\frac{\partial P_h}{\partial \alpha_N}$ or $\frac{\partial \eta}{\partial \alpha_N}$	The partial differential of the harvested power or efficiency with respect to the dimensionless force factor
$\int$	Integration
$W_i (i = 1, 2, \ldots.)$	The state variable
A	The "state matrix"
B	The "input matrix"
C	The "output matrix"
D	The "feed through matrix"
E	The system eigen matrix
F	The excitation inertia vector
k_e	The electromagnetic–mechanical coupling factor
ζ	The mechanical damping ratio
ζ_c	The load coefficient
t, t_0	The time instances
$\int_{t_0}^{t_0 + \frac{T}{2}}$	The integration from t_0 to $t_0 + \frac{T}{2}$
∞	The infinity
S	Switch
Q_i	The quality factor
I_s	The current from the total charge stored on the capacitor C_{st} before and after the voltage inversion
C_{st}	The external storage capacitor
C_0	The blocking capacitance of the piezoelectric insert

Subscripts

h	Harvested energy
in	Input
M	Amplitude or before the inversion process
max	Maximum value
m	After the inversion process
N	Dimensionless

Superscripts

-1	Inverse
$\underline{}$	Time average
$.$	The first differential
$..$	Second differential

Special function
| | Modulus or absolute value

Abbreviations
N/A	Not available
SDOF	Single degree of freedom
RMS	Root mean square
Standard	Standard interface circuit
SECE	Synchronous electric charge extraction
SSHI	Synchronous switch harvesting on inductor
SMFE	Synchronized flux extraction
EMVEH	Electromagnetic vibration energy harvester
RLC circuit	An electrical circuit consisting of a resistor, an inductor, and a capacitor connected in series or in parallel

Figure Legends
SECE	Simulated results of the single degree of freedom harvester connected to a synchronous electric charge extraction circuit using frequency domain analysis
Para SSHI	Simulated results of the single degree of freedom harvester connected to a parallel synchronous switch harvesting on inductor circuit using frequency domain analysis
Series SSHI	Simulated results of the single degree of freedom harvester connected to a series synchronous switch harvesting on inductor circuit using frequency domain analysis

REFERENCES

[1] Poulin G, Sarraute E, Costa F. Generation of electrical energy for portable devices. Sensors Actuators A Phys 2004;116(3):461−71.

[2] Priya S, Inman DJ. Energy harvesting technologies. London: Springer; 2008 [Limited].

[3] Rahimi A, Zorlu Ö, Muhtaroğlu A, Külah H. A compact electromagnetic vibration harvesting system with high performance interface electronics. Procedia Eng 2011; 25:215−8.

[4] Wang X, Koss LL. Frequency response function for power, power transfer ratio and coherence. J Sound Vib 1992;155(1):55−73.

[5] Shu YC, Lien IC. Analysis of power output for piezoelectric energy harvesting systems. Smart Mater Struct 2006;15(6):1499−512.

[6] Arroyoa E, Badela A, Formosaa F, Wu Y, Qiu J. Comparison of electromagnetic and piezoelectric vibration energy harvesters: model and experiments. Sensors Actuators A 2012;183:148−56.

[7] Gatti RR. Spatially varying multi degree of freedom electromagnetic energy harvesting [Ph.D. thesis]. Australia: Curtin University; 2013.

[8] Wickenheiser AM, Reissman T, Wu WJ, Garcia E. Modelling the effects of electromechanical coupling on energy storage through piezoelectric energy harvesting. Mechatronics, IEEE/ASME Trans 2010;15(3):400−11.

[9] Constantinou P, Mellor PH, Wilcox PD. A magnetically sprung generator for energy harvesting applications. Mechatronics, IEEE/ASME Trans 2012;17(3):415−24.

[10] Guyomar D, Sebald G, Pruvost S, Lallart M. Energy harvesting from ambient vibrations and heat. J Intelligent Material Syst Struct 2009;20:609−24.

[11] Giusa F, Giuffrida A, Trigona C, Andò B, Bulsara AR, Baglio S. "Random mechanical switching harvesting on inductor": a novel approach to collect and store energy from weak random vibrations with zero voltage threshold. Sensors Actuators A Phys 2013; 198:35−45.

[12] Renno JM, Daqaq MF, Inman DJ. On the optimal energy harvesting from a vibration source. J Sound Vib 2009;320(1−2):386−405.

[13] Wang X, Liang XY, Wei HQ. A study of electromagnetic vibration energy harvesters with different interface circuits. Mech Syst Signal Process 2015;58−59(2015):376−98.

[14] Bawahab M, Xiao H, Wang X. A study of linear regenerative electromagnetic shock absorber system. 2015. SAE 2015-01-0045.

[15] Wang X, Liang XY, Hao ZY, Du HP, Zhang N, Qian M. Comparison of electromagnetic and piezoelectric vibration energy harvesters with different interface circuits. Mech Syst Signal Process 2016;72−73:906−24.

[16] Wang X. Coupling loss factor of linear vibration energy harvesting systems in a framework of statistical energy analysis. J Sound Vib 2016;362(2016):125−41.

[17] Wang X, Liang XY, Shu GQ, Watkins S. Coupling analysis of linear vibration energy harvesting systems. Mech Syst Signal Process March 2016;70−71:428−44.

[18] Wang X, John S, Watkins S, Yu XH, Xiao H, Liang XY, et al. Similarity and duality of electromagnetic and piezoelectric vibration energy harvesters. Mech Syst Signal Process 2015;52−53(2015):672−84.

Similarity and Duality of Electromagnetic and Piezoelectric Vibration Energy Harvesters

5

CHAPTER OUTLINE

5.1 INTRODUCTION

The electromagnetic and piezoelectric vibration energy harvesters (VEHs) have been intensively studied in the literature where different piezoelectric VEHs or electromagnetic VEHs were connected to a single load resistor or connected to bridge rectification and storage interface circuits. The power output has been optimized and has been normalized in dimensionless forms.

Most research papers have discussed some optimizations of harvested power of the electromagnetic or piezoelectric VEHs, while limited works have been conducted to compare electromagnetic and the piezoelectric VEHs for the power harvesting performance.

The aim of this chapter is to develop an effective approach for analysis and optimization of ambient VEHs. The developed dimensionless analysis will be used to evaluate performance of those VEHs and maximize their output power and efficiency, therefore contributing to environmental sustainability.

5.2 DIMENSIONLESS COMPARISON OF SDOF PIEZOELECTRIC AND ELECTROMAGNETIC VIBRATION ENERGY HARVESTERS CONNECTED WITH A SINGLE LOAD RESISTANCE

For a single degree of freedom (SDOF) piezoelectric vibration harvester connected with a single load resistance, according to Eqs. (2.23) and (2.26) in Chapter 2, the normalized dimensionless harvested resonant power and energy harvesting efficiency are given by

$$\frac{P_h}{\left(\frac{M^2 \cdot A^2}{c}\right)} = \frac{1}{2} \cdot \frac{R_N \cdot \alpha_N^2}{\alpha_N^4 \cdot R_N^2 + 2 \cdot R_N \cdot \alpha_N^2 + \left(1 + R_N^2\right)} \tag{2.23}$$

where P_h is the harvested resonant power, M is the oscillator mass, c is the short circuit oscillator damping, A is the input excitation acceleration amplitude, R_N is the dimensionless resistance, α_N is the dimensionless force factor. R_N and α_N are defined by Eqn (2.21). The resonant energy harvesting efficiency is given by

$$\eta = \frac{R_N \cdot \alpha_N^2}{1 + R_N \cdot \alpha_N^2 + R_N^2} \tag{2.26}$$

where η is the resonant energy harvesting efficiency. For an SDOF electromagnetic vibration harvester connected with a single load resistance, according to Eqs. (4.25) and (4.30) in Chapter 4, the normalized dimensionless harvested resonant power and energy harvesting efficiency are given by

$$\frac{P_h}{\left(\frac{M^2 \cdot A^2}{c}\right)} = \frac{1}{2} \cdot \frac{R_N \cdot \alpha_N^2}{\left(\frac{1}{R_N}\right)^2 + \left[1 + R_N \cdot \alpha_N^2\right]^2} \tag{4.25}$$

and

$$\eta = \frac{R_N \cdot \alpha_N^2}{\frac{1}{R_N^2} + \left[1 + R_N \cdot \alpha_N^2\right]} \tag{4.30}$$

where R_N and α_N are defined by Eq. (4.23).

From Eqs. (2.23) and (4.25), it is seen that the formulas of the dimensionless harvested resonant power are interchangeable from the piezoelectric to the electromagnetic VEH, when only $(R_N)^2$ term is replaced by $(1/R_N)^2$ and the $R_N \cdot \alpha_N^2$ related term is not changed. From the pair of Eqs. (2.26) and (4.30), the formulas of the resonant energy harvesting efficiency are interchangeable from the piezoelectric to the electromagnetic VEH, when only the $(R_N)^2$ term is replaced by $(1/R_N)^2$ and the $R_N \cdot \alpha_N^2$ related term is not changed. This reflects the duality between the piezoelectric and electromagnetic VEHs, which supports the conclusions in Refs [1,3−14].

From the pair of Eqs. (2.23) and (4.25), it is seen that at the electric circuit oscillation resonance $R_N = R \cdot C_0 \cdot \omega = 1$ or $R_N = \frac{R}{L_e \cdot \omega} = 1$, the dimensionless harvested resonant power is same for both the piezoelectric and electromagnetic VEH. From the pair of Eqs. (2.26) and (4.30), it is seen that at the electric circuit oscillation resonance $R_N = R \cdot C_0 \cdot \omega = 1$ or $R_N = \frac{R}{L_e \cdot \omega} = 1$, the resonant energy harvesting efficiency is same for both the piezoelectric and electromagnetic VEH. The radial resonant frequency ω in radian/s is defined as $\omega = 2\pi f \approx \sqrt{\frac{K}{M}}$. f is the resonant frequency in Hz. When $R_N = \alpha_N = 1$, $\eta = 33.3\%$ and $P_h \left/ \left(\frac{M^2 \cdot A^2}{c} \right) \right. = 0.1$ for both the piezoelectric and electromagnetic VEH. From the pair of Eqs. (2.23) and (4.25) or Eqs. (2.26) and (4.30), for both the piezoelectric and electromagnetic VEH, there does not exist a pair of R_N and α_N to achieve a peak dimensionless harvested resonant power or energy harvesting efficiency. The maximum dimensionless harvested resonant power is less than the one-eighth of the squared input force magnitude divided by mechanical damping. The energy harvesting efficiency at the maximum harvested power is 50%.

If the dimensionless electric load resistance and dimensionless force factors change from 0.0 to 10.0 in a step size of 0.1, the energy harvesting efficiency as a function of the dimensionless electric load resistance and dimensionless force factors can be plotted in Fig. 5.1. It can be seen from Eqs. (2.26) and (4.30) and Fig. 5.1 that when the dimensionless force factor tends to be very large, the dimensionless resistance is nonzero constant, the efficiency tends to be 100%. When the dimensionless force factor tends to be zero, dimensionless resistance is nonzero constant, the efficiency tends to be zero. When the dimensionless electric load resistance tends to zero, dimensionless force factor is nonzero constant, the efficiency tends to be zero. Most of the above results of the efficiency limits are same for both the types of electromagnetic and piezoelectric harvesters. The difference between the electromagnetic and

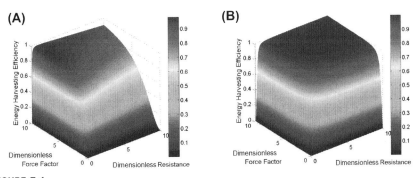

FIGURE 5.1

Dimensionless resonant energy harvesting efficiency of piezoelectric and electromagnetic vibration energy harvesters versus dimensionless resistance and force factors for the single degree of freedom harvesters connected to a load resistance; (A) piezoelectric; (B) electromagnetic.

piezoelectric harvesters is the efficiency limit of the very large dimensionless resistance. When the dimensionless resistance tends to be very large, the dimensionless force factor is not-zero constant, the efficiency for the electromagnetic harvester tends to be 100% while the efficiency for the piezoelectric harvester tends to be zero. The efficiency for the piezoelectric harvester tends to 100% when both the dimensionless resistance and dimensionless force factors tend to be large.

It is seen from Fig. 5.1 that when the dimensionless resistance R_N is very small, the efficiency of piezoelectric (A) VEH will be larger than that of electromagnetic (B) VEH. The efficiency of electromagnetic (B) VEH is very small, while the efficiency of piezoelectric (A) VEH may be large or small depending on the value of the dimensionless force factor. When the dimensionless resistance R_N is very large, the efficiency of electromagnetic (B) VEH will be larger than that of piezoelectric (A) VEH. It is seen from Fig. 5.1 that there is no local peak of the efficiency with respect to the dimensionless resistance factor R_N and force factor α_N. It can be seen that if the dimensionless resistance is nonzero constant, when the dimensionless force factor becomes larger, the efficiency goes higher. This is true for both the piezoelectric (A) and electromagnetic (B) VEHs. Physically, dimensionless force factor as defined in Eq. (2.21) or (4.23) is affected by various parameters. For example, small damping coefficient will result in large dimensionless force factor and higher energy efficiency. There is a special case in which $R_N \cdot \alpha_N^2 = \text{constant}$, when the dimensionless resistance becomes very large, while the dimensionless force factor becomes very small, the efficiency of the piezoelectric (A) VEH tends to be zero, the efficiency of the electromagnetic (B) VEH tends to be a fixed value. When the dimensionless resistance becomes very small while the dimensionless force factor becomes very large, the efficiency of the piezoelectric (A) VEH tends to be a fixed value, the efficiency of the electromagnetic (B) VEH tends to be zero.

The dimensionless harvested resonant power of the piezoelectric and electromagnetic harvesters can be calculated using Eqs. (2.23) and (4.25) where both the dimensionless electric load resistance and the dimensionless force factors are changed from 0.0 to 10.0 in a step size of 0.1. The dimensionless harvested resonant power versus the dimensionless resistance and the dimensionless force factors for the piezoelectric (A) and electromagnetic (B) harvesters is plotted in Fig. 5.2.

It is seen from Fig. 5.2 that when the dimensionless resistance R_N is very small, the dimensionless harvested resonant power of piezoelectric (A) VEH will be larger than that of electromagnetic (B) VEH. The dimensionless harvested resonant power of electromagnetic (B) VEH is very small, while the dimensionless harvested resonant power of piezoelectric (A) VEH may be large or small depending on the value of the dimensionless force factor. When the dimensionless resistance R_N is very large, the maximum dimensionless harvested resonant power of electromagnetic (B) VEH will be larger than that of piezoelectric (A) VEH.

It can be seen from Eqs. (2.23) and (4.25) and Fig. 5.2 that when the dimensionless force factor tends to be very large, while the dimensionless resistance is nonzero constant, the dimensionless harvested resonant power of both the piezoelectric (A) and electromagnetic VEHs tends to be zero. When the dimensionless force

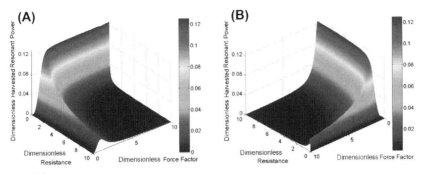

FIGURE 5.2

Dimensionless harvested resonant power versus dimensionless resistance and force factors for the single degree of freedom piezoelectric and electromagnetic harvesters connected to a load resistor; (A) piezoelectric; (B) electromagnetic.

factor tends to be very small, while the dimensionless resistance is nonzero constant, the dimensionless harvested resonant power of both the piezoelectric (A) and electromagnetic (B) VEHs tends to be zero. When the dimensionless resistance tends to be very small, while the dimensionless force factor is nonzero constant, the dimensionless harvested resonant power of the electromagnetic (B) VEH tends to be zero, the dimensionless harvested resonant power of the piezoelectric (A) VEH may be large or small, depending on the value of the dimensionless force factor. When the dimensionless resistance tends to be very large, while the dimensionless force factor is nonzero constant, the dimensionless harvested resonant power of the piezoelectric (A) VEH tends to be zero, the dimensionless harvested resonant power of the electromagnetic (B) VEH may be large or small, depending on the value of the dimensionless force factor. It is seen from Fig. 5.2 that the electromagnetic harvester harvests more power than the piezoelectric harvester in the specific range of a large dimensionless resistance and a small force factor. The piezoelectric harvester harvests more power than the electromagnetic harvester in the specific range of a small dimensionless resistance. There is a special case in which $R_N \cdot \alpha_N^2 = $ constant, when the dimensionless resistance becomes very large, while the dimensionless force factor becomes very small, the dimensionless harvested resonant power of the piezoelectric (A) VEH tends to be zero, the dimensionless harvested resonant power of the electromagnetic (B) VEH tends to be a fixed value. When the dimensionless resistance becomes very small while the dimensionless force factor becomes very large, the dimensionless harvested resonant power of the piezoelectric (A) VEH tends to be a fixed value, the dimensionless harvested resonant power of the electromagnetic (B) VEH tends to be zero.

It is seen from Eqs. (2.23) and (4.25) that the harvested resonant power is proportional to the squared oscillator mass and inversely proportional to the mechanical damping. As the mechanical damping increases, other parameters are kept constant, the harvested power decreases. It can be seen from Eqs. (2.23) and (4.25) that the

Table 5.1 Comparison of Electromagnetic and Piezoelectric Vibration Energy Harvesters Connected to a Single Load Resistor

Vibration Energy Harvester Type	Electromagnetic	Piezoelectric
Force factor	$\alpha = \frac{B \cdot l}{R}$	$\alpha = \frac{e_{33} \cdot S_0}{H}$
Dimensionless resistance and force factor	$\begin{cases} R_N = R \cdot C_0 \cdot \omega = \dfrac{R}{L_e \cdot \omega} \\[2ex] \alpha_N = \sqrt{\dfrac{\alpha^2}{c \cdot C_0 \cdot \omega}} = \sqrt{\dfrac{\alpha^2}{\left(\dfrac{c}{L_e \cdot \omega}\right)}} \end{cases}$	$\begin{cases} R_N = R \cdot C_0 \cdot \omega \\[2ex] \alpha_N = \sqrt{\dfrac{\alpha^2}{c \cdot C_0 \cdot \omega}} \end{cases}$
Equivalent capacity of material	$C_0 = \dfrac{R}{L_e \cdot \omega^2}$	$C_0 = \dfrac{\varepsilon_{33}^S \cdot S_0}{H}$
Dimensionless harvested power	$\dfrac{P_h}{\left(\frac{M^2 \cdot A^2}{c}\right)} \approx \dfrac{1}{2} \cdot \dfrac{R_N \cdot \alpha_N^2}{\left(\frac{1}{R_N}\right)^2 + \left[1 + R_N \cdot \alpha_N^2\right]^2}$	$\dfrac{P_h}{\left(\frac{M^2 \cdot A^2}{c}\right)} = \dfrac{1}{2} \cdot \dfrac{R_N \cdot \alpha_N^2}{\left(R_N \cdot \alpha_N^2 + 1\right)^2 + R_N^2}$
Efficiency	$\eta = \dfrac{R_N \cdot \alpha_N^2}{\frac{1}{R_N^2} + \left[1 + R_N \cdot \alpha_N^2\right]}$	$\eta = \dfrac{R_N \cdot \alpha_N^2}{1 + R_N \cdot \alpha_N^2 + R_N^2}$
Harvested resonant power limit	$\dfrac{M^2 \cdot A^2}{8 \cdot c}$	$\dfrac{M^2 \cdot A^2}{8 \cdot c}$
The energy harvesting efficiency at the peak harvested power	50%	50%

partial differentials of the dimensionless harvested resonant power with respect to mechanical damping are not equal to zero. When the mechanical damping changes, other parameters are kept constant, there does not exist an optimized mechanical damping value, which produces the peak dimensionless harvested resonant power. The main performance parameters of the electromagnetic and piezoelectric VEHs are compared and listed in Table 5.1.

If the dimensionless electric load resistance or the dimensionless force factor is fixed as 0.5 and 1.5, the dimensionless harvested resonant power of both the harvesters can be calculated using Eqs. (2.23) and (4.25) where the other variable is changed from 0.5 to 10.5 in a step size of 1.0. The dimensionless harvested resonant power versus the dimensionless force factor for both the harvesters are plotted in Figs. 5.3 and 5.4 for the fixed dimensionless resistances of 0.5 and 1.5. The dimensionless harvested resonant power curves versus the dimensionless resistance for both the harvesters are plotted in Figs. 5.5 and 5.6 for the fixed dimensionless force factors of 0.5 and 1.5.

Given the same oscillator mass, excitation acceleration/force amplitude, and mechanical damping, it is seen from Fig. 5.3 that when the dimensionless electric resistance is fixed to be less than one, the piezoelectric harvester harvests more power than the electromagnetic harvester. The difference is substantial when the

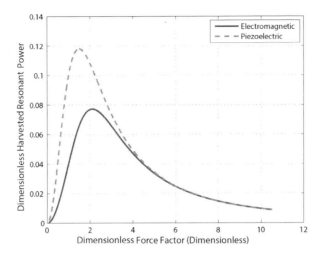

FIGURE 5.3

Dimensionless harvested resonant power versus the dimensionless force factor by fixing the dimensionless resistance of 0.5.

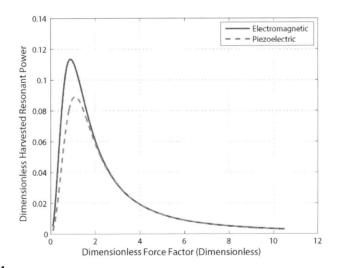

FIGURE 5.4

Dimensionless harvested resonant power versus the dimensionless force factor by fixing the dimensionless resistance of 1.5.

dimensionless force factor is less than 4. It is seen from Fig. 5.4 that when the dimensionless electric resistance is fixed to be larger than one, the piezoelectric harvester harvests less power than the electromagnetic harvester. The difference is substantial when the dimensionless force factor is less than 2. It is seen from

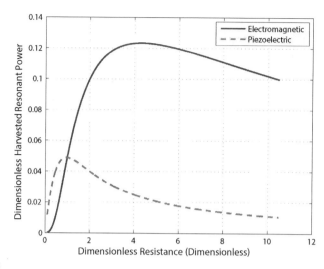

FIGURE 5.5

Dimensionless harvested resonant power versus the dimensionless resistance by fixing the dimensionless force factor of 0.5.

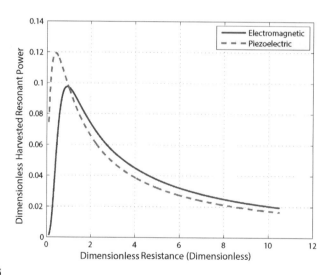

FIGURE 5.6

Dimensionless harvested resonant power versus the dimensionless resistance by fixing the dimensionless force factor of 1.5.

Eqs. (2.23) and (4.25) when the dimensionless electric resistance is set equal to one, the piezoelectric harvester harvests the same power as the electromagnetic harvester. It is seen from Figs. 5.3 and 5.4 that with dimensionless electric resistance fixed, when the dimensionless force factor varies, there exist the peak harvested resonant power for both the piezoelectric and the electromagnetic harvesters. Only an optimized design of piezoelectric material or magnetic and coil materials would produce the maximum harvested resonant power. Excessive or under design of the piezoelectric material or magnetic and coil material would decrease harvested resonant power. It is seen from Figs. 5.5 and 5.6 that with the dimensionless force factor fixed, when the dimensionless resistance changes to be larger than 1.5, the electromagnetic harvester is able to harvest more power than the piezoelectric harvester. The electromagnetic or piezoelectric harvester is able to harvest a peak power with a fixed dimensionless force factor larger than 0.3 and less than 0.8.

5.3 DIMENSIONLESS COMPARISON OF SDOF PIEZOELECTRIC AND ELECTROMAGNETIC VIBRATION ENERGY HARVESTERS CONNECTED WITH THE FOUR TYPES OF INTERFACE CIRCUITS

Four types of energy extraction and storage circuits are commonly adopted for energy harvesting devices as mentioned in the literature as shown in Fig. 3.1. They are the standard energy extraction and storage interface circuit shown in Fig. 3.1(A), the synchronous electric charge extraction (SECE) circuit shown in Fig. 3.1(B), the parallel synchronous switch harvesting on inductor (parallel SSHI) circuit shown in Fig. 3.1(C), and the series synchronous switch harvesting on inductor (series SSHI) circuit shown in Fig. 3.1(D) [2−14]. The dimensionless harvested power and efficiency and their optimized values for a piezoelectric vibration harvester connected to the four types of interface circuits have been derived in Chapters 2 and 3 according to those proposed in Refs. [2−14] and listed in Tables 5.2 and 5.3. The dimensionless harvested power and efficiency and their optimized values for an electromagnetic vibration harvester connected to the interface circuits have been derived in Chapter 4 in a similar way and listed in Tables 5.1−5.3 for comparison.

It is seen from Tables 5.1−5.3 that the dimensionless harvested resonant power peak for both the electromagnetic and piezoelectric VEHs connected to a single load resistor or the four types of interface circuits is limited to the one-eighth of the reference power or the one-eighth of squared input force amplitude divided by the damping coefficient. The energy harvesting efficiency at the peak harvested power is 50% which is independent of the types of the interface circuits or the load resistor. This shows a similarity of the electromagnetic and piezoelectric VEHs.

For the electromagnetic and piezoelectric VEHs connected to a load resistor or the standard SECE, parallel SSHI interface circuits, the calculation formulas of the energy harvesting efficiency and the dimensionless harvested resonant power are interchangeable between the piezoelectric harvester and the electromagnetic harvester with reciprocals of only individual $1/R_N$ or R_N related terms while $R_N \cdot \alpha_N^2$ related terms are kept with no changes. This shows a duality between the

Table 5.2 Comparison of Electromagnetic and Piezoelectric Vibration Energy Harvesters Connected to the Standard and SECE Interface Circuits

Vibration Energy Harvester Type	Electromagnetic + Standard Interface Circuit	Piezoelectric + Standard Interface Circuit	Electromagnetic + SECE Interface Circuit	Piezoelectric + SECE Interface Circuit
Equivalent force factor	$\alpha = \frac{B \cdot l}{R}$	$\alpha = \frac{e_{33} \cdot S_0}{H}$	$\alpha = \frac{B \cdot l}{R}$	$\alpha = \frac{e_{33} \cdot S_0}{H}$
Dimensionless resistance and force factors (load and coupling coefficients)	$\begin{cases} R_N = R \cdot C_0 \cdot \omega = \frac{R}{L_e \cdot \omega} \\ \alpha_N = \sqrt{\frac{\alpha^2}{c \cdot C_0 \omega}} = \sqrt{\left(\frac{c}{L_e \cdot \omega}\right)} \end{cases}$	$\begin{cases} R_N = R \cdot C_0 \cdot \omega \\ \alpha_N = \sqrt{\frac{\alpha^2}{c \cdot C_0 \cdot \omega}} \end{cases}$	$\begin{cases} R_N = R \cdot C_0 \cdot \omega = \frac{R}{L_e \cdot \omega} \\ \alpha_N = \sqrt{\frac{\alpha^2}{c \cdot C_0 \cdot \omega}} = \sqrt{\left(\frac{c}{L_e \cdot \omega}\right)} \end{cases}$	$\begin{cases} R_N = R \cdot C_0 \cdot \omega \\ \alpha_N = \sqrt{\frac{\alpha^2}{c \cdot C_0 \cdot \omega}} \end{cases}$
Equivalent capacity of material	$C_0 = \frac{R}{L_e \cdot \omega^2}$	$C_0 = \frac{e_{33}^S \cdot S_0}{H}$	$C_0 = \frac{R}{L_e \cdot \omega^2}$	$C_0 = \frac{e_{33}^S \cdot S_0}{H}$
Dimensionless harvested power	$\frac{P_h}{\left(\frac{M^2 \cdot A^2}{c}\right)} = \frac{\alpha_N^2 \cdot R_N \cdot \left(\frac{\pi}{2} + \frac{1}{R_N}\right)^2}{2 \cdot \left[\left(\frac{\pi}{2} + \frac{1}{R_N}\right)^2 + \alpha_N^2 \cdot R_N\right]^2}$	$\frac{P_h}{\left(\frac{M^2 \cdot A^2}{c}\right)} = \frac{R_N \cdot \alpha_N^2 \cdot \left(R_N + \frac{\pi}{2}\right)^2}{2 \cdot \left[\left(R_N + \frac{\pi}{2}\right)^2 + R_N \cdot \alpha_N^2\right]^2}$	$\frac{P_h}{\frac{M^2 \cdot A^2}{c}} = \frac{1}{2} \cdot \frac{\alpha_N^2 \cdot R_N^3}{\left(1 + \alpha_N^2 \cdot R_N^3\right)^2}$	$\frac{P_h}{\frac{(M \cdot A)^2}{c}} = \frac{1}{2} \cdot \frac{R_N \cdot \alpha_N^2}{\left(R_N + \alpha_N^2\right)^2}$
Efficiency	$\eta = \frac{\alpha_N^2 \cdot R_N}{\left(\frac{\pi}{2} + \frac{1}{R_N}\right)^2 + \alpha_N^2 \cdot R_N}$	$\eta = \frac{R_N \cdot \alpha_N^2}{\left(R_N + \frac{\pi}{2}\right)^2 + R_N \cdot \alpha_N^2}$	$\eta = \frac{\alpha_N^2 \cdot R_N^3}{1 + \alpha_N^2 \cdot R_N^3}$	$\eta = \frac{\alpha_N^2}{R_N + \alpha_N^2}$
Harvested resonant power limit	$\frac{M \cdot A^2}{8 \cdot c}$	$\frac{M \cdot A^2}{8 \cdot c}$	$\frac{M \cdot A^2}{8 \cdot c}$	$\frac{M \cdot A^2}{8 \cdot c}$
The energy harvesting efficiency at the peak harvested power	0.5	0.5	0.5	0.5

SECE, synchronous electric charge extraction.

Table 5.3 Comparison of Electromagnetic and Piezoelectric Vibration Energy Harvesters Connected to Series and Parallel SSHI Interface Circuits

Vibration Energy Harvester Type	Electromagnetic + Series SSHI Interface Circuit	Piezoelectric + series SSHI Interface Circuit	Electromagnetic + Parallel SSHI Interface Circuit	Piezoelectric + Parallel SSHI Interface Circuit
Equivalent force factor	$\alpha = \frac{B \cdot l}{R}$	$\alpha = \frac{e_{33} \cdot S_0}{H}$	$\alpha = \frac{B \cdot l}{R}$	$\alpha = \frac{e_{33} \cdot S_0}{H}$
Dimensionless resistance and force factors	$\begin{cases} R_N = R \cdot C_0 \cdot \omega = \frac{R}{L_e \cdot \omega} \\ \alpha_N = \sqrt{\frac{\alpha^2}{c \cdot C_0 \cdot \omega}} = \sqrt{\frac{\alpha^2}{\left(\frac{c}{L_e \cdot \omega}\right)}} \end{cases}$	$\begin{cases} R_N = R \cdot C_0 \cdot \omega \\ \alpha_N = \sqrt{\frac{\alpha^2}{c \cdot C_0 \cdot \omega}} \end{cases}$	$\begin{cases} R_N = R \cdot C_0 \cdot \omega = \frac{R}{L_e \cdot \omega} \\ \alpha_N = \sqrt{\frac{\alpha^2}{c \cdot C_0 \cdot \omega}} = \sqrt{\frac{\alpha^2}{\left(\frac{c}{L_e \cdot \omega}\right)}} \end{cases}$	$\begin{cases} R_N = R \cdot C_0 \cdot \omega \\ \alpha_N = \sqrt{\frac{\alpha^2}{c \cdot C_0 \cdot \omega}} \end{cases}$
Equivalent capacity of material	$C_0 = \frac{R}{L_e \cdot \omega^2}$	$C_0 = \frac{\varepsilon_{33}^S \cdot S_0}{H}$	$C_0 = \frac{R}{L_e \cdot \omega^2}$	$C_0 = \frac{\varepsilon_{33}^S \cdot S_0}{H}$
Dimensionless harvested power	$\frac{P_h}{\left(\frac{M^2 \cdot A^2}{c}\right)} = \frac{2 \cdot \pi^2 \cdot \alpha_N^2 \cdot R_N}{(\pi^2 + 4 \cdot \alpha_N^2 \cdot R_N)^2}$	$\frac{P_h}{\frac{\|MA\|^2}{c}} = \frac{2 \cdot \pi^2 \cdot \alpha_N^2 \cdot R_N}{(\pi^2 + 4 \cdot \alpha_N^2 \cdot R_N)^2}$	$\frac{P_h}{\left(\frac{M^2 \cdot A^2}{c}\right)} = \frac{1}{2} \cdot \frac{4 \cdot \alpha_N^2 \cdot R_N \cdot \left[\pi \cdot \frac{1}{R_N}\left(1 - e^{-\frac{\pi}{2 \cdot Q_i}}\right)\right]^2}{\left[\left(\pi + \frac{1}{R_N}\left(1 - e^{-\frac{\pi}{2 Q_i}}\right)\right)^2 + 4 \cdot \alpha_N^2 \cdot R_N\right]^2}$	$\frac{P_h}{\frac{\|MA\|^2}{c}} = \frac{1}{2} \cdot \frac{4 \cdot \alpha_N^2 \cdot R_N \cdot \left(R_N \cdot \left(1 - e^{-\frac{\pi}{2 Q_i}}\right) + \pi\right)^2}{\left[4 \cdot \alpha_N^2 \cdot R_N + \left(R_N \cdot \left(1 - e^{-\frac{\pi}{2 Q_i}}\right) + \pi\right)^2\right]^2}$
Efficiency	$\eta = \frac{4 R_N \alpha_N^2}{\pi^2 + 4 R_N \alpha_N^2}$	$\eta = \frac{4 \cdot R_N \cdot \alpha_N^2}{\pi^2 + 4 \cdot R_N \cdot \alpha_N^2}$	$\eta = \frac{4 \cdot \alpha_N^2 \cdot R_N}{\left[\pi + \frac{1}{R_N}\left(1 - e^{-\frac{\pi}{2 Q_i}}\right)\right]^2 + 4 \cdot \alpha_N^2 \cdot R_N}$	$\eta = \frac{4 \cdot R_N \cdot \alpha_N^2}{\left(R_N \cdot \left(1 - e^{-\frac{\pi}{2 Q_i}}\right) + \pi\right)^2 + 4 \cdot R_N \cdot \alpha_N^2}$
Harvested resonant power limit	$\frac{M \cdot A^2}{8 \cdot c}$	$\frac{M \cdot A^2}{8 \cdot c}$	$\frac{M \cdot A^2}{8 \cdot c}$	$\frac{M \cdot A^2}{8 \cdot c}$
The energy harvesting efficiency at the peak harvested power	0.5	0.5	0.5	0.5

SSHI, synchronous switch harvesting on inductor.

piezoelectric and electromagnetic VEHs. For the piezoelectric and electromagnetic harvesters connected to the series SSHI interface circuit, as shown in Table 5.3, the calculation formulas of the energy harvesting efficiency and the dimensionless harvested resonant power are identical, which shows the similarity of the two types of VEHs. It is seen from Tables 5.1–5.3 that when $R_N = 1$, the dimensionless harvested resonant power and energy efficiency of both the piezoelectric and electromagnetic VEHs are same, which also reflects the similarity of the two types of VEHs.

5.4 SUMMARY

Based on the Laplace and Fourier transform methods, an SDOF VEH connected to a single load resistor and the four types of energy extraction and storage interface circuits has been studied. The dimensionless efficiency and harvested resonant power of both the SDOF electromagnetic and piezoelectric VEHs have been analyzed and compared as shown in Tables 5.1–5.3. The normalization used points out for any type of the harvesters, only two characteristic normalized parameters of the dimensionless resistance and force factors are sufficient to describe the energy harvesting efficiency and harvested resonant power in a dimensionless form regardless of the size and configuration of the harvesters. There exists a duality between the piezoelectric and electromagnetic VEHs connected to a load resistor or the standard SECE and parallel SSHI interface circuits. The calculation formulas of the energy harvesting efficiency and the dimensionless harvested resonant power are interchangeable between the piezoelectric harvester and the electromagnetic harvester with reciprocals of only individual $1/R_N$ or R_N related terms while the $R_N \cdot \alpha_N^2$ related terms are not changed.

There exists a similarity of the harvested resonant power for the piezoelectric and electromagnetic VEHs. Both the electromagnetic and piezoelectric harvesters have a harvested resonant power limit which is one-eighth of the reference power or one-eighth of the squared input force amplitude divided by the damping coefficient. The peak power limit of the electromagnetic and piezoelectric harvesters occurs in the opposite range extremes of the dimensionless resistance. The energy harvesting efficiency at the peak harvested power is 50% for the two types of the harvesters connected to the load resistance or the four types of the interface circuits. For the electromagnetic and piezoelectric VEHs connected to the series SSHI interface circuits, the calculation formulas of the harvested resonant power and energy harvesting efficiency of the two types of the harvesters are identical. At the interface electric circuit resonance $R_N = 1$, the calculation formulas of the harvested resonant power and energy harvesting efficiency of the two types of the harvesters are identical. If the variable ranges of the dimensionless resistance and force factors are not limited, it is impossible to obtain an energy harvesting efficiency peak value at one pair of optimal dimensionless resistance and force factors for both the piezoelectric and electromagnetic harvesters. When the dimensionless electric load resistance tends to be very large, the dimensionless force factor is kept as constant, the harvesting efficiency tends to be 100% for the electromagnetic harvester while the harvesting efficiency tends to be zero for the piezoelectric harvester.

Given the same mass, excitation acceleration/force amplitude, and mechanical damping of the electromagnetic and piezoelectric harvesters, when only the dimensionless force factor varies, the dimensionless electric resistance is fixed to be less than one, the piezoelectric harvester harvests more power than the electromagnetic harvester. The difference of the harvested resonant power is large when the dimensionless force factor is less than 4. When the dimensionless electric resistance is fixed to be larger than one, the piezoelectric harvester harvests less power than the electromagnetic harvester. The difference of the harvested resonant power is large when the dimensionless force factor is less than 2. When the dimensionless electric resistance is fixed to be unity, the piezoelectric harvester harvests the same power as the electromagnetic harvester. There exist the peak harvested resonant power versus the dimensionless force factor for both the piezoelectric and the electromagnetic harvesters. Excessive or under design of piezoelectric material or magnetic and coil materials would decrease harvested resonant power.

Given the same mass, excitation acceleration/force amplitude, and mechanical damping of the electromagnetic and piezoelectric harvesters, when the dimensionless force factor is kept as constant, when only the dimensionless resistance changes to be larger than 1.5, the electromagnetic harvester is able to harvest more power than the piezoelectric harvester. The electromagnetic or piezoelectric harvester is able to harvest a peak power versus the dimensionless resistance with a fixed dimensionless force factor larger than 0.3 and less than 0.8.

There exists a similarity of the dimensionless harvested power between the piezoelectric and electromagnetic VEHs. If the variable ranges of the dimensionless resistance and force factors are not limited, it is impossible to obtain the peak dimensionless harvested power at one pair of optimal dimensionless resistance and force factors. For both the electromagnetic and piezoelectric harvesters, the harvested resonant power is proportional to the squared oscillator mass, the squared excitation acceleration, and inversely proportional to the mechanical damping. When the mechanical damping changes, other parameters are kept constant, there does not exist an optimized mechanical damping value, which produces the peak harvested resonant power. As the mechanical damping increases, other parameters are kept constant, the harvested resonant power will decrease.

NOMENCLATURE

S_0	The piezoelectric material insert disk surface area
A	The input excitation acceleration amplitude
c	The short circuit mechanical damping of the single degree of freedom system
C_0	The blocking capacity of the piezoelectric material insert
e_{33}	The piezoelectric constant

ε_{33}^{S}	The piezoelectric permittivity
f	The natural frequency in Hz
ω	The natural radial frequency in radian/s
K	The short circuit stiffness of the single degree of freedom (SDOF) system
B	The magnetic field constant, magnetic flux density
L	The length of the coil in the electromagnetic generator
L_e	The self inductance of the coil
R	The external load resistance
H	The thickness of the piezoelectric material insert disk thickness
M	The oscillator mass of the single degree of freedom system
P_h	The harvested resonant power
P_{in}	The input resonant power
$\frac{P_h}{\left(\frac{M^2 A^2}{c}\right)}$	The dimensionless harvested resonant power
$\frac{P_{h\ max}}{\left(\frac{M^2 A^2}{c}\right)}$	The peak dimensionless harvested resonant power
α	The force factor of the piezoelectrical material insert or equivalent force factor of the electromagnetic device
α_N	The dimensionless force factor
R_N	The dimensionless resistance
η	The resonant energy harvesting efficiency
η_{max}	The maximum resonant energy harvesting efficiency
π	3.1415928

Subscripts

0	The rectified or blocking capacity of the piezoelectric material insert
33	Piezoelectrical working mode having the same direction of loading and electric poles
h	Harvested energy
N	Normalized

Superscripts

.	The first differential
..	The second differential

Special function

| |	Modulus or absolute value

Abbreviations

N/A	Not available
SDOF	Single degree of freedom

REFERENCES

[1] Shu YC, Lien IC, Wu WJ. An improved analysis of the SSHI interface in piezoelectric energy harvesting. Smart Mater Struct 2007;16(6):2253−64.

[2] Guyomar D, Sebald G, Pruvost S, Lallart M. Energy harvesting from ambient vibrations and heat. J Intell Mater Syst Struct 2009;20:609−24.

[3] Wang X, Liang XY, Wei HQ. A study of electromagnetic vibration energy harvesters with different interface circuits. Mech Syst Signal Process 2015;58−59(2015):376−98.

[4] Bawahab M, Xiao H, Wang X. A study of linear regenerative electromagnetic shock absorber system. SAE 2015-01-0045. 2015.

[5] Xiao H, Wang X, John S. A dimensionless analysis of a 2DOF piezoelectric vibration energy harvester. Mech Syst Signal Process 2015;58−59(2015):355−75.

[6] Xiao H, Wang X, John S. A multi degree of freedom piezoelectric vibration energy harvester with piezoelectric elements inserted between two nearby oscillators. Mech Syst Signal Process February 2016;68−69:138−54.

[7] Xiao H, Wang X. A review of piezoelectric vibration energy harvesting techniques. Int Rev Mech Eng 2014;8(3). ISSN: 1970-8742:609−20.

[8] Wang X, Lin LW. Dimensionless optimization of piezoelectric vibration energy harvesters with different interface circuits'. Smart Mater Struct 2013;22. ISSN: 0964-1726:1−20.

[9] Wang X, Xiao H. Dimensionless analysis and optimization of piezoelectric vibration energy harvester. Int Rev Mech Eng 2013;7(4). ISSN: 1970-8742:607−24.

[10] Cojocariu B, Hill A, Escudero A, Xiao H, Wang X. Piezoelectric vibration energy harvester − design and prototype. In: Proceedings of ASME 2012 international mechanical engineering congress & exposition, ISBN 978-0-7918-4517-2.

[11] Wang X, Liang XY, Hao ZY, Du HP, Zhang N, Qian M. Comparison of electromagnetic and piezoelectric vibration energy harvesters with different interface circuits. Mech Syst Signal Process 2016;72−73:906−24. http://dx.doi.org/10.1016/j.ymssp.2015.10.016.

[12] Wang X. Coupling loss factor of linear vibration energy harvesting systems in a framework of statistical energy analysis. J Sound Vib 2016;362(2016):125−41.

[13] Wang X, Liang XY, Shu GQ, Watkins S. Coupling analysis of linear vibration energy harvesting systems. Mech Syst Signal Process March 2016;70−71:428−44.

[14] Wang X, John S, Watkins S, Yu XH, Xiao H, Liang XY, et al. Similarity and duality of electromagnetic and piezoelectric vibration energy harvesters. Mech Syst Signal Process 2015;52−53(2015):672−84.

A Study of a 2DOF Piezoelectric Vibration Energy Harvester and Its Application

H. Xiao, X. Wang

RMIT University, Melbourne, VIC, Australia

CHAPTER OUTLINE

6.1 INTRODUCTION

The vibration energy harvesting technique using piezoelectric materials has been intensively studied in the recent years. Conversion of ambient vibration energy into electric energy provides an attractive alternative energy source. Despite the power density of mechanical vibration (300 μm W/cm^3) is not as high as the power density of outdoor solar energy (15,000 μm W/cm^3), the vibration energy sources are potentially sustainable and perennial [1]. The piezoelectric vibration energy harvesting techniques have been well developed, numerous studies of the piezoelectric vibration energy harvesting have received the most attentions, because the piezoelectric vibration harvesters are able to operate in a wide frequency range and easy to fabricate [2,22–31]. Most of the research studies were focused on a cantilever beam attached with piezoelectric element [3–6], this is because the cantilever beam piezoelectric vibration energy harvester (PVEH) can be simplified as a single degree of freedom (SDOF) system for simulation and is easy to fabricate in microscale. The cantilever beam PVEH has many advantages such as distributing stress more evenly. However, the SDOF PVEH only works efficiently at a sole resonant frequency. Unfortunately, a majority of potential vibration energy sources are in the form of variable or random frequencies. Therefore, a major challenge is to improve the harvesting efficiency of PVEH under various excitation frequencies in a practical environment.

Several research studies were carried out to modify the structure of vibration energy harvesting device for tuning the resonant frequency to adapt the frequency of the ambient vibration energy source. One of the approaches was active self-tuning structures proposed by Wu and Roundy [7,8]. Though these techniques increase average harvested power by 30%, they require more power to activate the resonant frequency tuning structure than that the device can generate. On the other hand, passive or intermittent tuning techniques were studied by Cornwell [9] representing as "tuned auxiliary structure." However, it needs additional sensors or actuators to be added into harvested structure which has significantly increased the size of the device and increased the complexity of the mechanical structure. To widen the harvesting frequency bandwidth of the energy harvester is another research aspect. There are two major kinds of mechanical approaches. One is to attach multiple masses and springs to the harvesting device which converts the device into a multiple degree of freedom (MDOF) system with multiple resonant modes. Shahusz [10] proposed an MDOF PVEH which is constructed from many SDOF devices in a serial connection. Similarly, Erturk [11] demonstrated an L-shaped cantilever beam energy harvesting device which can operate in two modes of resonant frequency. In his research, two lumped masses are attached on the horizontal and vertical beam, respectively. Hence, to widen the frequency bandwidth of a vibration energy harvester, the second resonant frequency could be tuned not very far from the first natural frequency by changing the ratio of the two lumped masses. Another mechanical solution is to connect multiple cantilevered beams of different length. Sari [12] introduced a device consisting of an array of 40 cantilevered beams of variable length. It is useful as the vibration energy harvester works well in a wide frequency range of ambient vibration energy sources, nevertheless, not all the cantilevered beams are activated in a resonant frequency. However, the disadvantage of the array configuration is that the size of vibration energy device increases significantly which is not suitable for most of the microelectromechanical system (MEMS) applications. Wu [13] presented a novel two degrees of freedom (2DOF) PVEH which has the same size as an SDOF cantilever beam configuration but has two close resonant frequencies. The device can be easily converted from an SDOF cantilever beam energy harvester by cutting the inner beam inside and attaching another proof mass. It is a novel design concept which is extremely useful in practice and can be applied into a constrained space, especially in MEMS devices.

Despite of many solutions which are proposed to widen the harvesting frequency bandwidth, they are all focused on the small scale or microscale. As the piezoelectric vibration energy harvesting devices produce large output voltage or power under a large preload, it requires to increase the working stress. The working stress is limited by the material's mechanical strength in small MEMS systems [14]. It is easy to find large stresses in a large scale vibration energy harvesting environment, which can boost the power output range from 10—100 mW to 1 W—100 kW or more [15].

In this chapter, a dimensionless analysis method will be proposed for evaluating a 2DOF system, for example, it could be a quarter vehicle suspension model with built-in piezoelectric materials. It is also important to design a 2DOF vibration energy harvester for a designated ambient vibration energy source. As the proposed theoretical analysis method is in a dimensionless form, it can be used as a tool to design a 2DOF PVEH regardless of the geometries, size, or scale.

6.2 ANALYSIS AND SIMULATION OF A TWO DEGREE OF FREEDOM PIEZOELECTRIC VIBRATION ENERGY HARVESTER

A 2DOF piezoelectric vibration energy harvesting system model is built and shown in Fig. 6.1; the mechanical system governing equations are given by:

$$\begin{cases} m_1 \cdot \ddot{x}_1(t) = \alpha \cdot V(t) - k_2 \cdot [x_1(t) - x_2(t)] - c_2 \cdot [\dot{x}_1(t) - \dot{x}_2(t)] - k_1 \cdot [x_1(t) - y(t)] - c_1 \cdot [\dot{x}_1(t) - \dot{y}(t)] \\ m_2 \cdot \ddot{x}_2(t) = -k_2 \cdot [x_2(t) - x_1(t)] - c_2 \cdot [\dot{x}_2(t) - \dot{x}_1(t)] - \alpha \cdot V(t) \end{cases}$$

$$(6.1)$$

As well as the electrical system governing equation is given by

$$\alpha \cdot \left[\dot{x}_2(t) - \dot{x}_1(t)\right] = \frac{V(t)}{R} + C_0 \cdot \dot{V}(t) \qquad (6.2)$$

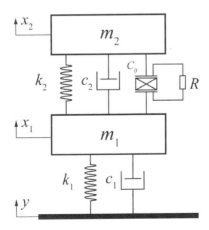

FIGURE 6.1

Schematic of a two degree of freedom piezoelectric vibration energy harvesting system model.

where the electrical energy generated by the piezoelectric element is the sum of the energy flow to the electric circuit and the electrostatic energy stored on the capacitance C_0 of the piezoelectric material [16].

For the 2DOF piezoelectric vibration energy harvesting system model, y is excitation displacement; m_1, k_1, and c_1 are the mass, stiffness, and damping coefficients of the primary oscillator; m_2, k_2, and c_2 are the mass, stiffness, and damping coefficients of the auxiliary oscillator; x_1 is the displacement of the primary oscillator; x_2 is the displacement of the auxiliary oscillator; V is the voltage generated by the piezoelectric element; R is the total resistance including the external load resistance and the internal resistance of the piezoelectric element insert; α and C_0 are the force factor and blocking capacitance of the piezoelectric insert, respectively, as defined in Eq. (6.3) [17]. The 2DOF PVEH system studied in this chapter has a piezo patch inserted between the primary and auxiliary oscillators. There is no piezo patch inserted between the primary oscillator and the base. The 2DOF PVEH system is assigned for the name of the conventional PVEH system.

$$\begin{cases} \alpha = -\dfrac{e_{33} \cdot S_0}{H} \\[3mm] C_0 = -\dfrac{\varepsilon^S \cdot S_0}{H} \end{cases} \tag{6.3}$$

where e_{33} and ε^S are the piezoelectric constant and permittivity of piezoelectric insert, respectively; S_0 and H are the surface area and thickness of piezoelectric insert, respectively.

Applying the Laplace transform to Eqs. (6.1) and (6.2), the dynamic equations of mechanical and electrical systems are given by:

$$\begin{cases} m_1 \cdot X_1 \cdot s^2 = \alpha \cdot \overline{V} - k_2 \cdot (X_1 - X_2) - c_2 \cdot s \cdot (X_1 - Y) - k_1 \cdot (X_1 - Y) - c_1 \cdot s \cdot (X_1 - Y) \\ m_2 \cdot X_2 \cdot s^2 = -k_2 \cdot (X_2 - X_1) - c_2 \cdot s \cdot (X_2 - X_1) - \alpha \cdot \overline{V} \end{cases}$$

$$\tag{6.4}$$

$$\frac{\overline{V}}{X_2 - X_1} = \frac{\alpha \cdot R \cdot s}{1 + R \cdot C_0 \cdot s} \tag{6.5}$$

where $s = i \cdot \omega$, X_i, Y, and $\overline{V}$ are the Laplace transform function of $x_i(t)$, $y(t)$ and $V(t)$, $i = 1, 2$. X_i, Y, and $\overline{V}$ in the Eq. (6.4) and onward equations are the short symbols of $X_i(s)$, $Y(s)$, $\overline{V}(s)$. The initial conditions are assumed that $x_i(0) = \frac{dx_i(0)}{dt} = 0$, $y(0) = \frac{dy(0)}{dt} = 0$, and $V(0) = \frac{dV(0)}{dt} = 0$ when $t = 0$.

The transfer function equations between the oscillator displacements and the excitation displacement are derived from Eqs. (6.4) and (6.5) and given by:

$$\begin{cases} \dfrac{X_1}{Y} = \dfrac{(k_1 + c_1 \cdot s) \cdot \left[(k_2 + c_2 \cdot s + m_2 \cdot s^2) \cdot (1 + R \cdot C_0 \cdot s) + \alpha^2 \cdot R \cdot s \right]}{(m_1 \cdot s^2 + k_1 + c_1 \cdot s) \cdot \left[(k_2 + c_2 \cdot s + m_2 \cdot s^2) \cdot (1 + R \cdot C_0 \cdot s) + \alpha^2 \cdot R \cdot s \right] + \left[(k_2 + c_2 \cdot s) \cdot (1 + R \cdot C_0 \cdot s) + \alpha^2 \cdot R \cdot s \right] \cdot m_2 \cdot s^2} \\[3mm] \dfrac{X_2}{Y} = \dfrac{(k_1 + c_1 \cdot s) \cdot \left[(k_2 + c_2 \cdot s) \cdot (1 + R \cdot C_0 \cdot s) + \alpha^2 \cdot R \cdot s \right]}{(m_1 \cdot s^2 + k_1 + c_1 \cdot s) \cdot \left[(k_2 + c_2 \cdot s + m_2 \cdot s^2) \cdot (1 + R \cdot C_0 \cdot s) + \alpha^2 \cdot R \cdot s \right] + \left[(k_2 + c_2 \cdot s) \cdot (1 + R \cdot C_0 \cdot s) + \alpha^2 \cdot R \cdot s \right] \cdot m_2 \cdot s^2} \end{cases} \tag{6.6}$$

Therefore, the transfer function equation between the output voltage and excitation displacement is given by:

$$
\frac{\bar{V}}{Y} = \frac{V}{X_2 - X_1} \cdot \frac{X_2 - X_1}{Y}
$$

$$
= \frac{-m_2 \cdot s^3 \cdot (k_1 + c_1 \cdot s) \cdot \alpha \cdot R}{(m_1 \cdot s^2 + k_1 + c_1 \cdot s) \cdot \begin{bmatrix} (k_2 + c_2 \cdot s + m_2 \cdot s^2) \cdot (1 + R \cdot C_0 \cdot s) \\ + \alpha^2 \cdot R \cdot s \end{bmatrix} + \begin{bmatrix} (k_2 + c_2 \cdot s) \cdot (1 + R \cdot C_0 \cdot s) \\ + \alpha^2 \cdot R \cdot s \end{bmatrix} \cdot m_2 \cdot s^2}
$$

(6.7)

For a harmonic excitation of $y(t) = Y(s) \cdot e^{s \cdot t}$, the relationship between the excitation displacement and the excitation acceleration can be described by:

$$
\ddot{y}(t) = s^2 \cdot Y(s) \cdot e^{s \cdot t} = s^2 \cdot y(t)
$$

(6.8)

As a result, the output voltage subjected to excitation acceleration is given by

$$
\frac{\bar{V}}{s^2 \cdot Y} = \frac{-m_2 \cdot (k_1 + c_1 \cdot s) \cdot (\alpha \cdot R \cdot s)}{m_2 \cdot s^2 \cdot \begin{bmatrix} (1 + R \cdot C_0 \cdot s) \cdot (k_2 + c_2 \cdot s) \\ + \alpha^2 \cdot R \cdot s \end{bmatrix} + (m_1 \cdot s^2 + k_1 + c_1 \cdot s) \cdot \begin{bmatrix} m_2 \cdot s^2 \cdot (1 + R \cdot C_0 \cdot s) + \alpha^2 \cdot R \cdot s \\ + (1 + R \cdot C_0 \cdot s) \cdot (k_2 + c_2 \cdot s) \end{bmatrix}}
$$

(6.9)

As well as the equations described the output voltage magnitude is given by

$$
\left| \frac{\bar{V}}{A} \right| = \frac{m_2 \cdot \alpha \cdot R \cdot \omega \cdot \sqrt{k_1^2 + c_1^2 \cdot \omega^2}}{\left\{ \begin{bmatrix} \left[(k_1 - m_1 \cdot \omega^2) \cdot (k_2 - m_2 \cdot \omega^2) - k_2 \cdot m_2 \cdot \omega^2 \right] - \begin{pmatrix} k_1 \cdot c_2 + k_2 \cdot c_1 - m_1 \cdot c_2 \cdot \omega^2 \\ -m_2 \cdot c_1 \cdot \omega^2 - m_2 \cdot c_2 \cdot \omega^2 \end{pmatrix} \cdot (R \cdot C_0 \cdot \omega^2) \\ -\omega^2 \cdot c_1 \cdot (c_2 + \alpha^2 \cdot R) \end{bmatrix}^2 \right.}{}
$$
$$
\left. + \begin{bmatrix} R \cdot C_0 \cdot \omega \cdot \left[(k_1 - m_1 \cdot \omega^2) \cdot (k_2 - m_2 \cdot \omega^2) - k_2 \cdot m_2 \cdot \omega^2 \right] \\ + \omega \cdot (c_2 + \alpha^2 \cdot R) \cdot \left[(k_1 - m_1 \cdot \omega^2) - m_2 \cdot \omega^2 \right] + c_1 \cdot \omega \cdot (k_2 - m_2 \cdot \omega^2 - R \cdot C_0 \cdot c_2 \cdot \omega^2) \end{bmatrix}^2 \right\}^{0.5}}
$$

(6.10)

And according to Ref. [21], the mean harvested power is given by

$$
\frac{P_h}{A^2} = \frac{1}{2} \cdot \frac{\left| \bar{V} \right|^2}{A^2}
$$

$$
= \frac{(\alpha^2 \cdot R \cdot \omega^2 \cdot m_2^2) \cdot (k_1^2 + c_1^2 \cdot \omega^2)}{2 \cdot \left[\left[(k_1 - m_1 \omega^2) \cdot (k_2 - m_2 \cdot \omega^2) - k_2 \cdot m_2 \cdot \omega^2 \right] - \begin{pmatrix} k_1 \cdot c_2 + k_2 \cdot c_1 - m_1 \cdot c_2 \cdot \omega^2 \\ -m_2 \cdot c_1 \cdot \omega^2 - m_2 \cdot c_2 \cdot \omega^2 \end{pmatrix} R \cdot C_0 \cdot \omega^2 - \omega^2 \cdot c_1 \cdot (c_2 + \alpha^2 \cdot R) \right]^2}
$$
$$
+ \begin{bmatrix} R \cdot C_0 \cdot \omega \cdot \left[(k_1 - m_1 \cdot \omega^2) \cdot (k_2 - m_2 \cdot \omega^2) - k_2 \cdot m_2 \cdot \omega^2 \right] + \omega \cdot (c_2 + \alpha^2 \cdot R) \cdot \left[(k_1 - m_1 \cdot \omega^2) - m_2 \cdot \omega^2 \right] \\ + c_1 \cdot \omega \cdot (k_2 - m_2 \cdot \omega^2 - R \cdot C_0 \cdot c_2 \cdot \omega^2) \end{bmatrix}^2
$$

(6.11)

where $A = \left| s^2 \cdot Y \right|$ is the input excitation acceleration amplitude. From Eqs. (6.10) and (6.11), the output voltage and harvested power of the 2DOF spring mass dashpot system can be simulated and calculated using a Matlab code.

6.3 DIMENSIONLESS ANALYSIS OF A WEAKLY COUPLED 2DOF PIEZOELECTRIC VIBRATION ENERGY HARVESTER MODEL

In this section, the dimensionless formulas of output voltage and harvested power are developed. These formulas will allow the performance comparison of vibration energy harvesters regardless the size or scale. The following dimensionless parameters are introduced for the analysis:

$$
\xi_{11} = \frac{c_1}{2 \cdot \sqrt{k_1 \cdot m_1}}, \quad \xi_{22} = \frac{c_2}{2 \cdot \sqrt{k_2 \cdot m_2}}
$$

$$
\alpha_{N_1} = \sqrt{\frac{\alpha^2}{k_1 \cdot C_0}}, \quad \alpha_{N_2} = \sqrt{\frac{\alpha^2}{k_2 \cdot C_0}}
$$

$$
R_N = R \cdot C_0 \cdot \omega_n
$$

$$
\omega_{11} = \sqrt{\frac{k_1}{m_1}}, \quad \omega_{22} = \sqrt{\frac{k_2}{m_2}}
\qquad (6.12)
$$

$$
\Phi_1 = \frac{\omega_{11}}{\omega_n}, \quad \Phi_2 = \frac{\omega_{22}}{\omega_n}
$$

$$
M_R = \frac{m_2}{m_1}, \quad K_R = \frac{k_2}{k_1}
$$

where R_N is the dimensionless resistance; α_{N_1} and α_{N_2} are the dimensionless force factors; ξ_{11} and ξ_{22} are the dimensionless damping coefficients; M_R is the mass ratio; and K_R is the stiffness ratio.

Consider the case of a weakly coupling at a resonance, from Eqs. (6.10) and (6.11), the natural frequency is solved from the roots of the following equation:

$$
\left(k_1 - m_1 \cdot \omega^2\right) \cdot \left(k_2 - m_2 \cdot \omega^2\right) - k_2 \cdot m_2 \cdot \omega^2 = 0
\qquad (6.13)
$$

where the damped resonant frequency is approximately equal to the natural frequency $\omega \approx \omega_n$.

Substituting Eqs. (6.12) and (6.13) into Eq. (6.10) gives the normalized dimensionless resonant output voltage as:

$$
\frac{|\overline{V}|}{\frac{m_2 \cdot A}{\alpha}} = \frac{\alpha_{N_2}^2 \cdot R_N \cdot \Phi_1 \cdot \Phi_2^2 \cdot \sqrt{\left(\Phi_1^2 + 4 \cdot \xi_{11}^2\right)}}{\left\{ \left[R_N \cdot \left(\begin{matrix} 2 \cdot \xi_{22} \cdot \Phi_2 \cdot \left(\Phi_1^2 - M_R - 1\right) \\ +2 \cdot \xi_{11} \cdot \Phi_1 \cdot \left(\Phi_2^2 + \alpha_{N_2}^2 \cdot \Phi_2^2 - M_R\right) \\ +4 \cdot \xi_{11} \cdot \xi_{22} \cdot \Phi_1 \cdot \Phi_2 \end{matrix} \right) \right]^2 + \left[\begin{matrix} \Phi_2 \cdot \left(2 \cdot \xi_{22} + \alpha_{N_2}^2 \cdot R_N \cdot \Phi_2\right) \cdot \left(\Phi_1^2 - M_R - 1\right) \\ +2 \cdot \xi_{11} \cdot \Phi_1 \cdot \left(\Phi_2^2 - 2 \cdot R_N \cdot \xi_{22} \cdot \Phi_2 - 1\right) \end{matrix} \right]^2 \right\}^{0.5}}
\qquad (6.14)
$$

As well as substituting Eqs. (6.12) and (6.13) into Eq. (6.11) gives the normalized dimensionless harvested resonant power as:

$$\frac{P_{\mathrm{h}}}{\frac{m_2^2 \cdot A^2}{\alpha^2 \cdot R}} = \frac{0.5 \cdot \alpha_{N_2}^4 \cdot R_N^2 \cdot \Phi_1^2 \cdot \Phi_2^4 \cdot \left(\Phi_1^2 + 4 \cdot \xi_{11}^2\right)}{\left[R_N \cdot \left(\begin{array}{c} 2 \cdot \xi_{22} \cdot \Phi_2 \cdot \left(\Phi_1^2 - M_R - 1\right) \\ +2 \cdot \xi_{11} \cdot \Phi_1 \cdot \left(\Phi_2^2 + \alpha_{N_2}^2 \cdot \Phi_2^2 - M_R\right) \end{array} \right) \\ +4 \cdot \xi_{11} \cdot \xi_{22} \cdot \Phi_1 \cdot \Phi_2 \right]^2 + \left[\begin{array}{c} \Phi_2 \cdot \left(2 \cdot \xi_{22} + \alpha_{N_2}^2 \cdot R_N \cdot \Phi_2\right) \cdot \left(\Phi_1^2 - M_R - 1\right) \\ +2 \cdot \xi_{11} \cdot \Phi_1 \cdot \left(\Phi_2^2 - 2 \cdot R_N \cdot \xi_{22} \cdot \Phi_2 - 1\right) \end{array} \right]^2} \tag{6.15}$$

The piezoelectric vibration energy harvesting efficiency is defined by:

$$\eta = \frac{P_h}{P_{in}} = \frac{\frac{P_h}{\frac{m_2^2 \cdot A^2}{\alpha^2 \cdot R}}}{\frac{P_{in}}{\frac{m_2^2 \cdot A^2}{\alpha^2 \cdot R}}} \tag{6.16}$$

To calculate the harvesting efficiency, according to Ref. [21], the equation representing mean input power is given by

$$P_{\mathrm{in}} = \frac{1}{2} \cdot \langle -m_1 \cdot \ddot{x}_1 \cdot \dot{y} \rangle + \frac{1}{2} \cdot \langle -m_2 \cdot \ddot{x}_2 \cdot \dot{y} \rangle \tag{6.17}$$

Considering the input is a harmonic excitation $y(t) = Y(s) \cdot e^{s \cdot t}$, and then the dimensionless input power is given by

$$\frac{P_{\mathrm{in}}}{\frac{m_2^2 \cdot A^2}{\alpha^2 \cdot R}} = \frac{\alpha^2 \cdot R}{2 \cdot \omega^2 \cdot m_2^2} \cdot \left\{ m_1 \cdot \mathrm{Re}\left[\frac{s \cdot X_1}{Y}\right] + m_2 \cdot \mathrm{Re}\left[\frac{s \cdot X_2}{Y}\right] \right\} \tag{6.18}$$

where $\mathrm{Re}\left[\frac{s \cdot X_1}{Y}\right]$ and $\mathrm{Re}\left[\frac{s \cdot X_2}{Y}\right]$ can be calculated from Eq. (6.6), substituting Eq. (6.6) into Eq. (6.18) gives:

$$\frac{P_{\mathrm{in}}}{\frac{m_2^2 \cdot A^2}{\alpha^2 \cdot R}} = \frac{\alpha^2 \cdot R}{2 \cdot m_2^2} \cdot \frac{\left\{ \begin{array}{c} \left[\begin{array}{c} c_1 \cdot \left(c_2 + \alpha^2 \cdot R\right) \\ +R \cdot C_0 \cdot \left[\begin{array}{c} c_1 \cdot \left(k_2 - m_2 \cdot \omega^2\right) \\ +c_2 \cdot \left(\begin{array}{c} k_1 - m_1 \cdot \omega^2 \\ -m_2 \cdot \omega^2 \end{array}\right) \end{array} \right] \end{array} \right] \cdot \left[\left(m_1 \cdot \omega^2 + m_2 \cdot \omega^2\right) \cdot \left[\begin{array}{c} k_1 \cdot \left(c_2 + \alpha^2 \cdot R\right) \\ +R \cdot C_0 \cdot \left(k_1 \cdot k_2 - c_1 \cdot c_2 \cdot \omega^2\right) \end{array} \right] \\ +m_1 \cdot \omega^2 \cdot c_1 \cdot \left(k_2 - m_2 \cdot \omega^2\right) + m_2 \cdot \omega^2 \cdot \left(\begin{array}{c} k_2 \cdot c_1 \\ -m_1 \cdot \omega^2 \cdot R \cdot C_0 \cdot k_1 \end{array}\right) \right] \\ + \left[\begin{array}{c} R \cdot C_0 \cdot c_1 \cdot c_2 \cdot \omega^2 - c_1 \cdot \left(k_2 - m_2 \cdot \omega^2\right) \\ -\left(\begin{array}{c} k_1 - m_1 \cdot \omega^2 \\ -m_2 \cdot \omega^2 \end{array}\right) \cdot \left(c_2 + \alpha^2 \cdot R\right) \end{array} \right] \cdot \left[\left(m_1 + m_2\right) \cdot \left[\begin{array}{c} c_1 \cdot \omega^2 \cdot \left(c_2 + \alpha^2 \cdot R\right) \\ +R \cdot C_0 \cdot \left(\begin{array}{c} k_1 \cdot c_2 \cdot \omega^2 \\ +c_1 \cdot k_2 \cdot \omega^2 \end{array}\right) \end{array} \right] - m_2 \cdot k_1 \cdot k_2 \\ -m_1 \cdot k_1 \cdot \left(k_2 - m_2 \cdot \omega^2\right) - m_1 \cdot R \cdot C_0 \cdot c_1 \cdot m_2 \cdot \omega^4 \right] \end{array} \right\}}{\left\{ \left[\begin{array}{c} \left[\begin{array}{c} \left(k_1 - m_1 \cdot \omega^2\right) \cdot \left(k_2 - m_2 \cdot \omega^2\right) \\ -k_2 \cdot m_2 \cdot \omega^2 \end{array} \right] - \left(\begin{array}{c} k_1 \cdot c_2 + k_2 \cdot c_1 - m_1 \cdot c_2 \cdot \omega^2 \\ -m_2 \cdot c_1 \cdot \omega^2 - m_2 \cdot c_2 \cdot \omega^2 \end{array}\right) \cdot \left(R \cdot C_0 \cdot \omega^2\right) \\ -\omega^2 \cdot c_1 \cdot \left(c_2 + \alpha^2 \cdot R\right) \end{array} \right]^2 \\ + \left[\begin{array}{c} R \cdot C_0 \cdot \omega \cdot \left[\left(k_1 - m_1 \cdot \omega^2\right) \cdot \left(k_2 - m_2 \cdot \omega^2\right) - k_2 \cdot m_2 \cdot \omega^2\right] \\ +\omega \cdot \left(c_2 + \alpha^2 \cdot R\right) \cdot \left[\left(k_1 - m_1 \cdot \omega^2\right) - m_2 \cdot \omega^2\right] \\ +c_1 \cdot \omega \cdot \left(k_2 - m_2 \cdot \omega^2 - R \cdot C_0 \cdot c_2 \cdot \omega^2\right) \end{array} \right]^2 \right\}} \tag{6.19}$$

The dimensionless input power is normalized from Eq. (6.19) using Eq. (6.12) and is given by:

$$\frac{P_{in}}{\frac{m_2^2 \cdot A^2}{\alpha^2 \cdot R}} = \frac{0.5 \cdot \alpha_{N_2}^2 \cdot R_N \cdot \Phi_1 \cdot \Phi_2^2 \cdot \left\{ \begin{array}{l} \left[\begin{array}{l} 2\cdot\xi_{11}\cdot\Phi_1\cdot\left[R_N\cdot\left(\Phi_2^2\cdot\alpha_{N_2}^2+\Phi_2^2-1\right)+\Phi_2\cdot\xi_{22}\right] \\ +2\cdot\xi_{22}\cdot\Phi_2\cdot\left[R_N\cdot\left(\Phi_1^2-1-M_R\right)+\Phi_1\cdot\xi_{11}\right] \end{array} \right]\cdot \left\{ \frac{1}{M_R}\cdot \left[\begin{array}{l} 2\cdot\xi_{22}\cdot\Phi_2\cdot\left(\Phi_1-\xi_{11}\cdot R_N\right) \\ +\left[2\cdot\xi_{11}\cdot\left(\Phi_2^2-1\right)-R_N\cdot\Phi_1\right] \\ +R_N\cdot\Phi_2\cdot\left[\begin{array}{l} \Phi_1\cdot\Phi_2\cdot\left(\alpha_{N_2}^2+1\right) \\ -2\cdot\xi_{11}\cdot\xi_{22}\cdot R_N \end{array} \right] \end{array} \right] + \Phi_2\cdot\left[\begin{array}{l} 2\cdot\xi_{22}\cdot\left(\Phi_1-\xi_{11}\cdot R_N\right) \\ +\Phi_1\cdot\Phi_2\cdot R_N\cdot\left(\alpha_{N_2}^2+1\right) \\ +2\cdot\xi_{11}\cdot\left(\Phi_2-\xi_{22}\cdot R_N\right) \end{array} \right] \right\} \\ + \left[\begin{array}{l} 2\cdot\xi_{22}\cdot\Phi_1\cdot\Phi_2\cdot\left(R_N\cdot\xi_{11}-\Phi_1\right) \\ -2\cdot\xi_{11}\cdot\Phi_1\cdot\left(\Phi_2^2-1\right) \\ +2\cdot R_N\cdot\Phi_1\cdot\Phi_2\cdot\left(2\cdot\xi_{11}\cdot\xi_{22}-\alpha_{N_2}^2\cdot\Phi_1\cdot\Phi_2\right) \\ +\left(2\cdot\xi_{22}\cdot\Phi_2+\alpha_{N_2}^2\cdot R_N\cdot\Phi_2^2\right)\cdot\left(M_R+1\right) \end{array} \right]\cdot \left[\begin{array}{l} \left(\frac{1}{M_R}\right)\cdot\left[\begin{array}{l} 2\cdot\xi_{11}\cdot\Phi_2\cdot\left(\xi_{22}+\alpha_{N_2}^2\cdot R_N\cdot\Phi_2\right) \\ -\left(\Phi_1-2\cdot\xi_{11}\cdot R_N\right)\cdot\left(\Phi_2^2-1\right) \\ +2\cdot\xi_{22}\cdot\Phi_2\cdot\left(R_N\cdot\Phi_1+\xi_{11}\right) \end{array} \right] \\ +\Phi_2\cdot\left[\begin{array}{l} 2\cdot\xi_{11}\cdot\left[\xi_{22}+R_N\cdot\Phi_2\cdot\left(\alpha_{N_2}^2+1\right)\right] \\ +2\cdot\xi_{22}\cdot\left(R_N\cdot\Phi_1+\xi_{11}\right)-\Phi_1\cdot\Phi_2 \end{array} \right] \end{array} \right] \end{array} \right\}}{\left[\begin{array}{l} R_N\cdot\left(\begin{array}{l} \left(M_R+\Phi_1^2\right)\cdot 2\cdot\xi_{22}\cdot\Phi_2 \\ -2\cdot\xi_{11}\cdot\Phi_1\cdot\left(\Phi_2^2+1\right)-2\cdot\xi_{22}\cdot\Phi_2 \end{array} \right) \\ -2\cdot\xi_{11}\cdot\Phi_1\cdot\Phi_2\cdot\left(2\cdot\xi_{22}+\alpha_{N_2}^2\cdot\Phi_2\cdot R_N\right) \end{array} \right]^2 + \left[\begin{array}{l} \left(\frac{\Phi_1^2-1}{\Phi_2}\right)\cdot\left(2\cdot\xi_{22}+\alpha_{N_2}^2\cdot\Phi_2\cdot R_N\right) \\ +2\cdot\xi_{11}\cdot\Phi_1\cdot\left(\Phi_2^2-1-2\cdot\xi_{22}\cdot\Phi_2\cdot R_N\right) \end{array} \right]^2} \tag{6.20}$$

Therefore from Eqs. (6.15), (6.16), and (6.20), the normalized resonant energy harvesting efficiency is written as:

$$\eta = \frac{\alpha_{N_2}^2 \cdot R_N \cdot \Phi_1 \cdot \Phi_2^2 \cdot \left(\Phi_1^2+4\cdot\xi_{11}^2\right)}{\left\{ \begin{array}{l} \left[\begin{array}{l} 2\cdot\xi_{11}\cdot\Phi_1\cdot\left[\begin{array}{l} R_N\cdot\left(\Phi_2^2\cdot\alpha_{N_2}^2+\Phi_2^2-1\right) \\ +\Phi_2\cdot\xi_{22} \end{array} \right] \\ +2\cdot\xi_{22}\cdot\Phi_2\cdot\left[\begin{array}{l} R_N\cdot\left(\Phi_1^2-1-M_R\right) \\ +\Phi_1\cdot\xi_{11} \end{array} \right] \end{array} \right]\cdot \left\{ \frac{1}{M_R}\cdot\left[\begin{array}{l} 2\cdot\xi_{22}\cdot\Phi_2\cdot\left(\Phi_1-\xi_{11}\cdot R_N\right) \\ +\left[2\cdot\xi_{11}\cdot\left(\Phi_2^2-1\right)-R_N\cdot\Phi_1\right] \\ +R_N\cdot\Phi_2\cdot\left[\begin{array}{l} \Phi_1\cdot\Phi_2\cdot\left(\alpha_{N_2}^2+1\right) \\ -2\cdot\xi_{11}\cdot\xi_{22}\cdot R_N \end{array} \right] \end{array} \right] +\Phi_2\cdot\left[\begin{array}{l} 2\cdot\xi_{22}\cdot\left(\Phi_1-\xi_{11}\cdot R_N\right) \\ +\Phi_1\cdot\Phi_2\cdot R_N\cdot\left(\alpha_{N_2}^2+1\right) \\ +2\cdot\xi_{11}\cdot\left(\Phi_2-\xi_{22}\cdot R_N\right) \end{array} \right] \right\} \\ +\left[\begin{array}{l} 2\cdot\xi_{22}\cdot\Phi_1\cdot\Phi_2\cdot\left(R_N\cdot\xi_{11}-\Phi_1\right) \\ -2\cdot\xi_{11}\cdot\Phi_1\cdot\left(\Phi_2^2-1\right) \\ +2\cdot R_N\cdot\Phi_1\cdot\Phi_2\cdot\left(2\cdot\xi_{11}\cdot\xi_{22}-\alpha_{N_2}^2\cdot\Phi_1\cdot\Phi_2\right) \\ +\left(2\cdot\xi_{22}\cdot\Phi_2+\alpha_{N_2}^2\cdot R_N\cdot\Phi_2^2\right)\cdot\left(M_R+1\right) \end{array} \right]\cdot \left[\begin{array}{l} \left(\frac{1}{M_R}\right)\cdot\left[\begin{array}{l} 2\cdot\xi_{11}\cdot\Phi_2\cdot\left(\xi_{22}+\alpha_{N_2}^2\cdot R_N\cdot\Phi_2\right) \\ -\left(\Phi_1-2\cdot\xi_{11}\cdot R_N\right)\cdot\left(\Phi_2^2-1\right) \\ +2\cdot\xi_{22}\cdot\Phi_2\cdot\left(R_N\cdot\Phi_1+\xi_{11}\right) \end{array} \right] \\ +\Phi_2\cdot\left[\begin{array}{l} 2\cdot\xi_{11}\cdot\left[\xi_{22}+R_N\cdot\Phi_2\cdot\left(\alpha_{N_2}^2+1\right)\right] \\ +2\cdot\xi_{22}\cdot\left(R_N\cdot\Phi_1+\xi_{11}\right)-\Phi_1\cdot\Phi_2 \end{array} \right] \end{array} \right] \end{array} \right\}} \tag{6.21}$$

Moreover, in a special condition, it is assumed that the 2DOF PVEH system operates at a resonant frequency and the damping of the system is ignored ($c_1 = c_2 = 0$). Therefore the dimensionless piezoelectric vibration energy harvesting power is given by:

$$\left(\frac{P_h}{\frac{m_2^2 \cdot A^2}{\alpha^2 \cdot R}} \right)_{c_1=c_2=0} = \frac{0.5 \cdot \Phi_1^4}{\left(\Phi_1^2 - M_R - 1 \right)^2} \tag{6.22}$$

As well as the piezoelectric vibration energy harvesting efficiency is given by:

$$\eta = \frac{M_R \cdot \Phi_1^2}{\left[M_R + 1 - 2 \cdot \Phi_1^2 \right] \cdot \left[1 - \Phi_2^2 \cdot (1 + M_R) \right]} \tag{6.23}$$

According to Eqs. (6.12) and (6.13), Eqs. (6.22) and (6.23) can also be written as

$$
\begin{cases}
\dfrac{P_h}{\frac{m_2^2 \cdot A^2}{R \cdot \alpha^2}} = \dfrac{1}{\left[2 \cdot \left(1 - \frac{1}{2} \cdot (1+M_R) \cdot \left[\frac{K_R}{M_R} + 1 + K_R - \sqrt{\left(\frac{K_R}{M_R} + 1 + K_R \right)^2 - 4 \cdot \frac{K_R}{M_R}} \right] \right) \right]^2} \\[3em]
\eta = \dfrac{2 \cdot \left[K_R + M_R + K_R \cdot M_R - \sqrt{\left(K_R + M_R + K_R \cdot M_R \right)^2 - 4 \cdot K_R \cdot M_R} \right]}{\left[(M_R + 1) \cdot \left(\frac{\frac{K_R}{M_R} + 1 + K_R}{-\sqrt{\left(\frac{K_R}{M_R} + 1 + K_R \right)^2 - 4 \frac{K_R}{M_R}}} \right) - 4 \right] \cdot \left[\left(\frac{\frac{K_R}{M_R} + 1 + K_R}{-\sqrt{\left(\frac{K_R}{M_R} + 1 + K_R \right)^2 - 4 \frac{K_R}{M_R}}} \right) - 4 \cdot \frac{K_R}{M_R} - 4 \cdot K_R \right]}
\end{cases} \tag{6.24}
$$

In Eq. (6.24), the stiffness ratio (K_R) is fixed and the mass ratio (M_R) is changed from 0.5 to 8, the dimensionless harvested resonant power $P_h \Big/ \left(\frac{m_2^2 \cdot A^2}{R \cdot \alpha^2} \right)$ and energy harvesting efficiency η are plotted in Fig. 6.2. As well as the mass ratio (M_R) is fixed and the stiffness (K_R) ratio is changed from 0.001 to 2 are plotted in Fig. 6.3. It is seen from Figs. 6.2 and 6.3 that the dimensionless harvested resonant power decreases when the mass ratio increases, but the harvested efficiency increases. The large magnitude of stiffness ratio could be beneficial for the dimensionless harvested resonant power, but sacrifice the harvested efficiency. It can also be concluded from Figs. 6.2 and 6.3 that the stiffness ratio is much more sensitive to both harvested efficiency and dimensionless harvested resonant power than the mass ratio. Therefore, there is more tuning space for the mass ratio.

It is seen from Eq. (6.24) that when the damping value of the harvesting system is small enough to be ignored, the resonant energy harvesting efficiency is not affected by piezoelectric physical material properties. In this case, the performance of the 2DOF PVEH is only related to the mass ratio and the stiffness ratio. Moreover, it is clearly shown in Eq. (6.24) that the excitation amplitude, force factor, and external load resistance have no influence on the energy harvesting efficiency.

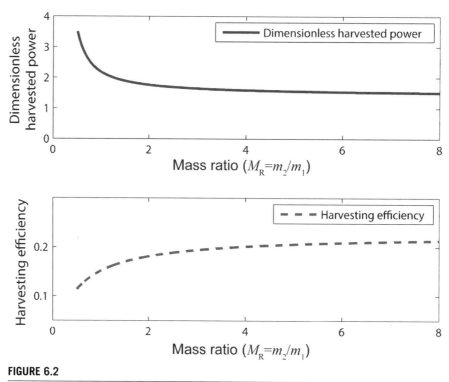

FIGURE 6.2

The dimensionless harvested resonant power and energy harvesting efficiency versus various mass ratio ($M_R = m_2/m_1$).

6.4 CASE STUDY OF A QUARTER VEHICLE SUSPENSION MODEL AND SIMULATION

In this section, a quarter vehicle suspension built with piezoelectric element inserter, as shown in Fig. 6.4, has been chosen for a case study to perform parameter studies and optimization. The piezoelectric material can be mounted under a specific preload at the shock tower between the body/chassis and suspension spring/shock absorber. The vibrations generated by tire–road interactions are transmitted through the suspension generating strains on the piezoelectric material insert, which could be partly converted into electrical energy. Without the piezoelectric material insert, the transmitted mechanical vibration energy is usually dissipated into heat energy which is wasted. It is because of that the quarter vehicle suspension can be modeled as a 2DOF PVEH as mentioned above. Furthermore, a quarter of the vehicle mass would be large enough and able to deliver a large amount of stress to the piezoelectric materials. The parameters of the quarter vehicle suspension model are given in Table 6.1 [18].

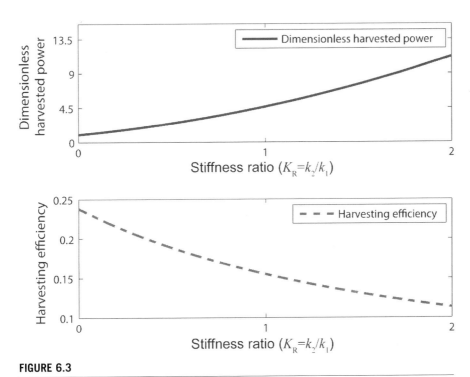

FIGURE 6.3

The dimensionless harvested resonant power and energy harvesting efficiency versus various stiffness ratio ($K_R = k_2/k_1$).

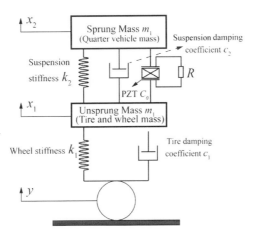

FIGURE 6.4

Case study of a quarter vehicle suspension model with piezoelectric element insert.

Table 6.1 Parameters of a Quarter Vehicle Suspension Model With Piezoelectric Insert

Parameter	Type	Units	Values
m_1	Vehicle wheel–tire mass	kg	40
m_2	Quarter vehicle mass	kg	260
c_1	Wheel–tire damping coefficient	N s/m	264.73
c_2	Suspension shock absorber damping coefficient	N s/m	520
C_0	Blocking capacitance of the piezoelectric inserter	F	1.89×10^{-8}
k_1	Wheel–tire stiffness	N/m	130,000
k_2	Suspension spring stiffness	N/m	26,000
α	Force factor	N/Volt	1.52×10^{-3}
f_n	Natural frequency	Hz	1.45
R	Electrical resistance	Ω	30,455.3

For a "vehicle quarter suspension model" with piezoelectric material insert in place of damping materials in a shake absorber, then y is the excitation displacement; m_1 is the unsprung mass or the mass of wheel and tire of a quarter vehicle; m_2 is the sprung mass or a quarter vehicle's mass; k_1 is the wheel–tire stiffness; k_2 is the suspension spring stiffness; c_1 is the wheel–tire damping coefficient; c_2 is the suspension damping coefficient; x_1 is the displacement of the unsprung mass m_1; x_2 is the displacement of the sprung mass m_2; V is the voltage generated by the piezoelectric insert.

To verify the output voltage and power calculated using the above analytical frequency response analysis, Matlab Simulink was applied to conduct time domain simulation for the performance of the harvesting system. The parameters in Table 6.1 were measured and identified from experiments. The system parameters are applied in the simulation schedule as shown in Fig. 6.5 and applied in Eqs. (6.1) and (6.2). The harvested power was calculated by the squared voltage divided by the resistance. In the simulation scheme, the excited acceleration was simulated by a sine wave acceleration of 1 g (9.80 m/s^2) amplitude and an excitation frequency of 1.45 Hz generated by a signal source block from the Matlab Simulink library. The excitation acceleration signal was passed through the Matlab Simulink wiring diagram which calculated the output voltage and harvested power. The predicted output voltage and harvested power using the time domain simulation program code are displayed by signal sink scopes of the Matlab Simulink library as shown in Figs. 6.6 and 6.7.

It is obtained from Figs. 6.6 and 6.7 that at the very beginning, the 2DOF PVEH has a transition response to the excitation acceleration. However, after a couple of seconds, the transition ends. At this stage, the peak output voltage and harvested power can be recorded, which are 274.62 V and 2.48 W, respectively. The root mean

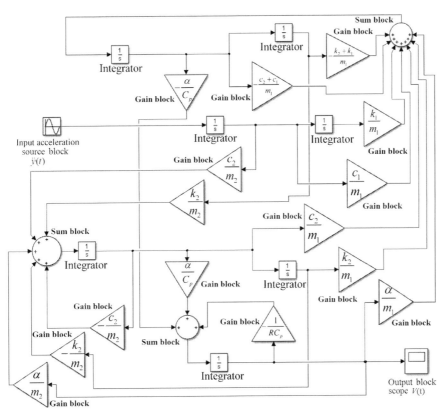

FIGURE 6.5

Simulation schematic for Eqs. (6.1) and (6.2) with a sine wave base excitation input and a sinusoidal voltage output at a frequency using the integration method.

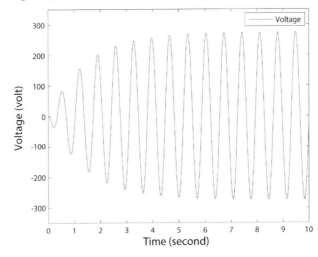

FIGURE 6.6

Output voltage for the input excitation acceleration amplitude of 1 g (9.80 m/s²) and excitation frequency of 1.45 Hz.

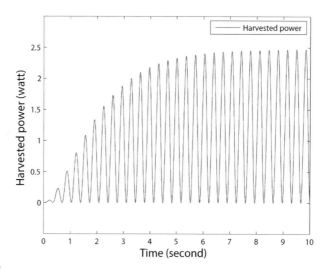

FIGURE 6.7

Output power for the input excitation acceleration amplitude of 1 g (9.80 m/s^2) and excitation frequency of 1.45 Hz.

square (RMS) value is calculated by the peak value divided by the square root 2 as the input excitation acceleration signal is assumed to be a sine wave. Hence, in this case, the RMS value of output voltage and harvested power are 194.14 V and 1.23 W, respectively. In the frequency response analysis, it is assumed that the frequency value varies but the other parameters are kept as constant. The relationships between the system oscillator displacement ratios and frequency are presented in Fig. 6.8. In Fig. 6.8, the $|X_1/Y|$ is the displacement amplitude of Mass 1 divided by the input displacement amplitude; and $|X_2/Y|$ is the displacement amplitude of Mass 2 divided by the input displacement amplitude. The two natural frequencies of the quarter vehicle suspension system with the built-in piezoelectric material can be identified from the peak frequencies of the displacement amplitude curves of $|X_1/Y|$ and $|X_2/Y|$ as demonstrated. As well as the maximum displacement ratio peaks of the sprung and unsprung masses can be identified. It is seen that there are two resonant peaks, the first mode of 1.45 Hz is related to suspension bouncing, or called the bouncing mode, the second mode of 9.7 Hz is related to wheel hopping, it can be called the suspension hop mode [19,20].

In addition, to compare the simulation results with those calculated using the analytical frequency response analysis, the RMS voltage and RMS harvested power data points obtained from the time domain simulation are presented by discrete triangle and discrete star marks in Figs. 6.9—6.14, respectively. To investigate the 2DOF piezoelectric vibration energy harvesting system performance versus various parameters such as the input excitation acceleration amplitude, electric resistance load, suspension damping, tire damping and force factor, it is assumed that one of the parameters in Table 6.1 is varied; the others are not changed, substituting the

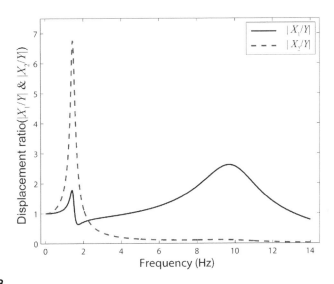

FIGURE 6.8

Displacement amplitude ratios of mass 1 and mass 2 with respect to the input displacement amplitude versus frequency.

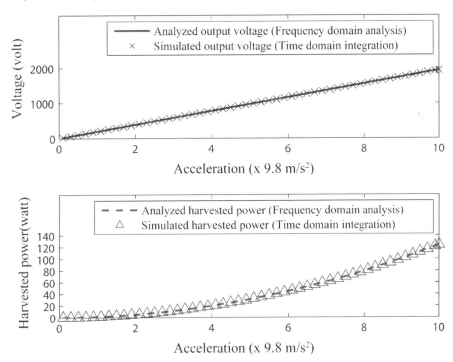

FIGURE 6.9

Output voltage and harvested resonant power versus the input excitation acceleration amplitude.

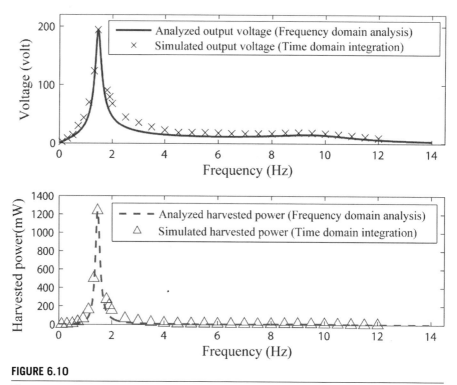

FIGURE 6.10

The output voltage and harvested resonant power versus frequency.

parameters into Eqs. (6.10) and (6.11) gives the peak value of resonant output voltage and harvested resonant power of the 2DOF system. The RMS values of output voltage and harvested power are calculated based on their peak values. For a better comparison, the output RMS voltage and harvested power calculated by the analytical frequency response analysis are plotted by solid curves as shown in Figs. 6.9−6.14, respectively.

As shown in Fig. 6.9, the base excitation acceleration amplitude increases from 0 times to 10 times of 1 g (9.80 m/s^2) in a step size of 1 times g; the output voltage linearly increases in proportion to the excitation acceleration amplitude. However, the harvested power quadratically increases with the excitation acceleration amplitude. It can be seen that the results from the time domain simulation and the analytical frequency response analysis methods are close to each other. It is important that the results given by the analytical frequency response analysis method are validated by the time domain simulation method.

The output voltage and harvested power versus frequency are shown in Fig. 6.10. Assuming the frequency changes from 0 to 14 Hz in a step size of 1 Hz, the other parameters in Table 6.1 remain constant. The results from the time domain simulation and analytical frequency response analysis methods both show the highest

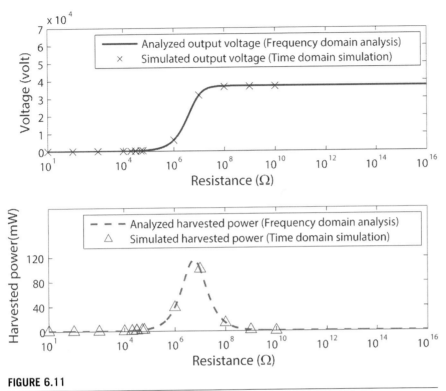

FIGURE 6.11

Output voltage and harvested resonant power versus electric load resistance.

output voltage and harvested power at around 1.45 Hz, which coincides with the first bouncing resonant frequency shown in Fig. 6.8. It is obvious that the 2DOF system power harvesting performance is much better at resonant frequencies than that at nonresonant frequencies, which has been validated by both the time domain simulation and the analytical frequency response analysis methods. It produces the highest RMS output voltage of 194.18 V and the highest RMS harvested power of 1238 mW. It should be noticed that there are some slight differences between the results from the time domain simulation and the analytical frequency response analysis methods around 2 Hz. The simulation errors might be caused by the different solver type of the Matlab Simulink software or by using a coarse step size of the Ruger-Kuta method.

It was assumed that the electric resistance load changes from 1000 to 1×10^{15} Ω in a step size of 10 times the resistance value of the last step, the other parameters in the Table 6.1 were fixed, the output voltage and harvested power were calculated by the time domain simulation and analytical frequency response analysis methods and are shown in Fig. 6.11. It is seen that the harvested power climbs to a peak, and then decreases. However, the voltage increases to a value, then maintains at this level

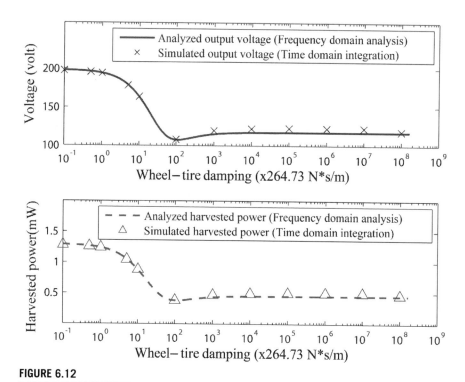

FIGURE 6.12

Output voltage and harvested resonant power versus the wheel−tire damping.

when the electric resistance load increases. There exists an optimal electric resistance load for achieving the maximum harvested power and output voltage for the 2DOF system. The time domain simulation results represented by discrete star and triangle marks are very close to the analytical frequency response analysis results represented by the solid curves. In other words, the results of the analytical frequency response analysis method have been validated by those of the time domain simulation method.

If the wheel−tire damping value was changed from 0.01 times to 1,000,000 times of the original value (264.73 N s/m) in a step size of 10 times the damping value of the last step, the other parameters were kept constant. It can be seen from Fig. 6.12 that the harvested power and output voltage decrease to a level as the wheel−tire damping increases. After reaching that bottom values, then, the harvested power and output voltage slightly increase and then maintain at a certain level when the wheel−tire damping value further increases.

If the suspension damping was changed from 0.0001 times to 1000 times of the original suspension damping (520 N s/m) in a step size of 10 times the damping value of the last step, the other parameters in Table 6.1 were fixed. The results of the harvested power and output voltage from the time domain simulation and

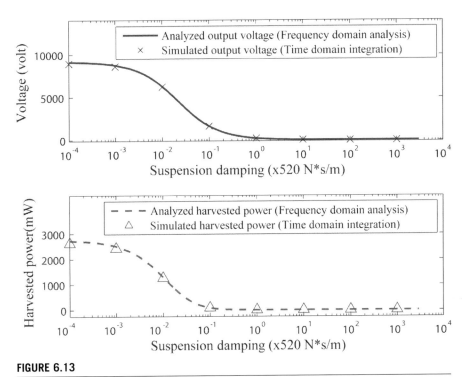

FIGURE 6.13

Output voltage and harvested resonant power versus the suspension damping.

analytical frequency response analysis methods are plotted in Fig. 6.13. It is seen that the harvested power and output voltage significantly decrease when the value of suspension damping increases. It is suggested that less suspension damping would allow for more stresses being applied into piezoelectric materials; therefore it should give high harvested power and voltage output. However, vehicle vibration isolation is very sensitive to the suspension damping; less suspension damping would produce better vibration energy harvesting performance, but worse vehicle vibration isolation, ride, and handling performance. The passengers would feel uncomfortable and experience harsh driving. A balance point between the energy harvesting performance and the vehicle vibration isolation, ride, and handling performance should be identified and reached.

According to Eq. (6.3), the force factor is determined by the ratio of material section area and thickness multiplying piezoelectric or permittivity constant of piezoelectric. The force factor was changed from 0 times to 600 times of the original force factor (1.52×10^{-3} N/Volt) in a step size of a 100 times the force factor of the last step and the other parameters were fixed, the results of harvested power and output voltage are plotted in Fig. 6.14. It can be seen that there exists an optimized force factor which gives the highest harvested power and output voltage. In other words, if the piezoelectric constant and permittivity of piezoelectric insert

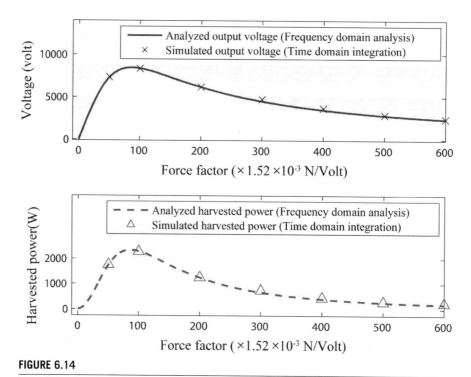

FIGURE 6.14

Output voltage and harvested resonant power versus the force factor.

are not changed, tuning the ratio of material surface area and thickness will help to achieve the optimum force factor. It is seen from Fig. 6.14 that the harvested power and output voltage increase and reach a peak, then decrease when the force factor increases. The output voltage and harvested power obtained from the time domain simulation method and represented by discrete triangle and star marks are close to the results obtained from the analytical frequency response analysis method. The results obtained from the analytical frequency response analysis method are represented by solid curves. The time domain simulation results have validated the analytical frequency response analysis method.

To predict the output voltage and harvested power by the frequency response analysis of the Laplace or Fourier transform, Eq. (6.1) and Eq. (6.2) could also be written as:

$$
\begin{bmatrix}
m_1 s^2 + (c_1 + c_2)\cdot s + k_1 + k_2 & -c_2 s - k_2 & -\alpha \\
-c_2 s - k_2 & m_2 s^2 + c_2 s + k_2 & \alpha \\
\alpha s & -\alpha s & C_0 s + \dfrac{1}{R}
\end{bmatrix}
\begin{Bmatrix} X_1 \\ X_2 \\ V \end{Bmatrix}
=
\begin{Bmatrix} c_1 s + k_1 \\ 0 \\ 0 \end{Bmatrix}
\cdot Y
$$

$$(6.25)$$

Eq. (6.25) can be simulated by the Matlab program code where Y is the Laplace or Fourier transform of input signal y and the X_1, X_2, and $\overline{V}$ are the Laplace or Fourier transforms of the x_1, x_2, and V. The output voltage, harvested power, dimensionless harvested voltage and dimensionless harvested power can be predicted and analyzed.

To better understand the effects of the system parameters on the performance of 2DOF piezoelectric vibration energy harvesting system, the simulation of a quarter vehicle suspension model with various selected parameters were carried out in a frequency domain based on Eq. (6.25) using the Matlab software code similar to that in Fig. 2.13 and the results are plotted in Figs. 6.15–6.22. It is seen from Fig. 6.16 that when the wheel–tire mass increases, the voltage magnitude of the bouncing resonant mode will increase, although the bouncing resonant frequency is rarely shifted. In the contrast, while the wheel–tire mass increases, the hopping resonant frequency is shifted to lower frequency but the voltage magnitude of the hopping resonant mode increases. In other words, the wheel–tire mass has very little influence on the bouncing resonant frequency but has some influences on the bouncing and hopping harvested resonant power magnitudes and the overall harvesting frequency bandwidth. A larger wheel–tire mass contributes to a larger bouncing harvested resonant power and a larger harvesting frequency bandwidth.

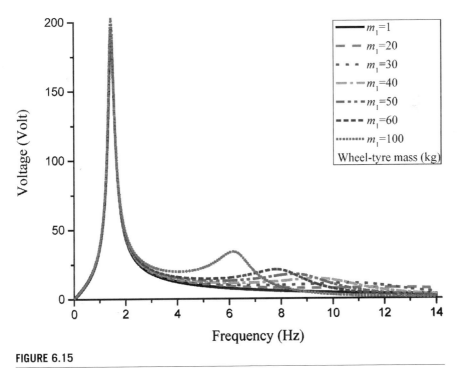

FIGURE 6.15

Output voltage versus frequency for various wheel–tire mass.

If the vehicle mass is not changed, as the wheel–tire mass increases, the mass ratio ($M_R = m_2/m_1$) decreases, according to Fig. 6.2, the dimensionless harvested resonant power $P_h \Big/ \left(\frac{m_2^2 \cdot A^2}{R \cdot \alpha^2}\right)$ increases, and therefore the harvested resonant power P_h increases. The result in Fig. 6.2 has verified to that shown in Fig. 6.15.

It is seen from Fig. 6.16 that when the vehicle mass increases, the bouncing resonant frequency decreases, the bouncing resonant voltage magnitude or harvested power increases. This result coincides with that in Fig. 6.2 where the wheel–tire mass is assumed not to be changed. When the vehicle mass increases, the mass ratio ($M_R = m_2/m_1$) increases, the dimensionless harvested power $P_h \Big/ \left(\frac{m_2^2 \cdot A^2}{R \cdot \alpha^2}\right)$ in Fig. 6.2 therefore decreases. However, the harvested power P_h is proportional to the dimensionless harvested power $P_h \Big/ \left(\frac{m_2^2 \cdot A^2}{R \cdot \alpha^2}\right)$ multiplied by the squared quarter vehicle mass (m_2). When the vehicle mass increases, the dimensionless harvested power $P_h \Big/ \left(\frac{m_2^2 \cdot A^2}{R \cdot \alpha^2}\right)$ decreases, the harvested power P_h will increase, which corresponds to the increased output voltage amplitude in Fig. 6.16.

The results in Figs. 6.2, 6.15, and 6.16 reveal the effect of the vehicle mass (m_2) and wheel–tire mass (m_1) on the harvested power or voltage and energy harvesting efficiency which is not clearly shown in Eqs. (6.24) and (6.25). The vehicle mass (m_2) has very little influence on the voltage magnitude of the hopping resonant mode. A smaller vehicle mass contributes to a larger harvesting frequency bandwidth. As the bouncing resonant voltage magnitude is much larger than the hopping

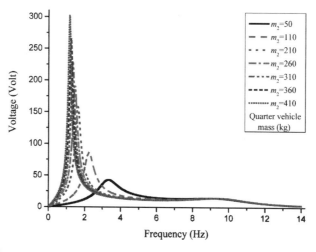

FIGURE 6.16

Output voltage versus frequency for various quarter vehicle masses.

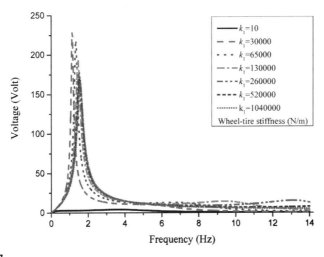

FIGURE 6.17

Output voltage versus frequency for various wheel—tire stiffness.

resonant voltage magnitude, the vehicle mass plays a more important role than the wheel—tire mass for the resonant output voltage magnitude.

The similar results can be obtained from different wheel—tire stiffness and suspension stiffness as shown in Figs. 6.17 and 6.18. It is seen from Fig. 6.17 that when the wheel—tire stiffness increases, the bouncing resonant frequency increases, the bouncing resonant voltage magnitude or harvested power decreases. The effect of

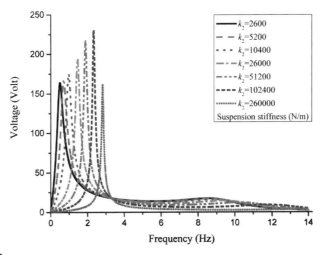

FIGURE 6.18

Output voltage versus frequency for various suspension stiffness.

the tire stiffness on the bouncing resonant voltage magnitude is much larger than that on the hopping resonant voltage magnitude. This result coincides with that in Fig. 6.3 where the suspension spring stiffness is assumed not to be changed. When the wheel–tire stiffness increases, the stiffness ratio k_2/k_1 decreases, the dimensionless harvested power $P_h \Big/ \left(\frac{m_2^2 \cdot A^2}{R \cdot \alpha^2} \right)$ in Fig. 6.3 or the harvested power P_h decreases, which corresponds to the decreased output voltage amplitude in Fig. 6.17.

It is seen from Fig. 6.18 that when the suspension spring stiffness increases, the bouncing resonant frequency increases; however, the bouncing resonant voltage magnitude first increases when the suspension spring stiffness increases from 2.6 kN/m to 102.4 kN/m, then decreases after the suspension spring stiffness is larger than 102.4 kN/m. The result coincides with that in Fig. 6.3 when the suspension spring stiffness is less than 102.4 kN/m where the wheel–tire stiffness is assumed not to be changed. When the suspension spring stiffness increases, the stiffness ratio k_2/k_1 increases, the dimensionless harvested power $P_h \Big/ \left(\frac{m_2^2 \cdot A^2}{R \cdot \alpha^2} \right)$ in Fig. 6.3 or the harvested power P_h increases. This corresponds to the increased output voltage amplitude until the suspension spring stiffness reaches 102.4 kN/m as shown in Fig. 6.18. When the suspension spring stiffness is larger than 102.4 kN/m, the weak couple assumption for Fig. 6.3 or Eq. (6.24) is not valid any more. The suspension system becomes a strong coupling system which can only be modeled using Eq. (6.25). This explains why when the suspension spring stiffness is larger than 102.4 kN/m, the result of Fig. 6.3 or Eq. (6.24) does not coincide with that in Fig. 6.18 or Eq. (6.25).

The results in Figs. 6.3, 6.17, and 6.18 reveal the effect of the suspension stiffness (k_2) and wheel–tire stiffness (k_1) on the harvested power and energy harvesting efficiency which is not clearly shown in Eqs. (6.24) and (6.25). It is seen from Fig. 6.18 that the effect of the suspension spring stiffness on the bouncing resonant voltage magnitude is much larger than that on the hopping resonant voltage magnitude. The smaller suspension spring stiffness would increase the hopping resonant voltage magnitude as well as increasing the harvesting frequency bandwidth. It is seen from Figs. 6.17 and 6.18 that the effect of increasing the suspension spring stiffness is larger than that of increasing the wheel–tire stiffness in regard to the bouncing resonant voltage magnitude or harvested power.

Moreover, the effect of stiffness ratio on dimensionless harvested power considering the effect of damping has been studied and the result is presented in Fig. 6.19. The result is different from that in Fig. 6.3 where the effect of damping is neglected. The vibration energy harvesting system is always set to be operated at the resonant frequency as the natural frequency of the system varies with the stiffness ratio. The optimal value of the stiffness ratio is found to be 0.73 that maximizes the dimensionless harvested resonant power.

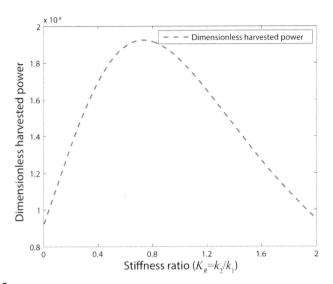

FIGURE 6.19

The dimensionless harvested resonant power versus the stiffness ratio (k_2/k_1).

In overall, both the wheel—tire stiffness and suspension spring stiffness play an important role to the voltage magnitude and harvested power of the bouncing resonant mode. The system harvesting resonant performance is more sensitive to the suspension stiffness rather than to the wheel—tire stiffness.

It is shown from Figs. 6.20 and 6.21 that the bouncing resonant voltage magnitude or harvested resonant power is nearly independent of the wheel—tire damping coefficient. On the other hand, the bouncing resonant output voltage magnitude or harvested resonant power is very sensitive to the suspension shock absorber damping coefficient. It clearly points out that the smallest suspension damping coefficient produces the largest bouncing resonant output voltage magnitude or harvested resonant power, which is preferred for the piezoelectric vibration energy harvesting system. The reason of the different trends of Figs. 6.20 and 6.21 from those of Figs. 6.12 and 6.13 is that only a small range of damping coefficient is chosen in Figs. 6.20 and 6.21. The small range of damping coefficient in Figs. 6.20 and 6.21 may not reflect the whole picture of the harvested power versus damping variation in Figs. 6.12 and 6.13. However, it allows us to compare the effects of suspension damping coefficient and wheel—tire damping coefficient on the output voltage or harvested resonant power.

Furthermore, the effect of the damping ratio on the dimensionless harvested resonant power is studied and plotted in Fig. 6.22. It is seen from Fig. 6.22 that when the damping ratio (c_1/c_2) is larger than 0.25, the dimensionless harvested resonant power will significantly decrease. Physically, when the damping ratio is larger than 1/4, the amplitude of the relative displacement between the wheel and vehicle body will become smaller which benefits the vehicle handling and comfort.

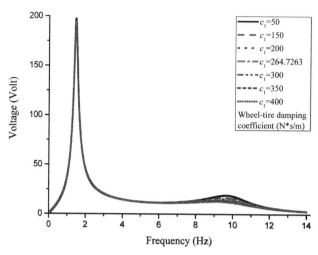

FIGURE 6.20

Output voltage versus frequency for various wheel–tire damping coefficients.

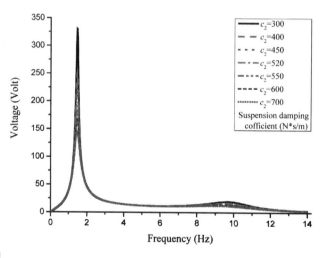

FIGURE 6.21

Output voltage versus frequency for various suspension damping coefficients.

6.5 SUMMARY

In this chapter, a dimensionless analysis method based on the Laplace transform is studied. It could provide accurate and reliable evaluation and analysis of the 2DOF piezoelectric vibration energy harvesting system performance as the results from the time domain simulation and analytical frequency response analysis methods

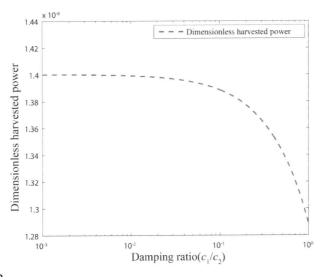

FIGURE 6.22

Dimensionless harvested resonant power versus the damping ratio (c_1/c_2).

are able to verify each other. The system parametric study has been conducted in the analysis approach.

Under the case of a small damping and a weak coupling, it has been proved that the dimensionless harvested resonant power and efficiency only depends on the stiffness and mass of the two oscillators and have nothing to do with the piezoelectric material property such as the force factor. When the mass ratio m_2/m_1 increases, the dimensionless harvested resonant power decreases, the resonant energy harvesting efficiency increases. When the stiffness ratio k_2/k_1 increases, the dimensionless harvested resonant power increases, the energy harvesting efficiency decreases. However, when the damping effect is considered, the optimal stiffness ratio is found to be 0.73 for the maximum dimensionless harvested resonant power. When the damping ratio (c_1/c_2) is greater than 0.25, the harvested resonant voltage or power P_h will significantly decrease.

If a vehicle quarter suspension system is simulated using the Laplace or Fourier transform−based frequency response analysis method for the 2DOF system model, physically, when the wheel mass is fixed, increasing the vehicle mass will increase the output resonant voltage or harvested resonant power P_h and increase the energy harvesting efficiency. When the vehicle mass is fixed, increasing the wheel−tire mass would decrease the mass ratio and therefore increase the harvested resonant power P_h and decrease the energy harvesting efficiency. When the wheel−tire stiffness is fixed, increasing the suspension spring stiffness would increase the harvested resonant power P_h and decrease the resonant energy harvesting efficiency. When the suspension spring stiffness is fixed, increasing the wheel—tire stiffness would decrease the stiffness ratio, therefore decrease the harvested power P_h, and increase the resonant energy harvesting efficiency.

The simulation results from Eq. (6.24) under the condition of a small damping and weak coupling have been verified by those from Eq. (6.25) under the condition of general damping and coupling. The approach could be applied as a tool to design and to optimize the 2DOF vibration energy harvester performance regardless of its configuration and dimension. The effective frequency bandwidth of the 2DOF vibration energy harvester has been studied and discussed in this chapter. The vibration energy harvesting frequency bandwidth can be widened through design and optimization of the mass and stiffness ratios of the oscillators. Increasing the energy harvesting frequency bandwidth and improving the vehicle vibration isolation can be achieved by optimizing the damping c_2 which is the suspension damping coefficient in the case study but at the cost of scarifying the output power and energy harvesting efficiency.

NOMENCLATURE

S_0	The piezoelectric material insert disc surface area
A	The acceleration amplitude
H	The thickness of the piezoelectric materials insert disc
e	2.718281828
e_{33}	The piezoelectric constant
ε^S	The piezoelectric permittivity
Q_i	The quality factor
f	The excitation or damped resonant frequency in Hz
f_n	The natural frequency in Hz
ω	The excitation or damped resonant frequency in radian/s
ω_n	The natural frequency in radian/s
π	3.1415926
$y, y(t)$	The base excitation displacement
$\dot{y}, \dot{y}(t)$	The base excitation velocity
$\ddot{y}, \ddot{y}(t)$	The base excitation acceleration
$Y, Y(s)$	The Laplace transform of y or $y(t)$
$x_1, x_1(t)$	The displacement of the primary oscillator mass (m_1)
$\dot{x}_1, \dot{x}_1(t)$	The velocity of the primary oscillator mass (m_1)
$\ddot{x}_1, \ddot{x}_1(t)$	The acceleration of the primary oscillator mass (m_1)
$X_1, X_1(s)$	The Laplace transform of x_1 or $x_1(t)$
$x_2, x_2(t)$	The displacement of the auxiliary oscillator mass (m_2)
$\dot{x}_2, \dot{x}_2(t)$	The velocity of the auxiliary oscillator mass (m_2)
$\ddot{x}_2, \ddot{x}_2(t)$	The acceleration of the auxiliary oscillator mass (m_2)
$X_2, X_2(s)$	The Laplace transform of x_2 or $x_2(t)$
m_1	The primary oscillator mass
m_2	|The auxiliary oscillator mass
k_1	The short circuit stiffness between the base and the primary oscillator (m_1)

k_2	The short circuit stiffness between the primary oscillator (m_1) and the auxiliary oscillator (m_2)
c_1	The short circuit mechanical damping between the base and the primary oscillator (m_1)
c_2	The short circuit mechanical damping between the primary oscillator (m_1) and the auxiliary oscillator (m_2)
C_0	The blocking capacity of the piezoelectric patch element
R	The sum of the external load resistance and the piezoelectric patch element resistance
R_N	The dimensionless resistance
α	The force factor of the piezoelectric material
α_{N_1}	The dimensionless force factor related to the primary oscillator 1
α_{N_2}	The dimensionless force factor related to the auxiliary oscillator 2
ξ_{11}	The dimensionless damping coefficient related to the primary oscillator 1
ξ_{22}	The dimensionless damping coefficient related to the auxiliary oscillator 2
Φ_1	The frequency ratio related to the primary oscillator
Φ_2	The frequency ratio related to the auxiliary oscillator
M_R	The primary and auxiliary mass ratio
K_R	The stiffness ratio of k_1 and k_2
$V, V(t)$	The output voltage of the 2DOF system
$\dot{V}$	The change rate of the output voltage
$|V|$	The amplitude of the output voltage
$\overline{V}, \overline{V}(s)$	The Laplace transform of V or $V(t)$
P_h	The harvested resonant power
P_{in}	The input resonant power
s	The Laplace variable
t	The variable of the integration or time variable
i	The square root of -1
H	The resonant energy harvesting efficiency

Subscripts

33	Piezoelectric working mode have the same direction of the loading and electric poles
c	The damping dissipated
e	Extracted vibration energy
h	Harvested energy
max	The maximum
N	Normalized

Superscripts

$\cdot$	The first differential
$\cdot\cdot$	The second differential
$-$	Time average

Special Function

$\|\ \|$	The modulus or absolute value
$\langle\rangle$	The time average, expectation, or mean value
Re[ ]	The real part of a complex variable

Abbreviations

SDOF   Single degree of freedom
2DOF   Two degrees of freedom
MDOF   Multiple degree of freedom
MEMS   Microelectromechanical system

REFERENCES

[1] Roundy S, Wright PK, Rabaey J. A study of low level vibrations as a power source for wireless sensor nodes. Comput Commun 2003;26(11):1131–44.

[2] Saadon S, Sidek O. A review of vibration-based MEMS piezoelectric energy harvesters. Energy Convers Manag 2011;52:500–4.

[3] Beeby SP, Tudor MJ, White NM. Energy harvesting vibration sources for microsystems applications. Meas Sci Technol 2006;17:R175–95.

[4] Choi WJ, Jeon Y, Jeong JH, Sood R, Kim SG. Energy harvesting MEMS device based on thin film piezoelectric cantilevers. J Electroceramics 2006;17:543–8.

[5] Cook-Chennault K, Thambi AN, Sastry AM. Powering MEMS portable devices—a review of non-regenerative and regenerative power supply systems with special emphasis on piezoelectric energy harvesting systems. Smart Mater Struct 2008;17: 043001.

[6] Ralib AA, Nordin MAN, Salleh H. A comparative study on MEMS piezoelectric microgenerators. Microsyst Technol 2010;16:1673–81.

[7] Roundy S, Zhang Y. Toward self-tuning adaptive vibration based micro-generators. In: Smart materials, nano- and micro-smart systems, Sydney; 2005. p. 373–84.

[8] Wu W-J, Chen Y-Y, Lee B-S, He J-J, Peng Y-T. Tunable resonant frequency power harvesting devices. In: Proceedings of smart structures and materials Conference; 2006. 61690A–8A.

[9] Cornwell P, Goethal J, Kowko JJ, Damianakis M. Enhancing power harvesting using a tuned auxiliary structure. J Intell Mater Syst Struct 2005;16:825–34.

[10] Shahruz SM. Design of mechanical band-pass filters for energy scavenging. J Sound Vib 2006;292:987–98.

[11] Erturk A, Renno JM, Inman DJ. Modeling of piezoelectric energy harvesting from an L-shaped beam-mass structure with an application to UAVs. J Intell Mater Syst Struct 2008;20:529–44.

[12] Sari I, Balkan T, Kulah H. An electromagnetic micro power generator for wideband environmental vibrations. Sens Actuators A Phys July–August 2008;145–146:405–13.

[13] Wu H, Tang L, Yang Y, Soh CK. A novel two-degrees-of-freedom piezoelectric energy harvester. J Intell Mater Syst Struct 2012;24:357–68.

[14] Tadesse Y, Shujun Z, Priya S. Multimodal energy harvesting system: piezoelectric and electromagnetic. J Intell Mater Syst Struct 2008;20:625–32.

[15] Zuo L, Tang X. Large-scale vibration energy harvesting. J Intell Mater Syst Struct 2013; 24:1405–30.

[16] Lefeuvre E, Badel A, Richard C, Petit L, Guyomar D. A comparison between several vibration-powered piezoelectric generators for standalone systems. Sens Actuators A Phys 2006;126:405–16.

[17] Guyomar D, Sebald G, Pruvost S, Lallart M, Khodayari A, Richard C. Energy harvesting from ambient vibrations and heat. J Intell Mater Syst Struct 2008;20:609—24.

[18] Happian-Smith J. An introduction to modern vehicle design. New York: Butterworth-Heinemann; 2001.

[19] Yagiz N, Hacioglu Y, Taskin Y. Fuzzy sliding-mode control of active suspensions. IEEE Trans Ind Electron 2008;55.

[20] Thite A. Development of a refined quarter car model for the analysis of discomfort due to vibration. Adv Acoust Vib 2012;2012.

[21] Wang X, Koss LL. Frequency response function for power, power transfer ratio and coherence. J Sound Vib 1992;155(1):55—73.

[22] Xiao H, Wang X, John S. A dimensionless analysis of a 2DOF piezoelectric vibration energy harvester. Mech Syst Signal Process 2015;58—59(2015):355—75.

[23] Xiao H, Wang X, John S. A multi degree of freedom piezoelectric vibration energy harvester with piezoelectric elements inserted between two nearby oscillators. Mech Syst Signal Process February 2016;68—69:138—54.

[24] Xiao H, Wang X. A review of piezoelectric vibration energy harvesting techniques. Int Rev Mech Eng 2014;8(3). ISSN: 1970-8742:609—20.

[25] Wang X, Lin LW. Dimensionless optimization of piezoelectric vibration energy harvesters with different interface circuits. Smart Mater Struct 2013;22. ISSN: 0964-1726:1—20.

[26] Wang X, Xiao H. Dimensionless analysis and optimization of piezoelectric vibration energy harvester. Int Rev Mech Eng 2013;7(4). ISSN: 1970-8742:607—24.

[27] Cojocariu B, Hill A, Escudero A, Xiao H, Wang X. Piezoelectric vibration energy harvester — design and prototype. In: Proceedings of ASME 2012. International Mechanical Engineering Congress & Exposition, ISBN 978-0-7918-4517-2.

[28] Wang X, Liang XY, Hao ZY, Du HP, Zhang N, Qian M. Comparison of electromagnetic and piezoelectric vibration energy harvesters with different interface circuits. Mech Syst Signal Process May 2016;72—73:906—24. [accepted by October 29, 2015].

[29] Wang X. Coupling loss factor of linear vibration energy harvesting systems in a framework of statistical energy analysis. J Sound Vib 2016;362(2016):125—41.

[30] Wang X, Liang XY, Shu GQ, Watkins S. Coupling analysis of linear vibration energy harvesting systems. Mech Syst Signal Process March 2016;70—71:428—44.

[31] Wang X, John S, Watkins S, Yu XH, Xiao H, Liang XY, et al. Similarity and duality of electromagnetic and piezoelectric vibration energy harvesters. Mech Syst Signal Process 2015;52—53:672—84.

A Study of Multiple Degree of Freedom Piezoelectric Vibration Energy Harvester

H. Xiao, X. Wang

RMIT University, Melbourne, VIC, Australia

CHAPTER OUTLINE

7.1 INTRODUCTION

The biggest motivation behind energy harvesting is to provide the promising energy for self-powered wireless sensors or devices and to overcome the limitations imposed by the traditional power sources such as batteries and the electrical grid. The most common configuration of piezoelectric vibration energy harvester (PVEH) is the one degree of freedom (1DOF) cantilever beam structure as illustrated in Refs. [1,2,19–28] and in Chapters 2 and 3. It is easy and efficient in converting vibration energy into electrical energy in some scenarios, such as industry motors, or machines with known sufficient vibration levels and repeatable and consistent vibration frequency ranges. Thus, the harvested power falls significantly when ambient excitation frequency is different from the resonant frequency because the vibration energy harvester is only efficient in a particular resonant frequency. Unfortunately, potential ambient vibration energy sources exist in a wide range of frequencies and in a random form, which is a major challenge for the energy harvesting technology. As a result, a number of approaches have been pursued to overcome this limitation. The approaches include multifrequency arrays [3–5], multiple degree of freedom

(MDOF) energy harvester which is also known as multifunctional energy harvesting technology [6–8], passive and active self-resonant tuning technologies [9–12].

For the multifrequency arrays, the recent studies are focused on the effects of the harvesting electrical circuits interfaced with the array configuration of the energy harvesters to increase the harvested power. The principle of the MDOF energy harvesting technique is to achieve wider harvesting frequency bandwidth through tuning two or multiple resonant frequencies close to each other where the resonant response magnitudes are significant. Kim et al. [13] developed the concept of a two degree of freedom (2DOF) piezoelectric energy harvesting device which could gain two close resonant frequencies thus increasing the harvesting frequency bandwidth. This is achieved by adopting two cantilever beams connected with one proof mass, as this configuration took account in both translational and rotational degrees of freedom (DOF). Ou et al. [14] presented an experimental study of a 2DOF piezoelectric vibration energy harvesting system attached with two masses on one cantilever beam to achieve two close resonant frequencies. Zhou et al. [15] presented a multimode piezoelectric energy harvester which comprised a tip mass called "dynamic magnifier." Liu et al. [16] proposed a piezoelectric cantilever beam energy harvester attached with a spring and a mass as the oscillator. This type of the vibration energy harvester increased almost four times harvesting efficiency compared with a conventional type of the vibration energy harvester while operating at the first resonant frequency. However, according to the experimental results, the harvesting frequency bandwidth did not increase because the two resonant frequencies of the harvester were not tuned close to each other. The harvester may require further tuning such as increasing the mass of the oscillator to achieve the preset goal but it may result in a size increase. In most of the above reported researches, the tuning strategy to obtain two or multiple resonant frequencies has not been studied. Thus, in this chapter, a tuning strategy to achieve a wide harvesting frequency bandwidth will be studied. Besides, a novel PVEH model comprised of the multiple inserted piezoelectric elements is proposed and analyzed to enhance the harvesting performance without increasing the size or the weight of a conventional PVEH. At last, a generalized MDOF PVEH model with multiple pieces of piezoelectric elements will be introduced and analyzed. By using the generalized PVEH model, the harvesting performance comparison is conducted for the piezoelectric energy harvester from 1DOF to 5DOF where the total mass and the mass ratio of oscillators of the harvester system are kept constant.

7.2 A TWO DEGREE OF FREEDOM PIEZOELECTRIC VIBRATION ENERGY HARVESTER INSERTED WITH TWO PIEZOELECTRIC PATCH ELEMENTS

A conventional 2DOF PVEH is often designed based on a 1DOF primary oscillator attached with an auxiliary oscillator, which contributes a second modal peak. This configuration could widen the harvesting frequency bandwidth by tuning the two

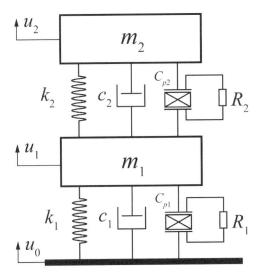

FIGURE 7.1

Schematic of a two degree of freedom piezoelectric vibration energy harvester inserted with two piezoelectric patch elements.

mode response peaks to be close to each other. The study of the proposed 2DOF PVEH model which is shown in Fig. 7.1 will provide the basis for analysis of a MDOF PEVH model inserted with multiple piezoelectric elements.

The governing equations of the 2DOF PVEH system are given by:

$$
\begin{cases}
m_1 \cdot \ddot{u}_1 = -k_1 \cdot (u_1 - u_0) - k_2 \cdot (u_1 - u_2) - c_1 \cdot (\dot{u}_1 - \dot{u}_0) \\
\qquad\qquad -c_2 \cdot (\dot{u}_1 - \dot{u}_2) + \alpha_2 \cdot V_2 - \alpha_1 \cdot V_1 \\
m_2 \cdot \ddot{u}_2 = -k_2 \cdot (u_2 - u_1) - c_2 \cdot (\dot{u}_2 - \dot{u}_1) - \alpha_2 \cdot V_2 \\
\dfrac{V_1}{R_1} = \alpha_1 \cdot (\dot{u}_1 - \dot{u}_0) - C_{p1} \cdot V_1 \\
\dfrac{V_2}{R_2} = \alpha_2 \cdot (\dot{u}_2 - \dot{u}_1) - C_{p2} \cdot V_2
\end{cases}
\tag{7.1}
$$

where m_1 and m_2 are the lumped masses; c_1 and c_2 are the mechanical damping coefficients of the system; k_1 and k_2 are the stiffness coefficients of the system; C_{p1} is the blocking capacitance of the first piezoelectric insert between the base and mass m_1; and the C_{p2} is the blocking capacitance of the second piezoelectric insert between the mass m_1 and mass m_2. α_1 and α_2 are the force factors of the first and second piezoelectric inserts, respectively. R_1 and R_2 are the external electric load resistances of the two piezoelectric elements, respectively. V_1 and V_2 are the voltages across R_1 and R_2, respectively. u_0, u_1, and u_2 are the displacements of the base, the mass m_1, and the mass m_2, respectively. In Eq. (7.1), u_0, u_1, u_2, V_1, and V_2 are functions of time t and can also be written as $u_m(t)$ and $V_j(t)$, $m = 0, 1, 2$ and $j = 1, 2$. By applying the Laplace transform to Eq. (7.1), it gives

$$
\begin{cases}
\left(m_1 \cdot s^2 + k_1 + k_2 + c_1 \cdot s + c_2 \cdot s\right) \cdot U_1 - \left(k_2 + c_2 \cdot s\right) \cdot U_2 \\
\quad + \alpha_1 \cdot \overline{V}_1 - a_2 \cdot \overline{V}_2 = \left(k_1 + c_1 \cdot s\right) \cdot U_0 \\[2mm]
-\left(k_2 + c_2 \cdot s\right) \cdot U_1 + \left(m_2 \cdot s^2 + k_2 + c_2 \cdot s\right) \cdot U_2 + \alpha_2 \cdot \overline{V}_2 = 0 \\[2mm]
-\alpha_1 \cdot U_1 \cdot s + \left(\dfrac{1}{R_1} + C_{p1} \cdot s\right) \cdot \overline{V}_1 = -\alpha_1 \cdot s \cdot U_0 \\[2mm]
\alpha_2 \cdot U_1 \cdot s - \alpha_2 \cdot U_2 \cdot s + \left(\dfrac{1}{R_2} + C_{p2} \cdot s\right) \cdot \overline{V}_2 = 0
\end{cases}
\tag{7.2}
$$

where s is the Laplace variable. U_m and $\overline{V}_j$ in Eq. (7.2) are the short symbols of $U_m(s)$ and $\overline{V}_j(s)$ which are the Laplace transform functions of $u_m(t)$ and $V_j(t)$, $m = 0, 1, 2$ and $j = 1, 2$. It is assumed that when $t = 0$, $u_m(0) = \frac{du_m(0)}{dt} = 0$, and $V_j(0) = \frac{dV_j(0)}{dt} = 0$. If $s = i \cdot \omega$, Eq. (7.2) can be written as:

$$
\begin{pmatrix}
-m_1 \cdot \omega^2 + c_1 \cdot \omega \cdot i + c_2 \cdot \omega \cdot i + k_1 + k_2 & -c_2 \cdot \omega \cdot i - k_2 & \alpha_1 & -\alpha_2 \\
-c_2 \cdot \omega \cdot i - k_2 & -m_2 \cdot \omega^2 + c_2 \cdot \omega \cdot i + k_2 & 0 & \alpha_2 \\
-\alpha_1 \cdot \omega \cdot i & 0 & C_{p1} \cdot \omega \cdot i + 1/R_1 & 0 \\
\alpha_2 \cdot \omega \cdot i & -\alpha_2 \cdot \omega \cdot i & 0 & C_{p2} \cdot \omega \cdot i + 1/R_2
\end{pmatrix}
$$

$$
\cdot
\begin{bmatrix} U_1 \\ U_2 \\ \overline{V}_1 \\ \overline{V}_2 \end{bmatrix}
=
\begin{pmatrix} k_1 + c_1 \cdot \omega \cdot i \\ 0 \\ -\alpha_1 \cdot \omega \cdot i \\ 0 \end{pmatrix}
\cdot U_0
\tag{7.3}
$$

To conduct the dimensionless analysis, all the parameters are normalized by:

$$
\omega_1 = \sqrt{\frac{k_1}{m_1}}, \qquad \omega_2 = \sqrt{\frac{k_2}{m_2}}
$$

$$
\zeta_1 = \frac{c_1}{2\sqrt{k_1 \cdot m_1}}, \qquad \zeta_2 = \frac{c_2}{2\sqrt{k_2 \cdot m_2}}
$$

$$
M_1 = \frac{m_2}{m_1}
$$

$$
\theta_1 = R_1 \cdot C_{p1} \cdot \omega_1, \qquad \theta_2 = R_2 \cdot C_{p2} \cdot \omega_2
\tag{7.4}
$$

$$
\Lambda_1^2 = \frac{\alpha_1^2}{C_{p1} \cdot k_1}, \qquad \Lambda_2^2 = \frac{\alpha_2^2}{C_{p2} \cdot k_2}
$$

$$
\Omega_1 = \frac{\omega_2}{\omega_1}, \qquad \Phi = \frac{\omega}{\omega_1}
$$

where ω_1 is the natural resonant frequency of the primary oscillator system with the mass m_2 removed and ω_2 is the natural resonant frequency of the auxiliary oscillator system with the mass m_1 clamped still. By substituting Eq. (7.4) into Eq. (7.3), the dimensionless voltages across R_1 and R_2 can be given by:

$$\frac{|\bar{V}_1|}{\frac{m_1 \cdot A}{\alpha_1}} = \left| \frac{\frac{\Lambda_1^2}{\Phi^2} \cdot \left[\left(1 - \left(\frac{\Omega_1^2}{\Phi^2} + \frac{2 \cdot \zeta_2 \cdot \Omega_1}{\Phi} \cdot i \right) \cdot (M_1 + 1) \right) \cdot \frac{\Omega_1 + \theta_2 \cdot \Phi \cdot i}{\theta_2 \cdot \Phi \cdot i} - \frac{\Lambda_2^2 \cdot \Omega_1^2}{\Phi^2} \cdot (M_1 + 1) \right]}{\left\{ \frac{1 + \theta_1 \cdot \Phi \cdot i}{\theta_1 \cdot \Phi \cdot i} \cdot \left[\left(1 - \left(\frac{\Omega_1^2}{\Phi^2} + \frac{2 \cdot \zeta_2 \cdot \Omega_1}{\Phi} \cdot i \right) \cdot (M_1 + 1) \right) \cdot \frac{\Omega_1 + \theta_2 \cdot \Phi \cdot i}{\theta_2 \cdot \Phi \cdot i} - \frac{\Lambda_2^2 \cdot \Omega_1^2}{\Phi^2} \cdot (M_1 + 1) \right] \atop - \left[\left(1 - \frac{\Omega_1^2}{\Phi^2} - \frac{2 \cdot \zeta_2 \cdot \Omega_1}{\Phi} \cdot i \right) \cdot \frac{\Omega_1 + \theta_2 \cdot \Phi \cdot i}{\theta_2 \cdot \Phi \cdot i} - \frac{\Lambda_2^2 \cdot \Omega_1^2}{\Phi^2} \right] \cdot \left[\left(\frac{1}{\Phi^2} + \frac{2 \cdot \zeta_1}{\Phi} \cdot i \right) \cdot \frac{1 + \theta_1 \cdot \Phi \cdot i}{\theta_1 \cdot \Phi \cdot i} + \frac{\Lambda_1^2}{\Phi^2} \right] \right\}} \right| \quad (7.5)$$

$$\left[1 - \left(\frac{\Omega_1^2}{\Phi^2} + \frac{\Lambda_2^2 \cdot \Omega_1^2}{\Phi^2} \right) \cdot (M_1 + 1) \right] - \left(1 - \frac{\Omega_1^2}{\Phi^2} - \frac{\Lambda_2^2 \cdot \Omega_1^2}{\Phi^2} \right) \cdot \left(\frac{1}{\Phi^2} + \frac{\Lambda_1^2}{\Phi^2} \right) = 0$$

$$\frac{|\bar{V}_2|}{\frac{m_2 \cdot A}{\alpha_2}} = \left| \frac{\frac{\Lambda_2^2 \cdot \Omega_1^2}{\Phi^2} \cdot \left[\left(\frac{1}{\Phi^2} + \frac{2 \cdot \zeta_1}{\Phi} \cdot i \right) \cdot \frac{1 + \theta_1 \cdot \Phi \cdot i}{\theta_1 \cdot \Phi \cdot i} + \frac{\Lambda_1^2}{\Phi^2} \right]}{\left\{ \frac{1 + \theta_1 \cdot \Phi \cdot i}{\theta_1 \cdot \Phi \cdot i} \cdot \left[\left(1 - \left(\frac{\Omega_1^2}{\Phi^2} + \frac{2 \cdot \zeta_2 \cdot \Omega_1}{\Phi} \cdot i \right) \cdot (M_1 + 1) \right) \cdot \frac{\Omega_1 + \theta_2 \cdot \Phi \cdot i}{\theta_2 \cdot \Phi \cdot i} - \frac{\Lambda_2^2 \cdot \Omega_1^2}{\Phi^2} \cdot (M_1 + 1) \right] \atop - \left[\left(1 - \frac{\Omega_1^2}{\Phi^2} - \frac{2 \cdot \zeta_2 \cdot \Omega_1}{\Phi} \cdot i \right) \cdot \frac{\Omega_1 + \theta_2 \cdot \Phi \cdot i}{\theta_2 \cdot \Phi \cdot i} - \frac{\Lambda_2^2 \cdot \Omega_1^2}{\Phi^2} \right] \cdot \left[\left(\frac{1}{\Phi^2} + \frac{2 \cdot \zeta_1}{\Phi} \cdot i \right) \cdot \frac{1 + \theta_1 \cdot \Phi \cdot i}{\theta_1 \cdot \Phi \cdot i} + \frac{\Lambda_1^2}{\Phi^2} \right] \right\}} \right| \quad (7.6)$$

where $A = |\omega^2 \cdot U_0|$. Hence, according to Ref. [17], the dimensionless harvested powers of the first and the second piezoelectric patches could be obtained from Eqs. (7.5) and (7.6) and are given by

$$\frac{P_1}{\frac{m_1 \cdot A^2}{\omega_1}} = \frac{1}{2 \cdot \Lambda_1^2 \cdot \theta_1} \cdot \left| \frac{\frac{\Lambda_1^2}{\Phi^2} \cdot \left[\left(1 - \left(\frac{\Omega_1^2}{\Phi^2} + \frac{2 \cdot \zeta_2 \cdot \Omega_1}{\Phi} \cdot i \right) \cdot (M_1 + 1) \right) \cdot \frac{\Omega_1 + \theta_2 \cdot \Phi \cdot i}{\theta_2 \cdot \Phi \cdot i} - \frac{\Lambda_2^2 \cdot \Omega_1^2}{\Phi^2} \cdot (M_1 + 1) \right]}{\left\{ \frac{1 + \theta_1 \cdot \Phi \cdot i}{\theta_1 \cdot \Phi \cdot i} \cdot \left[\left(1 - \left(\frac{\Omega_1^2}{\Phi^2} + \frac{2 \cdot \zeta_2 \cdot \Omega_1}{\Phi} \cdot i \right) \cdot (M_1 + 1) \right) \cdot \frac{\Omega_1 + \theta_2 \cdot \Phi \cdot i}{\theta_2 \cdot \Phi \cdot i} - \frac{\Lambda_2^2 \cdot \Omega_1^2}{\Phi^2} \cdot (M_1 + 1) \right] \atop - \left[\left(1 - \frac{\Omega_1^2}{\Phi^2} - \frac{2 \cdot \zeta_2 \cdot \Omega_1}{\Phi} \cdot i \right) \cdot \frac{\Omega_1 + \theta_2 \cdot \Phi \cdot i}{\theta_2 \cdot \Phi \cdot i} - \frac{\Lambda_2^2 \cdot \Omega_1^2}{\Phi^2} \right] \cdot \left[\left(\frac{1}{\Phi^2} + \frac{2 \cdot \zeta_1}{\Phi} \cdot i \right) \cdot \frac{1 + \theta_1 \cdot \Phi \cdot i}{\theta_1 \cdot \Phi \cdot i} + \frac{\Lambda_1^2}{\Phi^2} \right] \right\}} \right|^2$$

$$(7.7)$$

$$\frac{P_2}{\frac{m_2 \cdot A^2}{\omega_2}} = \frac{1}{2 \cdot \Lambda_2^2 \cdot \theta_2} \cdot \left| \frac{\frac{\Lambda_2^2 \cdot \Omega_1^2}{\Phi^2} \cdot \left(\left(\frac{1}{\Phi^2} + \frac{2 \cdot \zeta_1}{\Phi} \cdot i \right) \cdot \frac{1 + \theta_1 \cdot \Phi \cdot i}{\theta_1 \cdot \Phi \cdot i} + \frac{\Lambda_1^2}{\Phi^2} \right)}{\left\{ \frac{1 + \theta_1 \cdot \Phi \cdot i}{\theta_1 \cdot \Phi \cdot i} \cdot \left[\left(1 - \left(\frac{\Omega_1^2}{\Phi^2} + \frac{2 \cdot \zeta_2 \cdot \Omega_1}{\Phi} \cdot i \right) \cdot (M_1 + 1) \right) \cdot \frac{\Omega_1 + \theta_2 \cdot \Phi \cdot i}{\theta_2 \cdot \Phi \cdot i} - \frac{\Lambda_2^2 \cdot \Omega_1^2}{\Phi^2} \cdot (M_1 + 1) \right] \right.}{\left. - \left[\left(1 - \frac{\Omega_1^2}{\Phi^2} - \frac{2 \cdot \zeta_2 \cdot \Omega_1}{\Phi} \cdot i \right) \cdot \frac{\Omega_1 + \theta_2 \cdot \Phi \cdot i}{\theta_2 \cdot \Phi \cdot i} - \frac{\Lambda_2^2 \cdot \Omega_1^2}{\Phi^2} \right] \cdot \left[\left(\frac{1}{\Phi^2} + \frac{2 \cdot \zeta_1}{\Phi} \cdot i \right) \cdot \frac{1 + \theta_1 \cdot \Phi \cdot i}{\theta_1 \cdot \Phi \cdot i} + \frac{\Lambda_1^2}{\Phi^2} \right] \right\}} \right|^2 \tag{7.8}$$

To predict the harvesting efficiency, according to Ref. [17], the total mean input power is given by

$$P_{input} = -m_1 \cdot \langle \ddot{u}_0 \cdot \dot{u}_1 \rangle - m_2 \cdot \langle \ddot{u}_0 \cdot \dot{u}_2 \rangle$$

$$\frac{P_{input}}{A^2} = \frac{1}{2} \cdot \frac{m_1}{\omega} \cdot \text{Re} \left[\frac{i \cdot U_1}{U_0} \right] + \frac{1}{2} \cdot \frac{m_2}{\omega} \cdot \text{Re} \left[\frac{i \cdot U_2}{U_0} \right] \tag{7.9}$$

Therefore, the harvesting efficiencies of the two piezoelectric inserts are given by

$$\eta_1 = \frac{P_1}{P_{input}} = \frac{\dfrac{\dfrac{P_1}{m_1 \cdot A^2}}{\omega_1}}{\dfrac{P_{input}}{A^2} \cdot \dfrac{\omega_1}{m_1}} = \frac{\dfrac{\dfrac{P_1}{m_1 \cdot A^2}}{\omega_1}}{\dfrac{\omega_1}{\omega} \cdot \text{Re} \left[\dfrac{i \cdot U_1}{U_0} \right] + \dfrac{m_2}{m_1} \dfrac{\omega_1}{\omega} \cdot \text{Re} \left[\dfrac{i \cdot U_2}{U_0} \right]}$$

$$\eta_2 = \frac{P_2}{P_{input}} = \frac{\dfrac{\dfrac{P_2}{m_2 \cdot A^2}}{\omega_2}}{\dfrac{P_{input}}{A^2} \cdot \dfrac{\omega_2}{m_2}} = \frac{\dfrac{\dfrac{P_2}{m_2 \cdot A^2}}{\omega_2}}{\dfrac{m_1}{m_2} \dfrac{\omega_2}{\omega} \cdot \text{Re} \left[\dfrac{i \cdot U_1}{U_0} \right] + \dfrac{\omega_2}{\omega} \cdot \text{Re} \left[\dfrac{i \cdot U_2}{U_0} \right]} \tag{7.10}$$

where $\text{Re} \left[\frac{i \cdot U_1}{U_0} \right]$ and $\text{Re} \left[\frac{i \cdot U_1}{U_0} \right]$ are solved from Eq. (7.3); $\frac{P_1}{\frac{m_1 \cdot A^2}{\omega_1}}$ and $\frac{P_2}{\frac{m_2 \cdot A^2}{\omega_2}}$ are calculated from Eqs. (7.7) and (7.8).

To evaluate and compare the harvesting performance of the proposed PVEH with that of a conventional one, the parameters of the system as shown in Table 7.1 are taken from Tang's model [7] where the position effects of the piezoelectric patch on the harvesting performance were studied. The conventional 2DOF PVEH only has one piezoelectric patch inserted between the primary (m_1) and auxiliary (m_2) oscillators as illustrated in Chapter 6. The proposed 2DOF PVEH has two piezoelectric patches inserted between the primary (m_1) and auxiliary (m_2) oscillators and between the primary oscillator (m_1) and the base.

The harvested power could be calculated by substituting the values in Table 7.1 into Eqs. (7.7) and (7.8). It is assumed that the output voltage signals from the first and second piezoelectric elements have been compensated for

Table 7.1 System Parameters of a 2DOF Piezoelectric Vibration Energy Harvester [7]

Parameter	Type	Values	Units
m_1	Primary oscillator	0.04	kg
m_2	Auxiliary oscillator mass	0.008	kg
k_1	Primary oscillator stiffness	100	N/m
k_2	Auxiliary oscillator stiffness	14.45	N/m
c_1	Primary oscillator damping coefficient	0.08	N·s/m
c_2	Auxiliary oscillator stiffness damping coefficient	0.00272	N·s/m
α_1	First piezo-insert force factor	3.1623e-5	N/V
α_2	Second piezo-insert force factor	3.1623e-5	N/V
C_{p1}	Blocking capacitance of the first piezo patch	2.5e-8	F
C_{p2}	Blocking capacitance of the second piezo patch	2.5e-8	F
R_1	Electrical resistance of the first piezo patch	1.0e6	Ω
R_2	Electrical resistance of the second piezo patch	1.0e6	Ω

2DOF, *two degrees of freedom.*

their phase difference so that the two voltage signals are in phase and additive with each other. As a result, the harvesting performance of the proposed 2DOF PVEH model is predicted to have the power output of 2.497 mW and the power density of 51.0291 mW/kg. The harvesting performance of the proposed harvesting model is 9.78 times more than that of the original conventional 2DOF model reported in Ref. [7], whose power generation was 250.4 µW, and the power density was 5.22 mW/kg. The original conventional 2DOF model only had one piezoelectric element added between the primary and auxiliary oscillators. There was no piezoelectric element added between the primary and the base. In the proposed harvesting model, the entire system is not much changed, no extra mass is added or no structure complexity increases, only one additional piezoelectric element is added between the primary and the base to achieve this performance enhancement. The details of the parameter study will be presented in the following sections.

First of all, the principal advantage of the proposed 2DOF model is of a wider harvesting frequency bandwidth compared with that of the 1DOF model. To achieve that, the effects of the system parameters on the difference of the two resonant frequencies should be investigated, as the investigation will provide a useful method to tune the two resonant frequencies to be close to each other. Thus, from Eq. (7.5) or (7.6) under the nondamped or low damped and short-circuit condition, the two dimensionless resonant frequencies $\Phi_{1,2}$ are obtained from solving the following equation:

$$\left[1 - \left(\frac{\Omega_1^2}{\Phi^2} + \frac{\Lambda_2^2 \cdot \Omega_1^2}{\Phi^2}\right) \cdot (M_1 + 1)\right] - \left(1 - \frac{\Omega_1^2}{\Phi^2} - \frac{\Lambda_2^2 \cdot \Omega_1^2}{\Phi^2}\right) \cdot \left(\frac{1}{\Phi^2} + \frac{\Lambda_1^2}{\Phi^2}\right) = 0$$

$$(7.11)$$

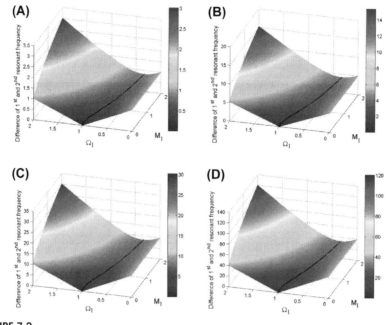

FIGURE 7.2

The difference of the two dimensionless resonant frequencies versus the mass ratio M_1 and dimensionless frequency ratio Ω_1 under the synchronous changes of the coupling strength of the piezoelectric patches. (A) $\frac{\Lambda_1^2}{\varsigma_1} = \frac{\Lambda_2^2}{\varsigma_2} = 0.02$; (B) $\frac{\Lambda_1^2}{\varsigma_1} = \frac{\Lambda_2^2}{\varsigma_2} = 5$; (C) $\frac{\Lambda_1^2}{\varsigma_1} = \frac{\Lambda_2^2}{\varsigma_2} = 10$; and (D) $\frac{\Lambda_1^2}{\varsigma_1} = \frac{\Lambda_2^2}{\varsigma_2} = 40$.

The discrepancy of the two dimensionless resonant frequencies versus the various mass ratio (M_1) and Ω_1 is shown in Fig. 7.2 where the coupling strengths of $\frac{\Lambda_1^2}{\varsigma_1}$ and $\frac{\Lambda_2^2}{\varsigma_2}$ are equal to 0.02, 5, 10, and 40, respectively. According to Ref. [18], the coupling strength values represent the coupling conditions of the weak, medium, strong, and very strong which influence the difference of the two dimensionless resonant frequencies. It is seen from Fig. 7.2 that the maximum dimensionless frequency difference under the strong coupling condition is larger than that under the weak coupling condition. The strong coupling condition requires more precise tuning of the optimal values of the Ω_1 and the mass ratio M_1 than the weak coupling condition. In general, the maximum dimensionless frequency difference occurs with a large number of mass ratio M_1 and the Ω_1, which is highlighted in red (dark grey in print version). In addition, the difference of the two dimensionless resonant frequencies increases when the coupling strength is increased from the weak to strong. In Fig. 7.2, there are boundary lines which pass the points of the optimal Ω_1 for the dimensionless resonant frequencies equal to one and the as small as possible mass ratio close to zero. The points reflect that the minimum

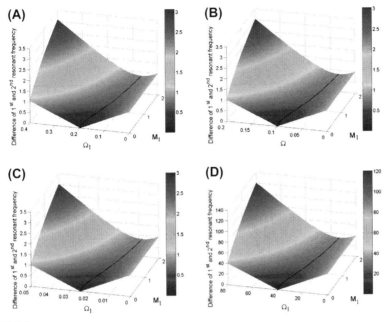

FIGURE 7.3

The difference of the two dimensionless resonant frequencies versus the mass ratio M_1 and the Ω_1 under the coupling strength changes of the primary and auxiliary oscillator systems. (A) $\frac{A_1^2}{\zeta_1} = 0.02 \frac{A_2^2}{\zeta_2} = 5$; (B) $\frac{A_1^2}{\zeta_1} = 0.02 \frac{A_2^2}{\zeta_2} = 10$; (C) $\frac{A_1^2}{\zeta_1} = 0.02 \frac{A_2^2}{\zeta_2} = 40$; and (D) $\frac{A_1^2}{\zeta_1} = 40 \frac{A_2^2}{\zeta_2} = 0.02$.

frequency difference is close to zero and that the 2DOF system degrades to the 1DOF system. On the left-hand side of the boundary lines, when the Ω_1 increases, the dimensionless frequency difference increases. When the mass ratio increases, the dimensionless frequency difference increases. On the right-hand side of the boundary lines, when the Ω_1 increases, the dimensionless frequency difference decreases. However, when the mass ratio increases, the dimensionless frequency difference does not change much and have a flat trend.

Fig. 7.3A–C shows the frequency difference $\Delta\Phi_{1,2}$ versus the dimensionless mass ratio M_1 and dimensionless frequency ratio Ω_1, when the primary oscillator system (with the mass m_2 removed) is under a weak coupling, and the auxiliary oscillator system (with the mass m_1 clamped) is changed from the weak to strong. Fig. 7.3D shows the frequency difference $\Delta\Phi_{1,2}$ versus the dimensionless mass ratio M_1 and the Ω_1, when the primary oscillator system is under the strong coupling, and the auxiliary oscillator system is under a weak coupling. The main trend of Fig. 7.3 is very similar to that of Fig. 7.2 as discussed previously. However, it is interesting to note that the maximum value of the frequency difference $\Delta\Phi_{1,2}$ is not changed much when only the auxiliary oscillator system changes the coupling strength from the

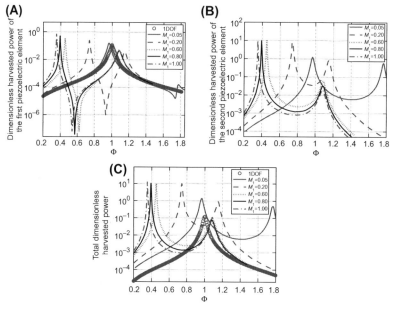

FIGURE 7.4

The dimensionless harvested power of the two degree of freedom piezoelectric vibration energy harvester versus the dimensionless frequency for different dimensionless mass ratio (M_1). (A) the dimensionless harvested power of the first piezo patch element; (B) the dimensionless harvested power of the second piezo patch element; and (C) the total dimensionless harvested power of the first and second piezo patch elements. *DOF*, degree of freedom.

weak to strong. In Fig. 7.3A and D, it clearly shows that the frequency difference $\Delta\Phi_{1,2}$ significantly increases, when the primary oscillator system is changed from the weak to strong coupling. It will prevent the tuning from widening the harvesting frequency bandwidth.

The effects of the mass ratio on the peak magnitude of dimensionless harvested power are illustrated in Fig. 7.4. The dimensionless harvested power of 1DOF system could be obtained when the mass ratio M_1 tends to be zero, and is plotted in Fig. 7.4 in the blue (grey in print version) circles. In this special case, for comparison, the mass of the 1DOF system is set to be equal to the sum of m_1 and m_2. For the first piezoelectric element which is located in the primary oscillator system, as shown in Fig. 7.4A, with a small value of the mass ratio M_1, the trend of the dimensionless harvested power of the 2DOF system is identical to that of the 1DOF system except for the second resonant peak of the 2DOF system. However, as the mass ratio increases, the magnitude of the first resonant peak increases and the first resonant frequency is shifted to lower frequency. However, the magnitude of the second peak increases first and is then remained at the same level as the mass ratio M_1

increases. For the second piezoelectric element which is placed in the auxiliary oscillator system, as shown in Fig. 7.4B, the magnitude of the first peak first increases and then remains same when the mass ratio M_1 increases. On the other hand, when the mass ratio increases, the magnitude of the second resonant peak increases slightly first, then decreases dramatically. The two resonant frequencies decrease as the mass ratio increases, which is similar for both the first and second piezoelectric patch elements.

The harvested energy is additive after the voltage signals are compensated for a phase delay and become in phase. The result in Fig. 7.4C shows that the total dimensionless harvested power of the 2DOF system can be tuned to achieve 85 times more than that of the 1DOF system.

If the mass ratio M_1 is fixed as the constant, the Ω_1 can represent the stiffness ratio (k_2/k_1). The effect of the stiffness ratio on the magnitude of the dimensionless power is demonstrated in Fig. 7.5. For the first piezoelectric element which is located in the primary oscillator system, as shown in Fig. 7.5A, as the stiffness ratio increases, the magnitude of the first resonant peak increases until the stiffness ratio equals to one, then the magnitude of the first resonant peak decreases. At the same time, the magnitude of the second resonant peak decreases when the stiffness

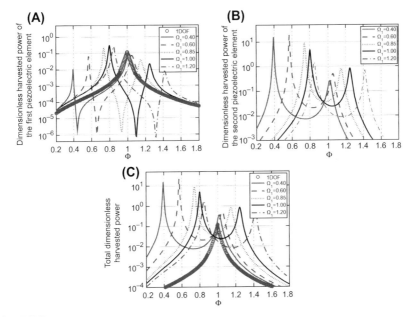

FIGURE 7.5

The dimensionless harvested power of the two degree of freedom piezoelectric vibration energy harvester versus the dimensionless frequency for different Ω_1. (A) the dimensionless harvested power of the first piezo patch element; (B) the dimensionless harvested power of the second piezo patch element; and (C) the total dimensionless harvested power. *DOF*, degree of freedom.

ratio increases. For the second piezoelectric element, as shown in Fig. 7.5B, the magnitude of the first resonant peak first increases slightly and then decreases as the stiffness ratio increases, as well as the magnitude of the second resonant peak. It is seen from Fig. 7.5 that the two harvested resonant power peak values of the second piezo patch element are larger than those of the first piezo patch element. It is seen from Figs. 7.4C and 7.5C that the first resonant peak value of the harvested power of the 2DOF system is larger than that of the 1DOF system described above.

Comparing Fig. 7.6 with Fig. 7.7, it is clearly shown that ζ_2 has less effect on the performance of the 2DOF PVEH than ζ_1 for both the first and second piezoelectric patch element. As a result, a small value of ζ_1 is much preferred to enhance the performance when a 2DOF PVEH is designed.

The effect of the mass ratio M_1 and dimensionless frequency Φ on harvested efficiency of the first and second piezoelectric patch element under different coupling strengths is shown in Fig. 7.8 and Fig. 7.9. High energy harvesting efficiency values could be achieved when the coupling strength increases. Furthermore, for the second piezoelectric patch element, a large mass ratio M_1 would result in a high energy harvesting efficiency with the optimal dimensionless frequency Φ. It is interesting to note

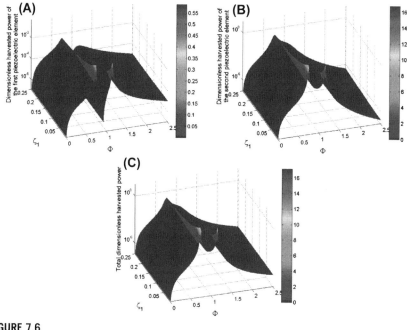

FIGURE 7.6

Dimensionless harvested power of the two degree of freedom piezoelectric vibration energy harvester versus Φ and ζ_1. (A) dimensionless harvested power of the first piezo patch element; (B) dimensionless harvested power of the second piezo patch element; and (C) total dimensionless harvested power.

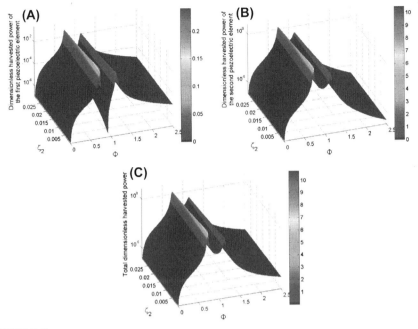

FIGURE 7.7

Dimensionless harvested power of the two degree of freedom piezoelectric vibration energy harvester versus Φ and ζ_2. (A) dimensionless harvested power of the first piezo patch element; (B) dimensionless harvested power of the second piezo patch element; and (C) total dimensionless harvested power.

that the first and second piezoelectric patch elements could not be tuned to most efficiently operate in the same parameter modes along with each other. As the second piezoelectric patch element has the maximum efficiency in certain values of Φ and M_1, the first piezoelectric patch element has the lowest efficiency. In contrast, when the first piezoelectric patch element has the maximum efficiency in certain values of Φ and M_1, the second piezoelectric patch element has a low efficiency. From the color scales of Figs. 7.8 and 7.9, it is seen that from an overall view, the first piezoelectric patch element can achieve a higher peak efficiency than the second one.

7.3 A THREE DEGREE OF FREEDOM PIEZOELECTRIC VIBRATION ENERGY HARVESTER INSERTED WITH THREE PIEZOELECTRIC PATCH ELEMENTS

As shown in Fig. 7.10, a 3DOF PVEH is built with the piezoelectric elements located between two nearby oscillators. In this study, the piezoelectric element type and the total oscillator mass are supposed to be exactly same as those of the 2DOF PVEH.

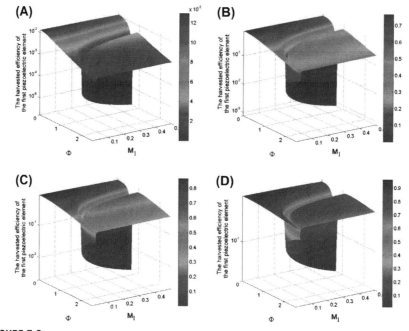

FIGURE 7.8

The energy harvesting efficiency of the first piezoelectric patch element versus Φ and M_1 for different coupling strengths. (A) $\frac{\Lambda_1^2}{\varsigma_1} = \frac{\Lambda_2^2}{\varsigma_2} = 0.02$; (B) $\frac{\Lambda_1^2}{\varsigma_1} = \frac{\Lambda_2^2}{\varsigma_2} = 5$; (C) $\frac{\Lambda_1^2}{\varsigma_1} = \frac{\Lambda_2^2}{\varsigma_2} = 10$; and (D) $\frac{\Lambda_1^2}{\varsigma_1} = \frac{\Lambda_2^2}{\varsigma_2} = 40$.

The governing equations of the 3DOF PVEH inserted with three piezoelectric patch elements are given by:

$$\begin{cases} m_1 \cdot \ddot{u}_1 = -k_1 \cdot (u_1 - u_0) - c_1 \cdot (\dot{u}_1 - \dot{u}_0) - k_2 \cdot (u_1 - u_2) - c_2 \cdot (\dot{u}_1 - \dot{u}_2) + \alpha_2 \cdot V_2 - \alpha_1 \cdot V_1 \\ m_2 \cdot \ddot{u}_2 = -k_2 \cdot (u_2 - u_1) - c_2 \cdot (\dot{u}_2 - \dot{u}_1) - k_3 \cdot (u_2 - u_3) - c_3 \cdot (\dot{u}_2 - \dot{u}_3) - \alpha_2 \cdot V_2 - \alpha_3 \cdot V_3 \\ \qquad m_3 \cdot \ddot{u}_3 = -k_3 \cdot (u_3 - u_2) - c_3 \cdot (\dot{u}_3 - \dot{u}_2) - \alpha_3 \cdot V_3 \\ \qquad\qquad \dfrac{V_1}{R_1} = \alpha_1 \cdot (\dot{u}_1 - \dot{u}_0) - C_{p1} \cdot \dot{V}_1 \\ \qquad\qquad \dfrac{V_2}{R_2} = \alpha_2 \cdot (\dot{u}_2 - \dot{u}_1) - C_{p2} \cdot \dot{V}_2 \\ \qquad\qquad \dfrac{V_3}{R_3} = \alpha_3 \cdot (\dot{u}_3 - \dot{u}_2) - C_{p3} \cdot \dot{V}_3 \end{cases}$$

(7.12)

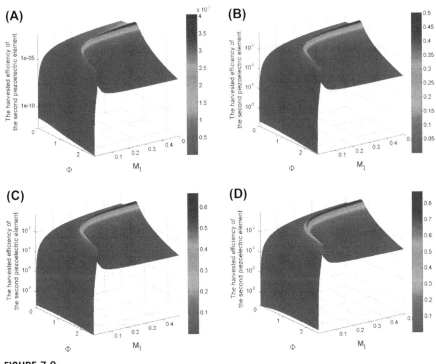

FIGURE 7.9

The energy harvesting efficiency of the second piezoelectric patch element versus Φ and M_1 for different coupling strengths. (A) $\frac{\Lambda_1^2}{\varsigma_1} = \frac{\Lambda_2^2}{\varsigma_2} = 0.02$; (B) $\frac{\Lambda_1^2}{\varsigma_1} = \frac{\Lambda_2^2}{\varsigma_2} = 5$; (C) $\frac{\Lambda_1^2}{\varsigma_1} = \frac{\Lambda_2^2}{\varsigma_2} = 10$; and (D) $\frac{\Lambda_1^2}{\varsigma_1} = \frac{\Lambda_2^2}{\varsigma_2} = 40$.

In Eq. (7.12), $u_0, u_1, u_2, u_3, V_1, V_2, V_3$ are functions of time t and can also be written as $u_m(t)$ and $V_j(t)$, $m = 0, 1, 2, 3$ and $j = 1, 2, 3$. By applying the Laplace transform to Eq. (7.1), it is obtained that

$$
\begin{cases}
(m_1 \cdot s^2 + k_1 + c_1 \cdot s + k_2 + c_2 \cdot s) \cdot U_1 - (k_2 + c_2 \cdot s) \cdot U_2 + \alpha_1 \cdot \overline{V}_1 - \alpha_2 \cdot \overline{V}_2 = (k_1 + c_1 \cdot s) \cdot U_0 \\
-(k_2 + c_2 \cdot s) \cdot U_1 + (m_2 \cdot s^2 + k_2 + c_2 \cdot s + k_3 + c_3 \cdot s) \cdot U_2 - (k_3 + c_3 \cdot s) \cdot U_3 + \alpha_2 \cdot \overline{V}_2 + \alpha_3 \cdot \overline{V}_3 = 0 \\
-(k_3 + c_3 \cdot s) \cdot U_2 + (m_3 \cdot s^2 + k_3 + c_3 \cdot s) \cdot U_3 + \alpha_3 \cdot \overline{V}_3 = 0 \\
-\alpha_1 \cdot U_1 \cdot s + \left(\dfrac{1}{R_1} + c_{p1} \cdot s \right) \cdot \overline{V}_1 = -\alpha_1 \cdot s \cdot U_0 \\
\alpha_2 \cdot U_1 \cdot s - \alpha_2 \cdot U_2 \cdot s + \left(\dfrac{1}{R_2} + C_{p2} \cdot s \right) \cdot \overline{V}_2 = 0 \\
\alpha_3 \cdot U_2 \cdot s - \alpha_3 \cdot U_3 \cdot s + \left(\dfrac{1}{R_3} + C_{p3} \cdot s \right) \cdot \overline{V}_3 = 0
\end{cases}
$$

$$(7.13)$$

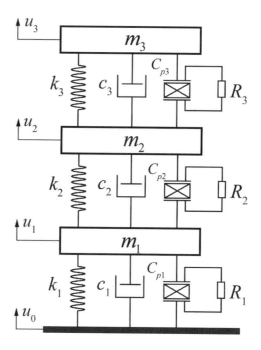

FIGURE 7.10

Schematic of a three degree of freedom piezoelectric vibration energy harvester inserted with three piezoelectric patch elements.

where s is the Laplace variable. U_m and $\overline{V}_j$ in Eq. (7.13) are the short symbols of $U_m(s)$ and $\overline{V}_j(s)$ and are the Laplace transform functions of $u_m(t)$ and $V_j(t)$, $m = 0$, 1, 2, 3 and $j = 1, 2, 3$. It is assumed that when $t = 0$, $u_m(0) = \frac{du_m(0)}{dt} = 0$ and $V_j(0) = \frac{dV_j(0)}{dt} = 0$. As $s = i \cdot \omega$, Eq. (7.13) can be written as:

$$
\begin{pmatrix}
\begin{pmatrix} m_1 \cdot s^2 \\ +k_1 + c_1 \cdot s \\ +k_2 + c_2 \cdot s \end{pmatrix} & -(k_2 + c_2 \cdot s) & 0 & \alpha_1 & -\alpha_2 & 0 \\[4pt]
-(k_2 + c_2 \cdot s) & \begin{pmatrix} m_2 \cdot s^2 \\ +k_2 + c_2 \cdot s \\ +k_3 + c_3 \cdot s \end{pmatrix} & -(k_3 + c_3 \cdot s) & 0 & \alpha_2 & \alpha_3 \\[4pt]
0 & -(k_3 + c_3 \cdot s) & \begin{pmatrix} m_3 \cdot s^2 \\ +k_3 + c_3 \cdot s \end{pmatrix} & 0 & 0 & \alpha_3 \\[4pt]
-\alpha_1 \cdot s & 0 & 0 & \begin{pmatrix} 1/R_1 \\ +C_{p1} \cdot s \end{pmatrix} & 0 & 0 \\[4pt]
\alpha_2 \cdot s & -\alpha_2 \cdot s & 0 & 0 & \begin{pmatrix} 1/R_2 \\ +C_{p2} \cdot s \end{pmatrix} & 0 \\[4pt]
0 & \alpha_3 \cdot s & -\alpha_3 \cdot s & 0 & 0 & \begin{pmatrix} 1/R_3 \\ +C_{p3} \cdot s \end{pmatrix}
\end{pmatrix}
\cdot
\begin{bmatrix} U_1 \\ U_2 \\ U_3 \\ \overline{V}_1 \\ \overline{V}_2 \\ \overline{V}_3 \end{bmatrix}
$$

$$
=
\begin{pmatrix} k_1 + c_1 \cdot s \\ 0 \\ 0 \\ -\alpha_1 \cdot s \\ 0 \\ 0 \end{pmatrix}
\cdot U_0
$$

(7.14)

The dimensionless analysis for the 3DOF vibration energy harvesting PVEH inserted with three piezoelectric elements can be extremely complex; therefore it is difficult to derive the analytical formulas or equations of the dimensionless analysis here. However, the dimensionless analysis could be conducted by Matlab using Eq. (7.14) and the following dimensionless parameters are defined as:

$$\omega_1 = \sqrt{\frac{k_1}{m_1}}, \qquad \omega_2 = \sqrt{\frac{k_2}{m_2}}, \qquad \omega_3 = \sqrt{\frac{k_3}{m_3}}$$

$$\zeta_1 = \frac{c_1}{2\sqrt{k_1 \cdot m_1}}, \qquad \zeta_2 = \frac{c_2}{2\sqrt{k_2 \cdot m_2}}, \qquad \zeta_3 = \frac{c_3}{2\sqrt{k_3 \cdot m_3}}$$

$$M_1 = \frac{m_2}{m_1}, \qquad M_2 = \frac{m_3}{m_2}$$

$$\theta_1 = R_1 \cdot C_{p1} \cdot \omega_1, \quad \theta_2 = R_2 \cdot C_{p2} \cdot \omega_2, \quad \theta_3 = R_3 \cdot C_{p3} \cdot \omega_3$$

$$\Lambda_1^2 = \frac{\alpha_1^2}{C_{p1} \cdot k_1}, \qquad \Lambda_2^2 = \frac{\alpha_2^2}{C_{p2} \cdot k_2}, \qquad \Lambda_3^2 = \frac{\alpha_3^2}{C_{p3} \cdot k_3}$$

$$\Omega_1 = \frac{\omega_2}{\omega_1}, \qquad \Omega_2 = \frac{\omega_3}{\omega_2}, \qquad \Phi = \frac{\omega}{\omega_1}$$

(7.15)

It is worth pointing out that the second auxiliary oscillator system (with the masses m_1 and m_2 clamped still) is identical to and duplicated form the first auxiliary oscillator system (with the mass m_1 clamped still and the mass m_3 removed). For the practical comparison of the 3DOF PVEH with the 2DOF PVEH, the mass, stiffness, and damping ratios between the primary and the first auxiliary oscillator systems are exactly same as those of the 2DOF system shown in Table 7.1. Furthermore, the total mass of 3DOF PVEH is set to be same as that of the 2DOF PVEH. In addition, the ratio (M_1) of the first auxiliary oscillator mass (m_2) over the primary oscillator mass (m_1) is equal to that of the second auxiliary oscillator mass (m_3) over the primary oscillator mass (m_1). In other words, the second auxiliary oscillator is identical to the first auxiliary oscillator.

The effects of the mass ratio M_1 on the harvested power of the 3DOF PVEH are demonstrated in Fig. 7.11. The blue (grey in print version) circle represented the dimensionless harvested power of the degraded 1DOF model described above. For the first piezoelectric element which is located in the primary oscillator system, the magnitude of the first resonant peak increases as the mass ratio M_1 increases. As well as the mass ratio increase will result in shifting the first resonant peak to a lower frequency. For the second and third piezoelectric elements, the magnitude of the first resonant frequency first slightly increases then stays at one level as the mass ratio M_1 increases. Furthermore, it is clearly shown in Fig. 7.11B and C that the mass ratio increases would reduce the discrepancy of the three resonant peaks and widen the effective harvesting frequency bandwidth.

If the mass ratio M_1 is kept as the constant, the Ω_1 can be considered as stiffness ratio. The effects of stiffness ratio on the dimensionless harvested power are illustrated in Fig. 7.12. Therefore, for the first piezoelectric patch element, the magnitude of first resonant peak increases when the stiffness ratio increases. For the second and third piezoelectric patch elements, the magnitude of the first resonant peak initially

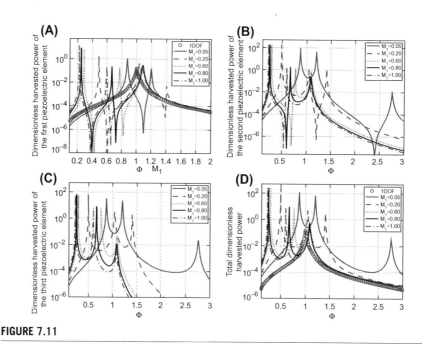

FIGURE 7.11

The dimensionless harvested power of the 3DOF piezoelectric vibration energy harvester versus the dimensionless frequency for different mass ratio M_1. (A) dimensionless harvested power of the first piezo patch element; (B) dimensionless harvested power of the second piezo patch element; (C) dimensionless harvested power of the third piezo patch element; and (D) total dimensionless harvested power. *DOF, degree of freedom.*

increases then decreases as the stiffness ratio increases. However, it is seen that the large values of the stiffness ratio results in large discrepancy of the three resonant peaks, which leads to a narrow effective harvesting frequency bandwidth as shown in Fig. 7.12B and C. Comparing Fig. 7.11D with Fig. 7.12D, overall, no matter how the mass ratio or the stiffness ratio changes, the first resonant peak value of the dimensionless harvested power of the 3DOF PVEH is much larger than that of the 1DOF PVEH. It is seen from Figs. 7.11 and 7.12 that the three harvested resonant power peak values of the second and third piezo patch elements are larger than those of the first piezo patch element.

Figs. 7.13 and 7.14 illustrate the effects of the dimensionless damping coefficient of ζ_1 and ζ_2 on the harvested power of the 3DOF PVEH. The conclusions from the 2DOF PVEH still hold for the 3DOF PVEH. However, the influence of ζ_2 on the harvested power in the 3DOF model is larger than that in the 2DOF PVEH. A small value of ζ_1 is more desirable than that of ζ_2 to improve the performance of the 3DOF PVEH.

The energy harvesting efficiency of 3DOF PVEH versus the mass ratio M_1 and frequency ratio Φ is illustrated in Fig. 7.15. For the first piezoelectric patch element, a small mass and frequency ratio is preferred to achieve a high energy harvesting

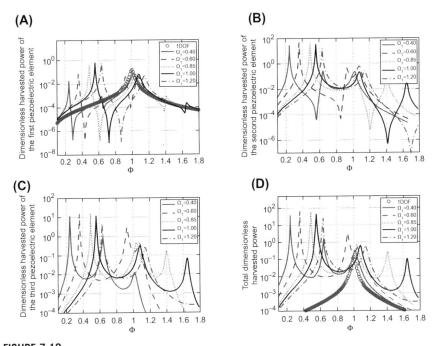

FIGURE 7.12

The dimensionless harvested power of the 3DOF system versus the dimensionless frequency for different Ω_1. (A) dimensionless harvested power of the first piezo patch element; (B) dimensionless harvested power of the second piezo patch element; (C) dimensionless harvested power of the third piezo patch element; and (D) total dimensionless harvested power. *DOF*, degree of freedom.

efficiency, but the mass ratio only has limited influence on the energy harvesting efficiency when the Φ is optimal for a high efficiency. On the other hand, for the second and third piezoelectric patch elements, when the Φ is optimal for a high efficiency, the energy harvesting efficiency increases when the mass ratio increases. However, the selection of optimal Φ is relatively flexible for the third piezoelectric element when the mass ratio M_1 is optimized for a high efficiency. Fig. 7.15D shows that the maximum harvesting efficiency occurs when the mass ratio is large but the frequency ratio is small.

7.4 A GENERALIZED MULTIPLE DEGREE OF FREEDOM PIEZOELECTRIC VIBRATION HARVESTER

Based on the above analysis of the 2DOF and 3DOF models, a versatile MDOF piezoelectric vibration harvester with piezoelectric elements inserted between all two nearby oscillators is developed and illustrated in Fig. 7.16.

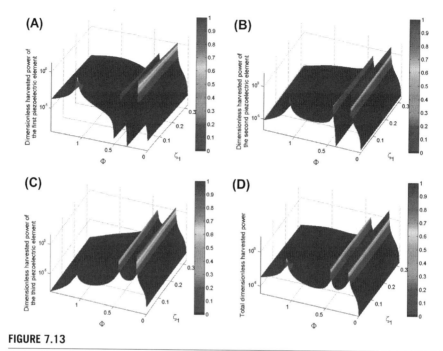

FIGURE 7.13

Dimensionless harvested power of the 3DOF piezoelectric vibration energy harvester versus Φ and ζ_1. (A) dimensionless harvested power of the first piezo patch element. (B) dimensionless harvested power of the second piezo patch element. (C) dimensionless harvested power of the third piezo patch element. (D) total dimensionless harvested power. *DOF*, degree of freedom.

The governing equations of the MDOF PVEH are given by:

$$
\begin{cases}
m_1 \cdot \ddot{u}_1 = -k_1 \cdot (u_1 - u_0) - c_1 \cdot (\dot{u}_1 - \dot{u}_0) - k_2 \cdot (u_1 - u_2) - c_2 \cdot (\dot{u}_1 - \dot{u}_2) + \alpha_2 \cdot V_2 - \alpha_1 \cdot V_1 \\[4pt]
m_2 \cdot \ddot{u}_2 = -k_2 \cdot (u_2 - u_1) - c_2 \cdot (\dot{u}_2 - \dot{u}_1) - k_3 \cdot (u_2 - u_3) - c_3 \cdot (\dot{u}_2 - \dot{u}_3) - \alpha_3 \cdot V_3 - \alpha_2 \cdot V_2 \\[4pt]
\qquad\qquad\qquad\qquad\qquad\qquad \vdots \\[4pt]
m_{n-1} \cdot \ddot{u}_{n-1} = -k_{n-1} \cdot (u_{n-1} - u_{n-2}) - c_{n-1} \cdot (\dot{u}_{n-1} - \dot{u}_{n-2}) \\[4pt]
\qquad\qquad\qquad\quad -k_n \cdot (u_{n-1} - u_n) - c_n \cdot (\dot{u}_{n-1} - \dot{u}_n) - \alpha_n \cdot V_n - \alpha_{(n-1)} \cdot V_{(n-1)} \\[4pt]
m_n \cdot \ddot{u}_n = -k_n \cdot (u_n - u_{n-1}) - c_n \cdot (\dot{u}_n - \dot{u}_{n-1}) - \alpha_n \cdot V_n \\[4pt]
\dfrac{V_1}{R_1} = \alpha_1 \cdot (\dot{u}_1 - \dot{u}_0) - C_{p1} \cdot \dot{V}_1 \\[8pt]
\dfrac{V_2}{R_2} = \alpha_2 \cdot (\dot{u}_2 - \dot{u}_1) - C_{p2} \cdot \dot{V}_2 \\[8pt]
\qquad\qquad\qquad \vdots \\[4pt]
\dfrac{V_n}{R_n} = \alpha_n \cdot (\dot{u}_n - \dot{u}_{n-1}) - C_{pn} \cdot \dot{V}_n
\end{cases}
$$

$$(7.16)$$

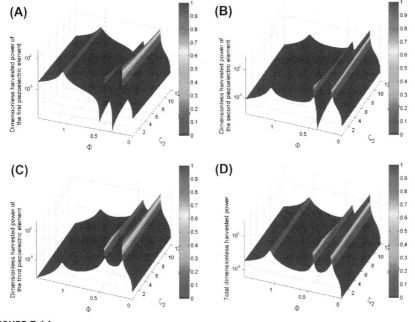

FIGURE 7.14

Dimensionless harvested power of the 3DOF piezoelectric vibration energy harvester versus Φ and ζ_2. (A) dimensionless harvested power of the first piezo patch element. (B) dimensionless harvested power of the second piezo patch element. (C) dimensionless harvested power of the third piezo patch element. (D) total dimensionless harvested power. *DOF*, degree of freedom.

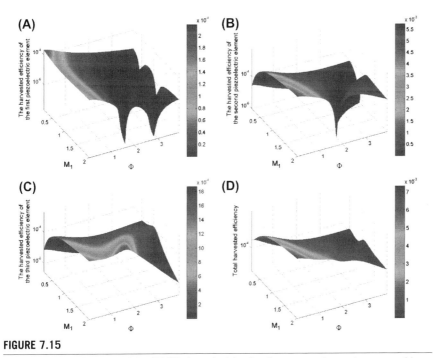

FIGURE 7.15

The harvested efficiency of the 3DOF piezoelectric vibration energy harvester versus M_1 and Φ. (A) the efficiency of the first piezo patch element. (B) the efficiency of the second piezo patch element. (C) the efficiency of the third piezo patch element. (D) total efficiency. *DOF*, degree of freedom.

In Eq. (7.16), u_1, u_2... u_n, and V_1, V_2... V_n are functions of time t and can also be written as $u_m(t)$ and $V_j(t)$, $m = 0, 1, 2...n$ and $j = 1, 2...n$. By applying the Laplace transform to Eq. (7.1), it gives

$$
\begin{pmatrix}
\begin{pmatrix} m_1 \cdot s^2 \\ +k_1+c_1\cdot s \\ +k_2+c_2\cdot s \end{pmatrix} & -(k_2+c_2\cdot s) & 0 & 0 & 0 & \alpha_1 & -\alpha_2 & \cdots & 0 \\
-(k_2+c_2\cdot s) & \begin{pmatrix} m_2 \cdot s^2 \\ +k_2+c_2\cdot s \\ +k_3+c_3\cdot s \end{pmatrix} & -(k_3+c_3\cdot s) & \ddots & \ddots & \ddots & \alpha_2 & \alpha_3 & 0 \\
0 & \ddots & \ddots & \ddots & \ddots & \ddots & \ddots & \ddots & 0 \\
\vdots & \ddots & -(k_{n-1}+c_{n-1}\cdot s) & \begin{pmatrix} m_{n-1}\cdot s^2 \\ +k_{n-1}+c_{n-1}\cdot s \\ k_n+c_n\cdot s \end{pmatrix} & -(k_n+c_n\cdot s) & \ddots & \ddots & \alpha_{n-1} & \alpha_n \\
0 & \ddots & \ddots & -(k_n+c_n\cdot s) & \begin{pmatrix} m_n\cdot s^2 \\ +k_n+c_n\cdot s \end{pmatrix} & 0 & \ddots & \ddots & \alpha_n \\
\alpha_1\cdot s & 0 & \ddots & \ddots & 0 & -\begin{pmatrix} 1/R_1 \\ +C_{p1}\cdot s \end{pmatrix} & \ddots & \ddots & 0 \\
0 & -\alpha_2\cdot s & \alpha_2\cdot s & \ddots & \ddots & \ddots & -\begin{pmatrix} 1/R_2 \\ +C_{p2}s \end{pmatrix} & \ddots & 0 \\
\vdots & \ddots & \ddots & \ddots & \ddots & \ddots & 0 & \ddots & 0 \\
0 & \cdots & 0 & -\alpha_n\cdot s & \alpha_n\cdot s & 0 & \cdots & \ddots & -\begin{pmatrix} 1/R_n \\ +C_{pn}\cdot s \end{pmatrix}
\end{pmatrix}
\cdot
\begin{bmatrix} U_1 \\ U_2 \\ \vdots \\ U_{n-1} \\ U_n \\ \overline{V}_1 \\ \overline{V}_2 \\ \vdots \\ \overline{V}_n \end{bmatrix}
$$

$$
=
\begin{pmatrix} \begin{pmatrix} k_1+c_1\cdot s \end{pmatrix} \\ 0 \\ \vdots \\ 0 \\ 0 \\ \alpha_1\cdot s \\ 0 \\ \vdots \\ 0 \end{pmatrix} \cdot U_0
\qquad (7.17)
$$

where s is the Laplace variable. U_m and $\overline{V}_j$ in Eq. (7.17) are the short symbols of $U_m(s)$ and $\overline{V}_j(s)$ and the functions of s. They are the Laplace transform functions of $u_m(t)$ and $V_j(t)$, $m = 0, 1, 2,...n$ and $j = 1, 2,...n$. It is assumed that when $t = 0$, $u_m(0) = \frac{du_m(0)}{dt} = 0$ and $V_j(0) = \frac{dV_j(0)}{dt} = 0$.

Eqs. (7.17) could be programmed into a Matlab code to predict the dimensionless harvested power and the harvested power density of a PVEH of a particular number

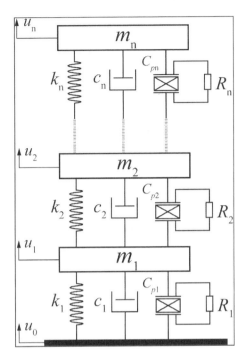

FIGURE 7.16

Schematic of a generalized multiple degree of freedom piezoelectric vibration energy harvester with piezoelectric elements between all two nearby oscillators.

of DOF. Therefore, the harvested power and the power density values of 4DOF and 5DOF PVEHs with the same total mass and the mass ratio of the auxiliary oscillator to the primary oscillator as those of the 2DOF and 3DOF PVEHs are summarized in Table 7.2. The harvested power and the power density values are plotted against the number of DOF in Fig. 7.17. It is seen from Fig. 7.17 that the dimensionless harvested power and power density increase as the number of DOF increases. However, the number of the DOF is limited by the system packaging volume, as a piezoelectric element has a certain dimension. It is found from Figs. 7.4D and 7.11D, Figs. 7.5D and 7.12D that the first resonant peak of the harvested power increases when the number of DOF increases. The first resonant frequency decreases as the number of DOF increases. Therefore increasing the number of DOF could be an alternative method to lower the resonant frequency without increasing the weight of the system. The proposed analysis method of an MDOF PVEH provides the guideline to improve the harvesting performance of a PVEH. That is to add more auxiliary oscillators or to increase the number of DOF of PVEH by inserting piezo patch elements between all two nearby oscillators. The method could be a useful tool to design and optimize a multiple DOF PVEH system.

Table 7.2 Comparison of Harvesting Performance From 1DOF to 5DOF Piezoelectric Vibration Energy Harvester

Number of Degree of PVEH Configuration	Dimensionless Harvested Power	Power Density (mW/kg)	First Resonant Frequency (Hz)
1DOF	0.0736	1.6114	7.264
2DOF	6.0443	25.5146	5.891
3DOF	37.7935	120.0293	4.232
4DOF	65.4518	168.6576	3.333
5DOF	109.9014	250.8898	2.626

DOF, *degree of freedom*; PVEH, *piezoelectric vibration energy harvester.*

7.5 MODAL ANALYSIS AND SIMULATION OF MULTIPLE DEGREE OF FREEDOM PIEZOELECTRIC VIBRATION ENERGY HARVESTER

To effectively integrate the piezoelectric material into plates and panels of the vehicle body structures for vibration energy harvesting purpose, a modal analysis and simulation model has to be adopted. The dynamic equation of the MDOF system is given by

$$\mathbf{M} \cdot \ddot{\mathbf{z}} + \mathbf{C} \cdot \dot{\mathbf{z}} + \mathbf{K} \cdot \mathbf{z} = -\mathbf{M} \cdot \ddot{\mathbf{y}} - \boldsymbol{\alpha} \cdot \mathbf{V} \tag{7.18}$$

The governing equation of the electrical system is given by

$$\mathbf{I} = \mathbf{R}^{-1} \cdot \mathbf{V} = \boldsymbol{\alpha} \cdot \dot{\mathbf{z}} - \mathbf{C_0} \cdot \dot{\mathbf{V}} \tag{7.19}$$

For a mode shape matrix of $\boldsymbol{\Phi}$, there exist the following coordinate transformations

$$\begin{cases} \mathbf{z} = \boldsymbol{\Phi} \cdot \mathbf{q_m} \\ \mathbf{y} = \boldsymbol{\Phi} \cdot \mathbf{y_m} \\ \mathbf{V} = \boldsymbol{\Phi} \cdot \mathbf{V_m} \end{cases} \tag{7.20}$$

Substituting Eq. (7.20) into Eq. (7.18) and Eq. (7.19) gives

$$\mathbf{M} \cdot \boldsymbol{\Phi} \cdot \ddot{\mathbf{q}}_{\mathbf{m}} + \mathbf{C} \cdot \boldsymbol{\Phi} \cdot \dot{\mathbf{q}}_{\mathbf{m}} + \mathbf{K} \cdot \boldsymbol{\Phi} \cdot \mathbf{q_m} = -\mathbf{M} \cdot \boldsymbol{\Phi} \cdot \ddot{\mathbf{y}}_{\mathbf{m}} - \boldsymbol{\alpha} \cdot \boldsymbol{\Phi} \cdot \mathbf{V_m} \tag{7.21}$$

$$\mathbf{R}^{-1} \cdot \boldsymbol{\Phi} \cdot \mathbf{V_m} = \boldsymbol{\alpha} \cdot \boldsymbol{\Phi} \cdot \dot{\mathbf{q}}_{\mathbf{m}} - \mathbf{C_0} \cdot \boldsymbol{\Phi} \cdot \dot{\mathbf{V}}_{\mathbf{m}} \tag{7.22}$$

Multiplying $\boldsymbol{\Phi}^{\mathrm{T}}$ on the both side of Eq. (7.21) gives

$$\boldsymbol{\Phi}^{\mathrm{T}} \mathbf{M} \boldsymbol{\Phi} \ddot{\mathbf{q}}_{\mathbf{m}} + \boldsymbol{\Phi}^{\mathrm{T}} \cdot \mathbf{C} \cdot \boldsymbol{\Phi} \cdot \dot{\mathbf{q}}_{\mathbf{m}} + \boldsymbol{\Phi}^{\mathrm{T}} \cdot \mathbf{K} \cdot \boldsymbol{\Phi} \cdot \mathbf{q_m} = \boldsymbol{\Phi}^{\mathrm{T}} \cdot \left(-\mathbf{M} \cdot \boldsymbol{\Phi} \cdot \ddot{\mathbf{y}}_{\mathbf{m}} - \boldsymbol{\alpha} \cdot \boldsymbol{\Phi} \cdot \mathbf{V_m} \right) \tag{7.23}$$

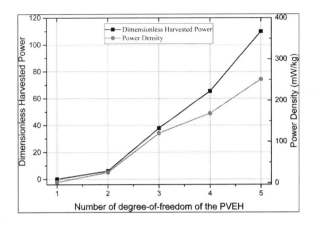

FIGURE 7.17

The dimensionless harvested power and the harvested power density versus the number of degrees of freedom of piezoelectric vibration energy harvester (PVEH).

Multiplying $\boldsymbol{\Phi}^{\mathrm{T}}$ on the both side of Eq. (7.22) gives

$$\boldsymbol{\Phi}^{\mathrm{T}} \cdot \mathbf{R}^{-1} \cdot \boldsymbol{\Phi} \cdot \mathbf{V_m} = \boldsymbol{\Phi}^{\mathrm{T}} \cdot \boldsymbol{\alpha} \cdot \boldsymbol{\Phi} \cdot \dot{\mathbf{q}}_{\mathbf{m}} - \boldsymbol{\Phi}^{\mathrm{T}} \cdot \mathbf{C_0} \cdot \boldsymbol{\Phi} \cdot \dot{\mathbf{V}}_{\mathbf{m}} \tag{7.24}$$

If the damping $\mathbf{C}$ is assumed as a proportional damping under the assumption of a weak coupling, there exist the following relationships:

$$\begin{cases} \boldsymbol{\Phi}^{\mathrm{T}} \cdot \mathbf{C} \cdot \boldsymbol{\Phi} = \mathbf{diag}[C_{\mathrm{m-i}}] \\ \boldsymbol{\Phi}^{\mathrm{T}} \cdot \mathbf{M} \cdot \boldsymbol{\Phi} = \mathbf{diag}[M_{\mathrm{m-i}}] \\ \boldsymbol{\Phi}^{\mathrm{T}} \cdot \mathbf{K} \cdot \boldsymbol{\Phi} = \mathbf{diag}[K_{\mathrm{m-i}}] \\ \boldsymbol{\Phi}^{\mathrm{T}} \cdot \mathbf{R}^{-1} \cdot \boldsymbol{\Phi} = \mathbf{diag}\left[\dfrac{1}{R_{\mathrm{m-i}}}\right] \\ \boldsymbol{\Phi}^{\mathrm{T}} \cdot \boldsymbol{\alpha} \cdot \boldsymbol{\Phi} = \mathbf{diag}[\alpha_{\mathrm{m-i}}] \\ \boldsymbol{\Phi}^{\mathrm{T}} \cdot \mathbf{C_0} \cdot \boldsymbol{\Phi} = \mathbf{diag}[C_{0\mathrm{m-i}}] \\ i = 1, 2, 3 \ldots N(\text{number of DOF}) \end{cases} \tag{7.25}$$

Then, from Eq. (7.25), Eqs. (7.23) and (7.24) are reduced to

$$M_{m-i}\ddot{q}_{m-i} + C_{m-i}\dot{q}_{m-i} + K_{m-i}q_{m-i} = -M_{m-i}\ddot{y}_{m-i} - \alpha_{m-i}V_{m-i} \tag{7.26}$$

$$\frac{1}{R_{m-i}}V_{m-i} = \alpha_{m-i}\dot{q}_{m-i} - C_{0m-i}\dot{V}_{m-i} \tag{7.27}$$

Eqs. (7.26) and (7.27) can be solved and simulated using the hybrid approach of the frequency analysis and time domain simulation methods as illustrated in

Chapters 2 and 3. Therefore displacement q_{m-i}, output voltage V_{m-i}, and power $P_{h(m-i)}$ of the ith oscillator/harvester can be calculated; z, $\dot{z}$, and V can then be calculated from Eq. (7.20). The total extracted power can be given by:

$$P_{e(total)} = \sum_{i=1}^{N} P_{e(m-i)} = -\dot{q}_m^T \cdot \mathbf{diag}[\alpha_{m-i}] \cdot \mathbf{V_m} = -\dot{q}_m^T \cdot \boldsymbol{\Phi}^T \cdot \boldsymbol{\alpha} \cdot \boldsymbol{\Phi} \cdot \mathbf{V_m} = \dot{z}^T \cdot \boldsymbol{\alpha}^T \cdot \mathbf{V}$$

(7.28)

The total harvested resonant power can be given by

$$P_{h(total)} = \sum_{i=1}^{N} \frac{1}{R_{m-i}} \cdot V_{m-i}^2 = \mathbf{V_m}^T \cdot \mathbf{diag}\left[\frac{1}{R_{m-i}}\right] \cdot \mathbf{V_m} = \mathbf{V_m}^T \cdot \boldsymbol{\Phi}^T \cdot \mathbf{R}^{-1} \cdot \boldsymbol{\Phi} \cdot \mathbf{V_m}$$

$$= \mathbf{V}^T \cdot \mathbf{R}^{-1} \cdot \mathbf{V}$$

(7.29)

The above analyses only consider a single electrical load resistor interface. To serve as a power source, the MDOF vibration energy harvester will have to be connected to specially designed energy extraction and storage interface circuits where the power loss of the circuits should be considered. This can be done by using the equivalent circuit and impedance methods. The analysis approach for MDOF vibration energy harvesters can be easily implemented into finite element software for analyzing, predicting, or evaluating the power generation performance of any continuous structures with built-in piezoelectric material.

7.6 SUMMARY

Following the study of a conventional 2DOF PVEH model in Chapter 6, in this chapter, a 2DOF PVEH model inserted with two piezoelectric elements and a 3DOF PVEH model inserted with three piezoelectric elements are proposed and investigated, the parameter effects on the performance of PVEH are analyzed. The results of the proposed 2DOF PVEH model show that as the coupling strength of the primary oscillator system increases, the maximum resonant frequency discrepancy increases. As the coupling strength increases, the energy harvesting efficiency increases.

The performance of the 2DOF PVEH is significantly improved with second piezoelectric element inserted into the auxiliary oscillator system. This does not increase the weight or the complexity of the entire energy harvesting system. Furthermore, the study of a 3DOF PVEH inserted with three piezoelectric patch elements and a generalized MDOF PVEH inserted with multiple piezoelectric patch elements between all two nearby oscillators has verified the conclusions from the study of the 2DOF PVEH. With an assistance of the MDOF PVEH analysis method, the harvested power and the power density of the PVEH from the 1DOF to 5DOF are compared. It is found that without additional weight being added into the system, the more DOF the system has, the more energy it can harvest.

The first resonant frequency decreases as the number of DOF of a PVEH system increases. As the number of DOF increases, the discrepancy of the modal frequencies decreases. This would result in a wide and effective harvesting frequency bandwidth. The harvested resonant power and efficiency from the piezo patch elements of the auxiliary oscillator systems are larger than those of the primary oscillator system. The maximum harvesting efficiencies of the piezo patch elements of the primary and auxiliary oscillator systems occur in the different system parameter ranges.

The analysis method presented in this paper enables to tune the PVEH toward the larger harvested power, higher harvesting efficiency, and wider harvesting frequency bandwidth.

NOMENCLATURE

$u_0, u_0(t)$	The base excitation displacement
$\dot{u}_0, \dot{u}_0(t)$	The base excitation velocity
$u_1, u_1(t)$	The displacement of the first oscillator mass (m_1)
$\dot{u}_1, \dot{u}_1(t)$	The velocity of the first oscillator mass (m_1)
$\ddot{u}_1, \ddot{u}_1(t)$	The acceleration of the first oscillator mass (m_1)
$u_2, u_2(t)$	The displacement of the second oscillator mass (m_2)
$\dot{u}_2, \dot{u}_2(t)$	The velocity of the second oscillator mass (m_2)
$\ddot{u}_2, \ddot{u}_2(t)$	The acceleration of the second oscillator mass (m_2)
$u_m, u_m(t)$	The displacement of the mth oscillator mass (m_m)
$\dot{u}_m, \dot{u}_m(t)$	The velocity of the mth oscillator mass (m_m)
$\ddot{u}_m, \ddot{u}_m(t)$	The acceleration of the mth oscillator mass (m_m)
$u_n, u_n(t)$	The displacement of the nth oscillator mass (m_n)
$\dot{u}_n, \dot{u}_n(t)$	The velocity of the nth oscillator mass (m_n)
$\ddot{u}_n, \ddot{u}_n(t)$	The acceleration of the nth oscillator mass (m_n)
$U_1, U_1(s)$	The Laplace transform of u_1 or $u_1(t)$
$U_2, U_2(s)$	The Laplace transform of u_2 or $u_2(t)$
$U_m, U_m(s)$	The Laplace transform of u_m or $u_m(t)$
$U_0, U_0(s)$	The Laplace transform of u_0 or $u_0(t)$
m_1	The first oscillator mass
m_2	The second oscillator mass
m_n	The nth oscillator mass
k_1	The short circuit stiffness between the base and the oscillator (m_1)
k_2	The short circuit stiffness between the m_1 and the m_2
k_n	The short circuit stiffness between the m_{n-1} and the m_n
c_1	The short circuit mechanical damping coefficient between the base and the oscillator (m_1)
c_2	The short circuit mechanical damping coefficient between the m_1 and the m_2
c_n	The short circuit mechanical damping coefficient between the m_{n-1} and the m_n

C_{p1}	The blocking capacity of the first piezoelectric patch element
C_{p2}	The blocking capacity of the second piezoelectric patch element
C_{pn}	The blocking capacity of the nth piezoelectric patch element
R_1	The external resistance connected with first piezoelectric patch element
R_2	The external resistance connected with second piezoelectric patch element
R_3	The external resistance connected with third piezoelectric patch element
R_n	The external resistance connected with nth piezoelectric patch element
α_1	The force factor of the first piezoelectric patch element
α_2	The force factor of the second piezoelectric patch element
α_n	The force factor of the nth piezoelectric patch element
$V_1, V_1(t)$	The output voltage of the first piezoelectric patch element
$V_2, V_2(t)$	The output voltage of the second piezoelectric patch element
$V_3, V_3(t)$	The output voltage of the third piezoelectric patch element
$V_j, V_j(t)$	The output voltage of the jth piezoelectric patch element
$V_n, V_n(t)$	The output voltage of the nth piezoelectric patch element
$\overline{V}_1, \overline{V}_1(s)$	The Laplace transform of V_1 or $V_1(t)$
$\overline{V}_2, \overline{V}_2(s)$	The Laplace transform of V_2 or $V_2(t)$
$\overline{V}_3, \overline{V}_3(s)$	The Laplace transform of V_3 or $V_3(t)$
$\overline{V}_j, \overline{V}_j(s)$	The Laplace transform of V_j or $V_j(t)$
$\overline{V}_n, \overline{V}_n(s)$	The Laplace transform of V_n or $V_n(t)$
A	The input excitation acceleration amplitude of the base
P_1	The harvested resonant power of the first piezoelectric patch element
P_2	The harvested resonant power of the second piezoelectric patch element
P_3	The harvested resonant power of the third piezoelectric patch element
P_{input}	The input resonant power from the base
t, t_0	Time instances
s	The Laplace variable
i	The square root of -1
ω_1	The natural frequency in rad/s of the first isolated oscillator mass (m_1)
ω_2	The natural frequency in rad/s of the second isolated oscillator mass (m_2)
ω_3	The natural frequency in rad/s of the third isolated oscillator mass (m_3)
ζ_1	The dimensionless damping coefficient of the first isolated oscillator mass (m_1)
ζ_2	The dimensionless damping coefficient of the second isolated oscillator mass (m_2)
ζ_3	The dimensionless damping coefficient of the third isolated oscillator mass (m_3)
Λ_1	The dimensionless force factor of the first piezoelectric patch element
Λ_2	The dimensionless force factor of the second piezoelectric patch element
Λ_3	The dimensionless force factor of the third piezoelectric patch element
θ_1	The dimensionless resistance of the first piezoelectric patch element
θ_2	The dimensionless resistance of the second piezoelectric patch element
θ_3	The dimensionless resistance of the third piezoelectric patch element
Ω_1	ω_2/ω_1, the ratio of the natural frequencies of the first and second isolated oscillators
Ω_2	$\Omega_2 = \frac{\omega_3}{\omega_2}$ the ratio of the natural frequencies of the second and third isolated oscillators
Φ	ω/ω_1, the ratio of the natural frequency of the first isolated oscillator and the natural resonant frequency of the MDOF oscillator system

Φ_1	ω_{n1}/ω_1, the ratio of the natural frequency of the first isolated oscillator and the first natural resonant frequency of the MDOF oscillator system
Φ_2	ω_{n2}/ω_1, the ratio of the natural frequency of the first isolated oscillator and the second natural resonant frequency of the MDOF oscillator system
$\Phi_{1,2}$	ω_{n1}/ω_1 or ω_{n2}/ω_1, the ratio of the natural frequency of the first isolated oscillator and the first natural resonant frequency of the MDOF oscillator system, or the ratio of the natural frequency of the first isolated oscillator and the second natural resonant frequency of the MDOF oscillator system
$\Delta\Phi_{1,2}$	$(\omega_{n2}-\omega_{n1})/\omega_1$, the difference of the two dimensionless resonant frequency ratios
M_1	m_2/m_1, the mass ratio of the first and second oscillators
M_2	m_3/m_2, the mass ratio of the second and third oscillators
η_1	The resonant energy harvesting efficiency of the first piezoelectric patch element
η_2	The resonant energy harvesting efficiency of the second piezoelectric patch element
η_3	The resonant energy harvesting efficiency of the third piezoelectric patch element
$\mathbf{M}$	The mass matrix for a multiple degree of freedom piezoelectric vibration energy harvester system
$M_{m\text{-}i}$	The ith oscillator mass under the mode coordinate
$\mathbf{K}$	The short circuit spring stiffness coefficient matrix for a multiple degree of freedom piezoelectric vibration energy harvester system
$K_{m\text{-}i}$	The ith oscillator short circuit spring stiffness coefficient under the mode coordinate
$\mathbf{z}$	The relative displacement vector array of a multiple degree of freedom piezoelectric vibration energy harvester system oscillators
$\dot{\mathbf{z}}$	The relative velocity vector array of a multiple degree of freedom piezoelectric vibration energy harvester system oscillators
$\ddot{\mathbf{z}}$	The relative acceleration vector array of a multiple degree of freedom piezoelectric vibration energy harvester system oscillators
$\mathbf{I}$	The output current vector array of a multiple degree of freedom piezoelectric vibration energy harvester system piezoelectric elements
$\boldsymbol{\Phi}$	The short-circuit mode shape matrix of a multiple degree of freedom piezoelectric vibration energy harvester system oscillators
$\mathbf{q_m}$	The modal relative displacement coordinate vector array of a multiple degree of freedom piezoelectric vibration energy harvester system oscillators
$q_{m\text{-}i}$	The ith oscillator modal relative displacement under the mode coordinate
$\dot{\mathbf{q}}_\mathbf{m}$	The modal relative velocity coordinate vector array of a multiple degree of freedom piezoelectric vibration energy harvester system oscillators
$\dot{q}_{m\text{-}i}$	The ith oscillator modal relative velocity under the mode coordinate
$\ddot{\mathbf{q}}_\mathbf{m}$	The modal relative acceleration coordinate vector array of a multiple degree of freedom piezoelectric vibration energy harvester system oscillators
$\ddot{q}_{m\text{-}i}$	The ith oscillator modal relative acceleration under the mode coordinate
$\mathbf{y_m}$	The base input excitation displacement vector array under the mode coordinate of a multiple degree of freedom piezoelectric vibration energy harvester system oscillators

$y_{m\text{-}i}$	The base input excitation displacement of the ith oscillator under the mode coordinate
$\ddot{y}_m$	The base input excitation acceleration vector array under the mode coordinate of a multiple degree of freedom piezoelectric vibration energy harvester system oscillators
$\ddot{y}_{m\text{-}i}$	The base input excitation acceleration of the ith oscillator under the mode coordinate
$\mathbf{V_m}$	The output voltage vector array under the mode coordinate of a multiple degree of freedom piezoelectric vibration energy harvester system oscillators
$V_{m\text{-}i}$	The output voltage of the ith oscillator under the mode coordinate
$P_{h(m\text{-}i)}$	The harvested resonant power of the ith oscillator under the mode coordinate
$P_{h(total)}$	The total harvested resonant power
$P_{e(total)}$	The total extracted resonant power
$\dot{\mathbf{V}}_\mathbf{m}$	The output voltage change rate vector array under the mode coordinate of a multiple degree of freedom piezoelectric vibration energy harvester system oscillators
$\dot{V}_m$	The output voltage change rate of the ith oscillator under the mode coordinate
C_0	The blocking capacity matrix of a multiple degree of freedom piezoelectric vibration energy harvester system piezoelectric elements
$C_{0m\text{-}i}$	The blocking capacity of the ith oscillator under the mode coordinate
$\mathbf{C}$	The short circuit damping coefficient matrix for a multiple degree of freedom piezoelectric vibration energy harvester system
$C_{m\text{-}i}$	The short circuit damping coefficient of the ith oscillator under the mode coordinate
$\mathbf{\Phi^T}$	The transpose of the short-circuit mode shape matrix $\mathbf{\Phi}$
$\mathbf{R}$	The external load resistance matrix of a multiple degree of freedom piezoelectric vibration energy harvester system piezoelectric elements
$R_{m\text{-}i}$	The external load resistance of the ith oscillator under the mode coordinate
α	The force factor matrix of a multiple degree of freedom piezoelectric vibration energy harvester system piezoelectric elements
$\alpha_{m\text{-}i}$	The force factor of the ith oscillator under the mode coordinate

Superscripts

$\cdot$	The first differential
$\cdot\cdot$	The second differential
$\mathbf{T}$	Transpose
-1	Inverse
$\underline{}$	The time average

Special function

$\lvert\rvert$	The modulus or absolute value
$\langle\rangle$	The time average, expectation, or mean value
$\mathrm{Re}[\ldots]$	The real part of a complex variable
$\mathrm{diag}[\ldots]$	Diagonal matrix operation

Abbreviations

DOF	Degree of freedom
MDOF	Multiple degree of freedom
PVEH	Piezoelectric vibration energy harvester

REFERENCES

[1] Calio RU, Rongala B, Camboni D, Milazzo M, Stefanini C, de Petris G, et al. Piezoelectric energy harvesting solutions. Sensors 2014;14:4755—90.

[2] Cook-Chennault KA, Thambi N, Sastry AM. Powering MEMS portable devices—a review of non-regenerative and regenerative power supply systems with special emphasis on piezoelectric energy harvesting systems. Smart Mater Struct 2008;17: 043001.

[3] Lin HC, Wu PH, Lien IC, Shu YC. Analysis of an array of piezoelectric energy harvesters connected in series. Smart Mater Struct 2013;22:094026.

[4] Lien IC, Shu YC. Array of piezoelectric energy harvesting by the equivalent impedance approach. Smart Mater Struct 2012;21:082001.

[5] Wen Z, Deng L, Zhao X, Shang Z, Yuan C, She Y. Improving voltage output with PZT beam array for MEMS-based vibration energy harvester: theory and experiment. Microsystem Technologies; 2014.

[6] Ashraf K, Md Khir MH, Dennis JO, Baharudin Z. Improved energy harvesting from low frequency vibrations by resonance amplification at multiple frequencies. Sens Actuators A Phys 2013;195:123—32.

[7] Tang L, Yang Y. A multiple-degree-of-freedom piezoelectric energy harvesting model. J Intell Mater Syst Struct 2012;23:1631—47.

[8] Sodano HA, Lien IC, Shu YC. In: Multiple piezoelectric energy harvesters connected to different interface circuitsvol. 8341; 2012. pp. 83410X-83410X-8.

[9] Takezawa A, Makihara K, Kogiso N, Kitamura M. Layout optimization methodology of piezoelectric transducers in energy-recycling semi-active vibration control systems. J Sound Vib 2014;333:327—44.

[10] Deterre M, Lefeuvre E, Dufour-Gergam E. An active piezoelectric energy extraction method for pressure energy harvesting. Smart Mater Struct 2012;21:085004.

[11] Acar MA, Yilmaz C. Design of an adaptive—passive dynamic vibration absorber composed of a string—mass system equipped with negative stiffness tension adjusting mechanism. J Sound Vib 2013;332:231—45.

[12] Wickenheiser AM, Garcia E. Power optimization of vibration energy harvesters utilizing passive and active circuits. J Intell Mater Syst Struct 2010;21:1343—61.

[13] Kim I-H, Jung H-J, Lee BM, Jang S-J. Broadband energy-harvesting using a two degree-of-freedom vibrating body. Appl Phys Lett 2011;98:214102.

[14] Ou Q, Chen X, Gutschmidt S, Wood A, Leigh N, Arrieta AF. An experimentally validated double-mass piezoelectric cantilever model for broadband vibration-based energy harvesting. J Intell Mater Syst Struct 2012;23:117—26.

[15] Zhou W, Penamalli GR, Zuo L. An efficient vibration energy harvester with a multi-mode dynamic magnifier. Smart Mater Struct 2012;21:015014.

[16] Liu H, Huang Z, Xu T, Chen D. Enhancing output power of a piezoelectric cantilever energy harvester using an oscillator. Smart Mater Struct 2012;21:065004.

[17] Wang X, Koss LL. Frequency response function for power, power transfer ratio and coherence. J Sound Vib 1992;155(1):55—73.

[18] Shu YC, Lien IC. Analysis of power output for piezoelectric energy harvesting systems. Smart Mater Struct 2006;15:1499—512.

[19] Xiao H, Wang X, John S. A dimensionless analysis of a 2DOF piezoelectric vibration energy harvester. Mech Syst Signal Process 2015;58—59(2015):355—75.

[20] Xiao H, Wang X, John S. A multi degree of freedom piezoelectric vibration energy harvester with piezoelectric elements inserted between two nearby oscillators. Mech Syst Signal Process February 2016;68−69:138−54.

[21] Xiao H, Wang X. A review of piezoelectric vibration energy harvesting techniques. Int Rev Mech Eng 2014;8(3). ISSN: 1970-8742:609−20.

[22] Wang X, Lin LW. Dimensionless optimization of piezoelectric vibration energy harvesters with different interface circuits. Smart Mater Struct 2013;22. ISSN: 0964-1726:1−20.

[23] Wang X, Xiao H. Dimensionless analysis and optimization of piezoelectric vibration energy harvester. Int Rev Mech Eng 2013;7(4). ISSN: 1970-8742:607−24.

[24] Cojocariu B, Hill A, Escudero A, Xiao H, Wang X. Piezoelectric vibration energy harvester − design and prototype. In: Proceedings of ASME 2012 International Mechanical Engineering Congress & Exposition, ISBN 978-0-7918-4517-2.

[25] Wang X, Liang XY, Hao ZY, Du HP, Zhang N, Qian M. Comparison of electromagnetic and piezoelectric vibration energy harvesters with different interface circuits. Mech Syst Signal Process 2016;72−73:906−24.

[26] Wang X. Coupling loss factor of linear vibration energy harvesting systems in a framework of statistical energy analysis. J Sound Vib 2016;362(2016):125−41.

[27] Wang X, Liang XY, Shu GQ, Watkins S. Coupling analysis of linear vibration energy harvesting systems. Mech Syst Signal Process March 2016;70−71:428−44.

[28] Wang X, John S, Watkins S, Yu XH, Xiao H, Liang XY, et al. Similarity and duality of electromagnetic and piezoelectric vibration energy harvesters. Mech Syst Signal Process 2015;52−53(2015):672−84.

Experimental Validation of Analytical Methods

H. Xiao, X. Wang

RMIT University, Melbourne, VIC, Australia

CHAPTER OUTLINE

8.1 INTRODUCTION

To validate the analytical methods illustrated in Chapters 2, 3, 6, and 7, the first experimental test is to test a cantilevered bimorph beryllium bronze beam with PZT-5H piezoelectric material coated on its top and bottom surfaces. The coated material on the beam forms a bimorph configuration. A tipped mass was added on the free end of the beam. The fixed end of the beam was excited by a shaker with a random noise signal or sinusoid signal from 0.5 to 100 Hz with a magnitude of 0.1 g. The outputs of the top and bottom surfaces are connected in series. A bridge rectifier of 1 A and 100 V was connected to the two output electrodes of the cantilevered beam to convert the AC voltage output to DC voltage output. The second and third experimental tests are to test two degrees of freedom (2DOF) vibration energy harvesters with one and two piezoelectric elements. The parameters of the 2DOF vibration energy harvesters were identified by testing the separate single degree of freedom (SDOF) primary and auxiliary oscillators. A white noise signal or sinusoid signal from 0.5 to 200 Hz was used to drive a shaker to excite the 2DOF vibration energy harvesters at the base. The excitation amplitudes were kept as 1.5 and 7.13 m/s^2 for the 2DOF vibration energy harvesters with one and two piezoelectric elements.

Frequency Analysis of Vibration Energy Harvesting Systems. http://dx.doi.org/10.1016/B978-0-12-802321-1.00008-X

8.2 EXPERIMENTAL RESULTS OF A SINGLE DEGREE OF FREEDOM VIBRATION ENERGY HARVESTER

To verify the analyses in Chapters 2 and 3, a cantilevered bimorph beryllium bronze beam was designed to have a length of 38.11 mm, width of 20 mm, and thickness of 0.21 mm. The PZT-5H piezoelectric material was coated on the top and bottom surfaces of the beam to form a bimorph configuration. The coated PZT-5H piezoelectric material has a length of 30 mm, width of 20 mm, and thickness of 0.45 mm on each side of the beam surface. A tip mass of 3 g was placed and glued on the beam at the free end, The other end of the beam was fixed and clamped by washers through bolt and nuts as shown in Fig. 8.1. The bolt was connected to a shaker push rod as shown in Fig. 8.2.

The property parameters of the bimorph cantilevered beam are listed in Table 8.1 where the natural frequency was calculated from the formula as below:

$$f_1 = \frac{1}{2\pi} \cdot 1.875^2 \cdot \sqrt{\frac{E \cdot J}{\rho \cdot A_b \cdot L_b^4}} \tag{8.1}$$

The short circuit stiffness is given by

$$K = \frac{3 \cdot E \cdot J}{L_b^3} \tag{8.2}$$

where E is the Young's modulus of the beam material, J is the moment of inertia for the cross section of the beam, $J = bh^3/12$; ρ is the mass density of the beam; b is the beam width, h is the beam thickness or height; A_b is the cross-section area of the beam; L_b is the beam length. The natural frequency and the short circuit stiffness of the equivalent SDOF vibration energy harvester can be calculated from Eqs. (8.1) and (8.2) and are given in Table 8.1. A cantilevered beam model with the property parameters in Table 8.1 was constructed in the ANSYS modal analysis module.

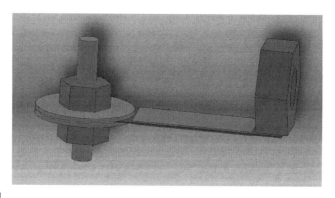

FIGURE 8.1

A cantilevered bimorph beam clamped by washers with a nut mass glued at the free end.

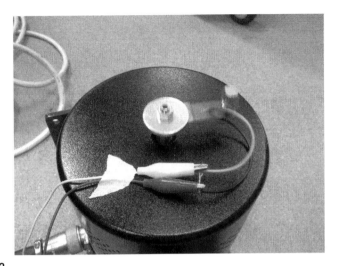

FIGURE 8.2

The bimorph cantilevered beam set up on the shaker for lab testing.

Table 8.1 Piezoelectric Vibration Energy Harvester Parameters

Parameters	Units	Value
Tip mass value, M_t	g	3
Total equivalent mass, M	g	5.3
The short circuit stiffness, K	N/m	125.5
Piezoelectric element PZT-5H length	mm	30
Piezoelectric element PZT-5H width	mm	20
Piezoelectric element PZT-5H thickness, t_p	mm	0.45
Beam material beryllium bronze mass density, ρ	kg/m³	8700
Piezoelectric element PZT-5H mass density, ρ_p	kg/m³	7500
Beryllium bronze Young's modulus, E	GPa	150
Piezoelectric element PZT-5H Young's modulus, E_p	GPa	76.5
Beryllium bronze Poisson ratio		0.334
Beam length, L_b	mm	38.11
Beam width, b	mm	20
Beam thickness, h	mm	0.21
Natural frequency of the beam, f_n	Hz	24.5
Mechanical damping coefficient, c	N·s/m	0.035
Piezoelectric blocking capacitance, C_0	F	1.39×10^{-8}
Force factor, α	N/V	1.88×10^{-4}
Electric load resistance, R	kΩ	434

The first modal natural frequency was obtained to be 26.192 Hz which is 1.692 Hz different from the calculated value of 24.5 Hz given in Table 8.1. The difference may be caused by the simplification of the bimorph beam structure into a cantilevered beam of mono beryllium bronze material of the same thickness.

To measure the harvested resonant voltage and power of the cantilevered vibration energy harvester, Polytec Laser Doppler Vibrometer system was used to drive the shaker and measure the beam surface vibration velocity according to the Laser Doppler principle. To improve the measurement accuracy and reduce the surface scattering to the laser beam, the beam surface was painted in red color, the laser beam was programmed to scan the painted surface following the blue (grey in print version) grid shown in Fig. 8.3.

A white noise random signal was generated to drive the shaker to excite the cantilevered beam piezoelectric vibration energy harvester (PVEH). The measured first natural frequency was shown to be 24.375 Hz in Fig. 8.4. It is seen that the differences of the calculated, simulated, and measured first natural frequencies are small.

8.2.1 FREQUENCY RESPONSE FUNCTION

After the first natural frequency was identified, a sinusoid signal was used to excite the same cantilevered beam vibration energy harvester at the identified natural frequency. The cantilevered beam system experienced a resonance with large displacement amplitude; the vibration energy of the beam was converted by the piezoelectric material into the electric energy carried by alternative current (AC) voltage. The electrodes of the bimorph cantilevered beam were connected in series. The top surface electrode was positive and connected in a red cable and the bottom electrode

FIGURE 8.3

Polytec Laser Doppler Vibrometer system display.

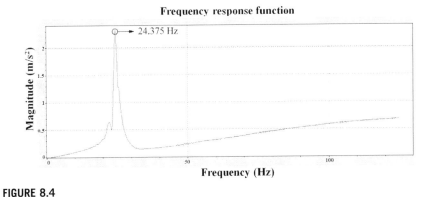

FIGURE 8.4

The measured vibration spectrum and first natural frequency of 24.375 Hz for the cantilevered beam under a white noise random force excitation.

was negative and connected in black cable as shown in Fig. 8.2. A sinusoid AC voltage was observed on the oscilloscope where the voltage amplitude increased with the shaker amplifier gain. As the shaker amplifier gain was linearly proportional to the excitation acceleration amplitude, the voltage amplitude linearly increased with the excitation acceleration amplitude, which has verified the simulation result of the output voltage linearly increasing with the excitation acceleration amplitude as illustrated in Chapters 2 and 3.

A shaker amplifier gain was chosen so that the cantilevered beam system vibrated largely and steadily without failures. As the sinusoid AC voltage was not able to be stored, to store the harvested vibration energy, a bridge rectifier of 1 A and 100 V was connected to the two output electrodes of the cantilevered beam. The open circuit output voltage generated from the cantilevered beam system was 2.262 V with an excitation acceleration amplitude of 0.1 g. The measured electric load resistance was 434 kΩ; therefore the harvested resonant power was 0.0118 mW. The output voltage and harvested resonant power predicted according to Eqs. (2.19) and (2.20) are 2.42 V and 0.0135 mW with the same electric load resistance. When the shaker excitation amplitude and external electric load resistance and other PZT-5H parameters were kept constant, only the sinusoidal excitation frequency was changed from 0.5 to 100 Hz, the output voltage, external electric load resistance, and excitation frequency of the PZT-5H beam were measured and recorded. The predicted and measured output voltage and harvested power under the same electrical load resistance of 434 kΩ at different excitation frequencies are plotted, compared, and shown in Figs. 8.5 and 8.6.

It is seen that the measured and predicted output voltages and harvested power are close at the resonance frequency. The measured and predicted output voltages and harvested power are different at nonresonant frequencies. This is because Eqs. (2.19) and (2.20) are derived only at the resonant frequency and the signal noise ratio is very low at nonresonant frequencies. Therefore, all the measurement results

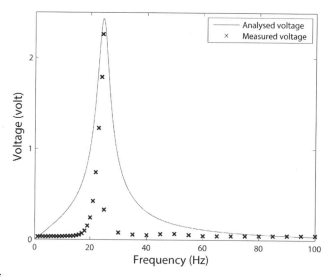

FIGURE 8.5

The predicted and measured voltage output versus excitation frequency for the PZT-5H cantilevered beam.

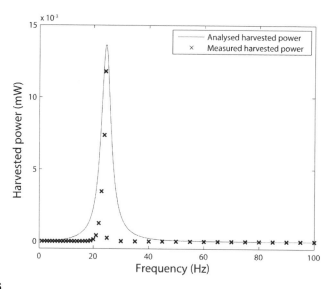

FIGURE 8.6

The predicted and measured harvested resonant power versus the excitation frequency for the PZT-5H cantilevered beam.

had better be taken only at the resonant frequency for the comparison of theory and experimental results. The other reason could be that the damping coefficient of the prediction could be underestimated. The damping coefficient of the prediction model could be much less than that of the experimental device.

When the excitation frequency was fixed at a resonant frequency of 24.4 Hz, the excitation acceleration amplitude was fixed at 0.1 g and other PZT-5H beam parameters were kept constant, only the external electric load resistance was changed from 434 kΩ to 10 MΩ, the output voltage and external electric load resistance of the PZT-5H cantilevered beam were measured and recorded. The predicted and measured resonant output voltages under different external load resistances are compared and shown in Fig. 8.7. It is seen from Fig. 8.7 that the measured output voltage at the resonant frequency is very close to the predicted output voltage under different external load resistances.

When the excitation frequency was fixed at a resonant frequency of 24.4 Hz, the external electric load resistance was kept as 434 kΩ, other PZT-5H beam parameters were kept constant, only the excitation acceleration amplitude was changed from 0.05 to 0.6 g (1 g = 9.8 m/s^2) in a step of 0.05 g, the output voltage and excitation acceleration amplitude of the PZT-5H cantilevered beam were measured and recorded. The predicted and measured resonant output voltages under different excitation acceleration amplitudes are compared and shown in Fig. 8.8. It is seen from Fig. 8.8 that the measured output voltages at the resonant frequency are very close to the predicted output voltages under the low excitation acceleration amplitudes. The difference between the measured and predicted output voltages becomes large under

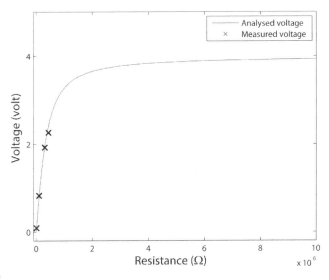

FIGURE 8.7

The predicted and measured resonant output voltage versus the external electric load resistance for the PZT-5H cantilevered beam.

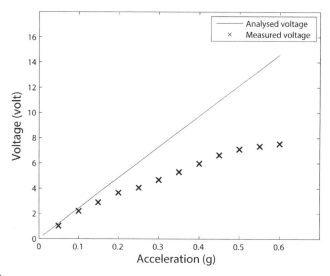

FIGURE 8.8

The predicted and measured resonant output voltages versus the excitation acceleration amplitude for the PZT-5H cantilevered beam.

the high excitation acceleration amplitudes, which may be caused by the large amplitude nonlinear effects.

From comparison of the predicted and measured output voltage and harvested power for variation of either the excitation frequencies, or the external electric load resistances or the excitation acceleration amplitudes, it is seen that the theoretical prediction results are very close to those of the experimental measurement. Therefore, the experimental results have validated the theoretical prediction and analysis as illustrated in Chapters 2 and 3. On the other hand, the hybrid analysis of both the frequency response analysis and time domain simulation methods illustrated in Chapters 2 and 3 has disclosed clear relationships between the performance of the SDOF piezoelectric vibration energy harvesting system and the selected system parameters. The hybrid analysis can provide accurate and reliable performance predictions of the SDOF PVEH as the time domain simulation and the frequency response analysis methods have all been validated.

8.3 EXPERIMENTAL RESULTS OF A TWO DEGREE OF FREEDOM VIBRATION ENERGY HARVESTERS WITH ONE AND TWO PIEZOELECTRIC ELEMENTS

To validate the theoretical analysis method of a 2DOF vibration energy harvester, a 2DOF PVEH with one piezoelectric element has been built and attached on the shaker for testing as shown in Fig. 8.9. There were three aluminum blocks which

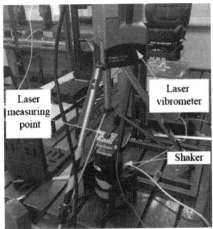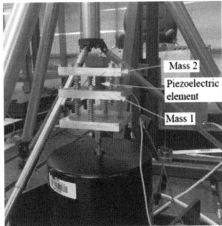

FIGURE 8.9

A two degree of freedom piezoelectric vibration energy harvester with one piezoelectric element was mounted on the shaker.

have a dimension of 83 mm × 83 mm × 10 mm and are connected by springs. The tipped mass was placed on the top aluminum block, and the piezoelectric stack was inserted between the top and the middle aluminum blocks.

The Laser Vibrometer system was used to measure vibration frequency spectrum and identify the resonant frequencies of the harvester device. The Laser Vibrometer system was also used to measure the velocity amplitude of the excitation that was generated by the shaker. The k_1 is the sum of the stiffness values of the four springs located below Mass 1, and k_2 is the sum of the stiffness values of the four springs located below Mass 2. Moreover, the parameters of the 2DOF piezoelectric vibration energy harvesting device were identified and summarized in Table 8.2. The first and the second resonant frequencies were calculated from Eq. (6.13) and the parameters in Table 8.2. The calculated results of the first and the second resonant frequencies are 37.42 and 101.8 Hz which are listed in Table 8.2. The calculated results of 37.42 and 101.8 Hz agree well with those of 38.58 and 102.34 Hz measured by the Laser Vibrometer system, respectively.

The predicted and experimentally measured voltage output values have been compared for different excitation frequencies and external electric load resistances. The excitation acceleration amplitude was kept as 1.5 m/s^2.

The measured voltage values presented by the scattered crosses are very close to the predicted voltage values presented by the solid curve as shown in Fig. 8.10. The maximum measured output voltage is 0.33 V at 38.58 Hz which is slightly higher than the predicted voltage. The maximum measured harvested power at that frequency is then 1.64 μW for the external resistance of 66,400 Ω. In this experiment, various external resistances ranging from 1 kΩ to 100 MΩ have been chosen to study the effect of the resistance on the harvested voltage of the 2DOF PVEH with one

Table 8.2 The Parameters of a Two Degree of Freedom Piezoelectric Vibration Energy Harvester With One Piezoelectric Element

Parameter	Type	Units	Values
m_1	Primary oscillator mass	kg	0.25
m_2	Auxiliary oscillator mass	kg	0.36
c_1	The short circuit damping coefficient	N·s/m	6.73
c_2	The short circuit damping coefficient	N·s/m	8.13
C_0	Blocking capacitance of the piezoelectric insert	F	7.2×10^{-6}
k_1	The short circuit spring stiffness	N/m	63,749.25
k_2	The short circuit spring stiffness	N/m	32,364.13
α	Force factor	N/Volt	5.14×10^{-3}
f_1	First natural frequency	Hz	37.42
f_2	Second natural frequency	Hz	101.8
R	Electrical resistance	Ω	66,400

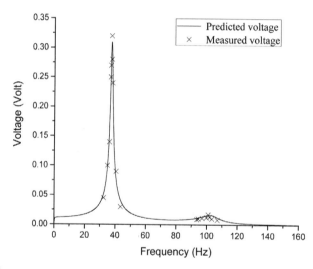

FIGURE 8.10

The predicted and experimentally measured voltage output values versus the excitation frequency for a two degree of freedom piezoelectric vibration energy harvester inserted with one piezoelectric element.

piezoelectric element. The experimentally measured output voltage results have been compared with the predicted results in Fig. 8.11. It is seen that the trend of the measured output voltage agrees with that of the predicted output voltage, although the measured voltage values are slightly higher than those of the predicted voltage values in the large resistance range. This is because the prediction is based

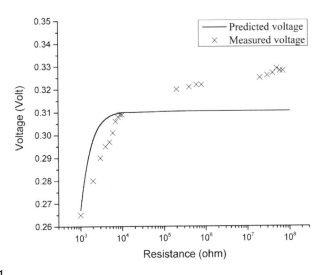

FIGURE 8.11

The predicted and experimentally measured voltage output values versus the external electric load resistance for a two degree of freedom piezoelectric vibration energy harvester inserted with one piezoelectric element.

on the assumption of a weak coupling. When the external load resistance increases and becomes very large, the electromechanical coupling becomes a strong coupling. Therefore, the prediction underestimates the output voltage of the harvester.

A 2DOF PVEH inserted with two piezoelectric elements was constructed and shown in Fig. 8.12 where the same three aluminum blocks as those shown in Fig. 8.9 were used. A tipped mass is attached on the top aluminum block. The first piezoelectric element was placed between the middle and bottom aluminum blocks, and the second piezoelectric element was placed between the top and middle aluminum blocks. If the bottom aluminum block, the springs, and their guide tubes between the middle and bottom aluminum blocks and the first piezoelectric patch element are removed, the middle aluminum block is fixed onto the push rod of the shaker, the top part of the 2DOF PVEH is formed and isolated as an auxiliary oscillator. If the top aluminum block, the springs and their guides between the top and middle aluminum blocks and the second piezoelectric patch element are removed, the bottom part of the 2DOF PVEH is formed and isolated as a primary oscillator. The top and bottom parts of the 2DOF PVEH are respectively tested to obtain the stiffness and damping coefficients of the auxiliary and primary oscillators as illustrated in Fig. 8.13 where the masses of the three aluminum blocks can be weighted by a scale. On the left-hand side of Fig. 8.13, the original top part of the 2DOF PVEH was turned up side down. The original top aluminum block became the bottom base mass and was connected to the shaker rod, and the original middle aluminum block with bolts became the top mass. The stiffness coefficients can be calculated from the measured masses and identified modal resonant frequencies

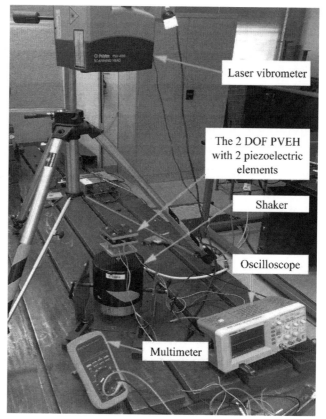

FIGURE 8.12

The experimental setup of the two degree of freedom (2DOF) piezoelectric vibration energy harvester (PVEH) built with two piezoelectric elements.

of the primary and auxiliary oscillators. The damping coefficients can be calculated from the measured masses, modal resonant frequencies, and the damping ratios identified using the half power bandwidth method. Therefore, the parameters of the 2DOF PVEH with two piezoelectric elements are summarized in Table 8.3.

In Table 8.3, m_1 is the mass of the middle aluminum block and the m_2 is the total mass of the tipped mass and the top aluminum block. The masses of the first and second piezo patch elements are small and neglected; k_1 and k_2 are the stiffness coefficients of the primary and auxiliary oscillators, respectively and are identified by the isolated tests; c_1 and c_2 are the damping coefficients of the primary and auxiliary oscillators, respectively and are measured from the isolated tests as well. The stiffness and damping coefficients of the primary and auxiliary oscillators should be same or very close to each other for the 2DOF PVEHs in Figs. 8.9 and 8.12. C_{p1} and C_{p2} are the blocking capacitances of the first and second piezoelectric patch

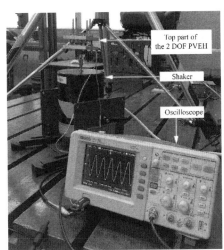

FIGURE 8.13

The isolated tests for the primary and auxiliary oscillators of the two degree of freedom (2DOF) piezoelectric vibration energy harvester (PVEH) built with two piezoelectric elements.

elements, respectively; R_1 and R_2 are the total external electrical resistances connected with the first and second piezoelectric patch elements, respectively; f_1 and f_2 are the first and second modal resonant frequencies predicted by Eq. (7.11).

The Laser Vibrometer was used to measure the velocity of the excitation generated by the shaker from which the vibration frequency response spectra and the resonant frequencies of the primary and auxiliary oscillators were measured. The amplitude of the excitation was set as 7.13 m/s² for the experiments.

The experimentally measured and theoretically predicted voltage outputs are compared under different excitation frequencies for the proposed 2DOF PVEH with two piezoelectric elements, as shown in Fig. 8.14. For the proposed 2DOF PVEH with two piezoelectric elements, the measured first and second modal resonant frequencies are 61.85 and 147.1 Hz, respectively, which are very close to the analytical results of 62.22 and 147.8 Hz in Table 8.3. The maximum measured voltage output values of the first and second piezoelectric elements are 0.98 and 1.04 V at 61.85 Hz, respectively, which are slightly lower than the analytical results as shown by the solid curve peaks in Fig. 8.14. The maximum measured voltage output values of 0.98 and 1.04 V at 61.85 Hz for the proposed 2DOF harvester with two piezoelectric elements are three times more than the maximum measured voltage output value of 0.33 V at 38.58 Hz for the above 2DOF harvester with one piezoelectric element. In the experimental tests, the measured voltage values depicted by the scattered crosses are very close to the predicted voltage values presented by the solid curve. The maximum harvested power of the proposed 2DOF PVEH with two piezoelectric elements is 204.02 µW for the two external load

Table 8.3 The Experimentally Identified Parameters of the 2DOF Piezoelectric Vibration Energy Harvester Built With Two Piezoelectric Elements

Parameters	Type	Values	Units
m_1	Primary oscillator mass	0.38	kg
m_2	Auxiliary oscillator mass	0.36	kg
k_1	The short circuit stiffness between the base and the primary oscillator mass (m_1)	1.79×10^5	N/m
k_2	The short circuit stiffness between the primary oscillator mass (m_1) and the auxiliary oscillator mass (m_1)	9.96×10^4	N/m
c_1	The short circuit mechanical damping coefficient between the base and the oscillator (m_1)	6.73	N·s/m
c_2	The short circuit mechanical damping coefficient between the primary oscillator (m_1) and the auxiliary oscillator mass (m_2)	8.13	N·s/m
α_1	First piezo-insert force factor	2.3×10^{-4}	N/V
α_2	Second piezo-insert force factor	2.1×10^{-4}	N/V
C_{p1}	Blocking capacitance of first piezo patch element	2.09×10^{-9}	F
C_{p2}	Blocking capacitance of second piezo patch element	2.09×10^{-9}	F
R_1	Electrical resistance of first piezo patch element	1.0×10^4	Ω
R_2	Electrical resistance of second piezo patch element	1.0×10^4	Ω
f_1	First modal resonant frequency	62.22	Hz
f_2	Second modal resonant frequency	147.8	Hz

resistances in Table 8.3. Therefore, the proposed 2DOF PVEH with two piezoelectric elements can harvest 204.02 µW which is 124 times more power than 1.64 µW of the 2DOF PVEH with one piezoelectric element.

The experimental results have proved that the proposed analytical methods in Chapters 6 and 7 are able to provide reliable performance predictions of the 2DOF PVEHs. Hence, the proposed analytical methods are useful in design and optimization of the 2DOF PVEHs to achieve the maximum harvested power and output voltage. On the other hand, the hybrid analysis of both the frequency response analysis and time domain simulation methods illustrated in Chapters 6 and 7 has disclosed clear relationships between the performance of the 2DOF piezoelectric vibration energy harvesting system and the selected system parameters. The hybrid analysis can provide accurate and reliable performance predictions as the time domain simulation and the frequency response analysis methods have all been validated.

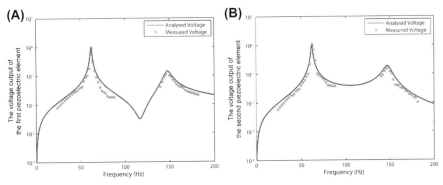

FIGURE 8.14

The analytically predicted and experimentally measured voltage outputs versus the excitation frequency for a two degree of freedom piezoelectric vibration energy harvester inserted with two piezoelectric elements. (A) The analytically predicted and experimentally measured voltage output of the first piezo patch element; (B) The analytically predicted and experimentally measured voltage output of the second piezo patch element.

NOMENCLATURE

M_t	The tip mass value
M	The total equivalent mass
t_p	The thickness of piezoelectric element PZT-5H
ρ	The material density of beam material beryllium bronze
ρ_p	The density of piezoelectric element PZT-5H
K	The short circuit stiffness of the SDOF system
E	The beryllium bronze Young's modulus
Ep	The piezoelectric element PZT-5H Young's modulus
A_b	The cross-section area of the cantilever beam
L_b	The length of the cantilever beam
b	The width of the cantilever beam
h	The thickness of the cantilever beam
f_n	Natural frequency
f_1	The first modal resonant frequency
f_2	The second modal resonant frequency
c	The short circuit mechanical damping coefficient
α	Force factor
R	The sum of the external load resistance and the piezoelectric patch element resistance
R_1	The sum of the external load resistance and the first piezoelectric patch element resistance
R_2	The sum of the external load resistance and the second piezoelectric patch element resistance
π	3.1415928
J	The moment of inertia for the cross section of the cantilever beam

m_1	The primary oscillator mass or mass of oscillator 1
m_2	The auxiliary oscillator mass or mass of oscillator 2
k_1	The short circuit stiffness between the base and the primary oscillator mass (m_1)
k_2	The short circuit stiffness between the primary oscillator mass (m_1) and the auxiliary oscillator mass (m_2)
c_1	The short circuit mechanical damping coefficient between the base and the oscillator (m_1)
c_2	The short circuit mechanical damping coefficient between the primary oscillator (m_1) and the auxiliary oscillator mass (m_2)
α_1	The force factor of the first piezo-insert
α_2	The force factor of the second piezo-insert
C_0	The blocking capacitance of the piezo-insert
C_{p1}	The blocking capacitance of the first piezoelectric patch element
C_{p2}	The blocking capacitance of the second piezoelectric patch element

Abbreviation

SDOF	Single degree of freedom
2DOF	Two degree of freedom
MDOF	Multiple degree of freedom

Coupling Analysis of Linear Vibration Energy Harvesting Systems

9

9.1 INTRODUCTION

Most of the previous vibration energy harvesting research work is focused on the resonant analysis of the harvesters of specially designed materials or structures with a single frequency harmonic excitation. The resonant analysis is often conducted by deterministic approach. In reality, ambient environment excitation is often random such as the excitation generated from the road surface to moving vehicles. The deterministic analysis is not sustainable. This is because harvested power and energy harvesting efficiency are much overestimated, when the excitation frequency is deviated from the resonant frequency.

Ambient environment vibrations or excitations are random, and they could be of a low or mid or high frequency. As a matter of fact, some of the piezoelectric vibration energy harvesters (PVEHs) work more efficiently in the middle and high frequency than in the low frequency [1–7]. In the middle or high frequency, the established methods are not able to reliably predict the vibration energy harvesting performance of a series of vibration energy harvesting systems of the same design. This is due to the system parameter variations which are caused by variations in materials and manufacturing processes. In this case, statistical energy analysis (SEA) of vibration energy harvesting systems needs to be developed. For this purpose,

Frequency Analysis of Vibration Energy Harvesting Systems. http://dx.doi.org/10.1016/B978-0-12-802321-1.00009-1

coupling analysis of mechanical and electrical subsystems needs to be established. SEA is a framework of study with primary variables of energy for the systems being studied from populations of similar design construction having known distributions of dynamic parameters. The procedures for SEA are model development, parameter evaluation, and calculation of response. The motivation of adopting the SEA framework is to avoid repeatedly resimulating a system with varied parameter adjustments, as variations of materials and manufacturing processes always exist even for vibration energy harvesters (VEHs) of the same design. This chapter focuses on establishing coupling loss factor which enables SEA of linear PVEH and electromagnetic vibration energy harvester (EMVEH).

Electromechanical coupling factor k_e of the EMVEH and PVEH was defined in Refs [2,8−20], which is a numerical measure of the conversion efficiency between electrical and mechanical energy. In those literatures, electromechanical coupling factor k_e^2 of piezoelectric materials is equal to squared piezoelectric coefficient of the piezoelectric disk divided by the product of the free stress permittivity and short-circuited elastic rigidity of the piezoelectric disk. The electromechanical coupling factor k_e^2 of piezoelectric materials is also equal to the squared force factor divided by the product of the short-circuited stiffness and the capacitance of the piezoelectric disk. Shu and Lien [21−23] defined critical coupling strength k_e^2/ξ_M, as the squared electromechanical coupling factor k_e^2 divided by the mechanical damping ratio ξ_M. They believed that when the critical coupling strength is much less than one, the harvester system is weakly coupled. When the critical coupling strength is larger than 1 and less than 10, the harvester system is moderately coupled. When the critical coupling strength is larger than 10, the harvester system is strongly coupled. As the previous research studies on vibration energy harvesting analysis have never been conducted in the framework of SEA, coupling loss factor has never been defined for vibration energy harvesting analysis. The coupling loss factor has never been related to the electromechanical coupling factor k_e^2, although they are two different concepts. In Chapters 2−5, dimensionless harvested power and energy harvesting efficiency are expressed as functions of only two dimensionless variables: dimensionless resistance R_N and dimensionless force factor α_N. However, physical meanings of the dimensionless resistance R_N and dimensionless force factor α_N are not illustrated. The dimensionless equivalent force factor α_N is not related to the critical coupling strength. Maidanik and Dickey [24] defined coupling quotient ν_{ME}^2 which is equal to the ratio of damping loss factor and coupling loss factor. The coupling was classified as weak, moderate, and strong according to a numerical range of the coupling quotient ν_{ME}^2. Based on the previous research work, following questions are raised: (1) How is the coupling loss factor defined for PVEH and EMVEH and associated with the electromechanical coupling factor k_e^2? (2) How is the coupling loss factor associated with the squared dimensionless force factor α_N^2? (3) How is the squared dimensionless force factor associated with the critical coupling strength k_e^2/ξ_M and the coupling quotient v_{ME}^2? The answers to these questions will enable SEA of linear vibration energy harvesting systems for reliable performance predictions and design optimization.

Starting from analyses of linear single degree of freedom (SDOF) PVEH and EMVEH systems, this paper classifies weak, moderate, and strong electromechanical coupling according to the numerical ranges of the dimensionless force factor α_N^2. Main contribution of this chapter is to determine in which frequency range, the vibration energy harvesting systems are in a weak coupling where the SEA is applicable.

This chapter establishes relationships between VEH performance and the dimensionless force factor α_N^2 and relationships of coupling loss factor with dimensionless force factor, critical coupling strength, coupling quotient, electromechanical coupling factor. The analysis results are also extended to linear multiple degrees of freedom (MDOH) PVEH and EMVEH systems.

Small damping and linear system are assumed in this paper. For an EMVEH system, the internal coil resistance is expected to be much less than the external electrical load resistance.

9.2 COUPLING ANALYSIS OF A LINEAR SINGLE DEGREE OF FREEDOM PIEZOELECTRIC VIBRATION ENERGY HARVESTING SYSTEM UNDER A HARMONIC EXCITATION

For a linear SDOF PVEH as shown in Fig. 2.1, governing equation of the mechanical subsystem is given by:

$$M \cdot \ddot{z} + c \cdot \dot{z} + K \cdot z = -M \cdot \ddot{y} - \alpha \cdot V \tag{2.1}$$

Governing equation of the electrical subsystem is given by:

$$I = \alpha \cdot \dot{z} - C_0 \cdot \dot{V} \tag{2.2}$$

where y is the excitation displacement; M is the mass; c is the short circuit mechanical damping coefficient; K is the short circuit stiffness of the SDOF PVEH system; z is the relative displacement of the mass with respect to the base; V is the voltage and I is the current. According to Guyomar et al. [10,11], the force factor α and the blocking capacitance of the piezoelectric insert C_0 are defined in Eq. (2.3). When the circuit in Fig. 2.1 is switched on and R is larger than zero, if the base excitation is harmonic, the relative oscillator displacement and output voltage are assumed to be:

$$\begin{cases} y = Y \cdot e^{s \cdot t} \\ z = Z \cdot e^{s \cdot t} \\ V = \overline{V} \cdot e^{s \cdot t} \end{cases} \tag{9.1}$$

Substituting Eq. (9.1) into Eq. (2.2) gives:

$$\left(C_0 \cdot s + \frac{1}{R}\right) \cdot \overline{V} = \alpha \cdot s \cdot Z \tag{9.2}$$

From Eq. (9.2), the harvested voltage response function was derived and given by

$$\frac{\overline{V}}{Z} = \frac{\alpha \cdot s \cdot R}{R \cdot C_0 \cdot s + 1} \tag{9.3}$$

Substitution of Eqs. (9.1) and (9.3) into Eq. (2.1) gives:

$$\frac{Z}{s^2 \cdot Y} = \frac{-M}{K + M \cdot s^2 + c \cdot s + \frac{\alpha^2 \cdot s \cdot R}{R \cdot C_0 \cdot s + 1}} \tag{9.4}$$

The eigen equation is

$$K + M \cdot s^2 + c \cdot s + \frac{\alpha^2 \cdot s \cdot R}{R \cdot C_0 \cdot s + 1} = 0 \tag{9.5}$$

When $s = i \cdot \omega$, consider a small damping and resistance R, at the mechanical resonance, $\omega \approx (K/M)^{0.5}$, which represents a weak coupling. According to the derivation in Chapter 2 or in Refs [1,3–7,26–28], the electrical load resistance and force factor are normalized into two dimensionless variables of R_N and α_N, which are given by:

$$\begin{cases} R_N = R \cdot C_0 \cdot \omega \\ \alpha_N^2 = \dfrac{\alpha^2}{c \cdot C_0 \cdot \omega} = \dfrac{\alpha^2}{M \cdot \eta_M \cdot C_0 \cdot \omega^2} \end{cases} \tag{2.21}$$

where $\eta_M = 2 \cdot \xi_M = \frac{c}{2 \cdot M \cdot \omega}$ is the mechanical damping loss factor; ξ_M is the mechanical damping ratio. According to Chapter 2, the dimensionless mean harvested resonant power is given by:

$$\frac{P_h}{\left(\frac{M^2 \cdot A^2}{c}\right)} = \frac{1}{2} \cdot \frac{R_N \cdot \alpha_N^2}{\alpha_N^4 \cdot R_N^2 + 2 \cdot R_N \cdot \alpha_N^2 + \left(1 + R_N^2\right)} \tag{2.23}$$

where $A = \left|\omega^2 \cdot Y\right|$ is the input excitation acceleration amplitude. Consider the circuit oscillation resonance, $R_N = 1$, Eq. (2.23) becomes

$$\frac{P_h}{\left(\frac{M \cdot A^2}{\omega \cdot \eta_M}\right)} = \frac{1}{2} \cdot \frac{\alpha_N^2}{\alpha_N^4 + 2 \cdot \alpha_N^2 + 2} \tag{9.6}$$

According to Chapter 2, the resonant energy harvesting efficiency is given by

$$\eta = \frac{R_N \alpha_N^2}{1 + R_N \alpha_N^2 + R_N^2} \tag{2.26}$$

Consider the circuit oscillation resonance, $R_N = 1$, Eq. (2.26) becomes

$$\eta = \frac{\alpha_N^2}{2 + \alpha_N^2} \tag{9.7}$$

When the external load resistance R tends to be zero, the circuit is short, Eq. (2.2) diminishes, while Eq. (2.1) becomes,

$$M \cdot \ddot{z} + C \cdot \dot{z} + K \cdot z = -M \cdot \ddot{y} \tag{9.8}$$

The eigen equation can be written as

$$M \cdot s^2 + c \cdot s + K = 0 \tag{9.9}$$

For a small damping, the eigen value is given by

$$\omega_n^2 \approx \frac{K}{M} \tag{9.10}$$

This represents a zero coupling. If $s = i\omega$, Eq. (9.4) becomes

$$\frac{Z}{i\omega \cdot Y} = \frac{-M \cdot i\omega}{K - M \cdot \omega^2 + c \cdot i \cdot \omega} \tag{9.11}$$

The modulus of Eq. (9.4) is given by

$$\frac{|Z|}{A} = \frac{M}{c \cdot \omega} \tag{9.12}$$

The real part of Eq. (9.11) is given by

$$\mathrm{Re}\left(\frac{Z}{i\omega \cdot Y}\right) = \frac{-M}{c} \tag{9.13}$$

From Eqs. (9.3) and (9.12), the modulus of the harvested voltage ratio is given by

$$\frac{|\overline{V}|}{A} = \frac{|\overline{V}|}{|Z|} \cdot \frac{|Z|}{A} = \left|\frac{\alpha \cdot i \cdot \omega \cdot R}{R \cdot C_0 \cdot i \cdot \omega + 1}\right| \cdot \frac{|Z|}{A} = \alpha \cdot R \cdot \frac{M}{c} \tag{9.14}$$

From Eqs. (9.14) and (2.21), according to Ref. [29] the mean harvested resonant power ratio is then given by

$$\frac{P_h}{A^2} = \frac{1}{2} \cdot \frac{|\overline{V}|^2}{R} = \frac{1}{2} \cdot \frac{M^2}{c} \cdot \frac{R \cdot \alpha^2}{c} = \frac{1}{2} \cdot \frac{M^2}{c} \cdot R_N \cdot \alpha_N^2 \tag{9.15}$$

The dimensionless mean harvested resonant power is given by

$$\frac{P_h}{\left(\frac{M^2 \cdot A^2}{c}\right)} = \frac{P_h}{\left(\frac{M \cdot A^2}{\omega \cdot \eta_M}\right)} = \frac{1}{2} \cdot R_N \cdot \alpha_N^2 \tag{9.16}$$

From Eq. (9.13), according to Ref. [29], the mean input power at the resonance frequency is given by

$$P_{in} = -\frac{1}{2} \cdot M \cdot A^2 \cdot \text{Re}\left(\frac{Z}{i\omega \cdot Y}\right) = \frac{1}{2} \cdot \frac{M^2}{c} \cdot A^2 \qquad (9.17)$$

From Eq. (2.21), the dimensionless mean input resonant power is given by

$$\frac{P_{in}}{\left(\frac{M^2 \cdot A^2}{c}\right)} = \frac{P_{in}}{\left(\frac{M \cdot A^2}{\omega \cdot \eta_M}\right)} = \frac{1}{2} \qquad (9.18)$$

From Eqs. (9.16) and (9.18), the resonant energy harvesting efficiency of the linear SDOF system connected with a single load resistor gives

$$\eta = \frac{\frac{P_h}{\left(\frac{M^2 \cdot A^2}{c}\right)}}{\frac{P_{in}}{\left(\frac{M^2 \cdot A^2}{c}\right)}} = R_N \cdot \alpha_N^2 \qquad (9.19)$$

Consider the circuit oscillation resonance $R_N = 1$, Eq. (9.16) becomes

$$\frac{P_h}{\left(\frac{M \cdot A^2}{\omega \cdot \eta_M}\right)} = \frac{1}{2} \cdot \alpha_N^2 \qquad (9.20)$$

Consider the circuit oscillation resonance $R_N = 1$, Eq. (9.19) becomes

$$\eta = \alpha_N^2 \qquad (9.21)$$

The dimensionless resistance R_N defined in Chapter 2 or in Refs [1,3−7,26−28,30] reflects how close the harvesting circuit oscillation is to the resonance. Physical meaning of α_N will be given in the following sections.

When the external load resistance R tends to be very large, the circuit is open. Eq. (9.5) becomes

$$K + M \cdot s^2 + c \cdot s + \frac{\alpha^2}{C_0} = 0$$

For the open circuit and non-damping system, the eigen value becomes

$$\omega^2 = \frac{K}{M} + \frac{\alpha^2}{M \cdot C_0} = \frac{K}{M} \cdot \left(1 + k_e^2\right) = \omega_n^2 \cdot \left(1 + k_e^2\right) \qquad (9.22)$$

where k_e^2 is the electromechanical coupling factor and given by

$$k_e^2 = \frac{\alpha^2}{K \cdot C_0} \qquad (9.23)$$

The frequency ratios $\Omega = \omega/\omega_n$ of the short circuit and open circuit are given by

$$\Omega_{sc} = 1; \quad \Omega_{oc} = \sqrt{1 + k_e^2}; \qquad (9.24)$$

where Ω_{sc} is the frequency ratio of the short circuit, and Ω_{oc} is the frequency ratio of the open circuit. Eqs. (9.23) and (9.24) are identical to those equations given in Refs [21−23]. Note that the resonant frequency shift is pronounced in Eq. (9.22) if the electromechanical coupling factor k_e is large. Consider a small damping and very large external load resistance R, the circuit is open, and the system has a strong electromechanical coupling at the resonance of $\omega = \sqrt{\frac{K}{M}\left(1 + k_e^2\right)}$. In the case of the strong coupling, Eq. (9.4) becomes

$$\frac{Z}{s^2 \cdot Y} = \frac{-M}{K + M \cdot s^2 + c \cdot s + \frac{\alpha^2}{C_0}} \tag{9.25}$$

If $s = i\omega$, Eq. (9.25) becomes

$$\frac{Z}{\omega^2 \cdot Y} = \frac{M}{K - M \cdot \omega^2 + c \cdot i \cdot \omega + \frac{\alpha^2}{C_0}} \tag{9.26}$$

At the resonance of $\omega = \sqrt{\frac{K}{M}\left(1 + k_e^2\right)}$, Eq. (9.26) becomes

$$\frac{Z}{\omega^2 \cdot Y} = \frac{M}{c \cdot i \cdot \omega} \tag{9.27}$$

According to Eq. (9.3), if the resistance is very large, the transfer function between the base excitation acceleration and output voltage is given by:

$$\frac{\overline{V}}{s^2 \cdot Y} = \frac{\overline{V}}{Z} \cdot \frac{Z}{s^2 \cdot Y} = \frac{\alpha}{C_0} \cdot \frac{Z}{s^2 \cdot Y} \tag{9.28}$$

Substitution of Eq. (9.27) into Eq. (9.28) gives its modulus of

$$\frac{|\overline{V}|}{A} = \frac{M \cdot \alpha}{c \cdot \omega \cdot C_0} \tag{9.29}$$

From Eq. (9.29), according to Ref. [29], the mean harvested resonant power ratio is then given by

$$\frac{P_h}{A^2} = \frac{1}{2} \cdot \frac{\frac{|\overline{V}|^2}{A^2}}{R} = \frac{1}{2} \cdot \frac{M^2}{c} \cdot \frac{\alpha^2}{c \cdot \omega \cdot C_0 \cdot (R \cdot \omega \cdot C_0)} \tag{9.30}$$

Substitution of Eq. (2.21) into Eq. (9.30) gives

$$\frac{P_h}{\left(\frac{M^2 \cdot A^2}{c}\right)} = \frac{P_h}{\left(\frac{M \cdot A^2}{\omega \cdot \eta_M}\right)} = \frac{1}{2} \cdot \frac{\alpha_N^2}{R_N} \tag{9.31}$$

where η_M is the mechanical damping loss factor. From Eq. (9.27), its real part is given by

$$\mathrm{Re}\left[\frac{Z}{i \cdot \omega \cdot Y}\right] = \frac{-M}{c} \tag{9.32}$$

According to Eq. (9.32) and Ref. [29], the mean input resonant power is given by

$$P_{in} = -\frac{1}{2} \cdot M \cdot A^2 \cdot Re\left[\frac{Z}{i \cdot \omega \cdot Y}\right] = \frac{1}{2} \cdot \frac{M^2}{c} \cdot A^2 \tag{9.33}$$

According to Ref. [29], the dimensionless mean input resonant power is given by

$$\frac{P_{in}}{\left(\frac{M^2 \cdot A^2}{c}\right)} = \frac{P_{in}}{\left(\frac{M \cdot A^2}{\omega \cdot \eta_M}\right)} = \frac{1}{2} \tag{9.34}$$

From Eqs. (9.31) and (9.34), the resonant energy harvesting efficiency of the linear SDOF system connected with a single load resistor is given by

$$\eta = \frac{\frac{P_h}{\left(\frac{M^2 \cdot A^2}{c}\right)}}{\frac{P_{in}}{\left(\frac{M^2 \cdot A^2}{c}\right)}} = \frac{\alpha_N^2}{R_N} \tag{9.35}$$

Consider the circuit oscillation resonance $R_N = 1$, Eq. (9.31) becomes

$$\frac{P_h}{\left(\frac{M \cdot A^2}{\omega \cdot \eta_M}\right)} = \frac{1}{2} \cdot \alpha_N^2 \tag{9.36}$$

Consider the circuit oscillation resonance, $R_N = 1$, Eq. (9.35) becomes

$$\eta = \alpha_N^2 \tag{9.37}$$

From Eqs. (9.16) and (9.31) in Table 9.2, the formulas of the mean harvested resonant power are interchangeable from the zero coupling to a strong coupling when the harvested resonant power is divided or multiplied by $(R_N)^2$. From Eqs. (9.19) and (9.35) in Table 9.4, the formulas of the resonant energy harvesting

Table 9.1 Dimensionless Mean Harvested Resonant Power of Piezoelectric and Electromagnetic Vibration Energy Harvesters at the Circuit Oscillation Resonance

Dimensionless Mean Harvested Resonant Power When $R_N = 1$	Zero Coupling $R \to 0$	Weak Coupling $R > 0$, $\omega \approx (K/M)^{0.5}$	Strong Coupling $R \to \infty$
Piezoelectric vibration energy harvester	$\frac{P_h}{\left(\frac{M \cdot A^2}{\omega \cdot \eta_M}\right)} = \frac{1}{2} \cdot \alpha_N^2$ (9.20)	$\frac{P_h}{\left(\frac{M \cdot A^2}{\omega \cdot \eta_M}\right)} = \frac{1}{2} \cdot \frac{\alpha_N^2}{\alpha_N^4 + 2 \cdot \alpha_N^2 + 2}$ (9.6)	$\frac{P_h}{\left(\frac{M \cdot A^2}{\omega \cdot \eta_M}\right)} = \frac{1}{2} \cdot \alpha_N^2$ (9.36)
Electromagnetic vibration energy harvester	$\frac{P_h}{\left(\frac{M \cdot A^2}{\omega \cdot \eta_M}\right)} = \frac{1}{2} \cdot \alpha_N^2$ (9.53)	$\frac{P_h}{\left(\frac{M \cdot A^2}{\omega \cdot \eta_M}\right)} = \frac{1}{2} \cdot \frac{\alpha_N^2}{\alpha_N^4 + 2 \cdot \alpha_N^2 + 2}$ (9.41)	$\frac{P_h}{\left(\frac{M \cdot A^2}{\omega \cdot \eta_M}\right)} = \frac{1}{2} \cdot \alpha_N^2$ (9.63)

Table 9.2 Dimensionless Mean Harvested Resonant Power of Piezoelectric and Electromagnetic Vibration Energy Harvesters Without the Circuit Oscillation Resonance

Dimensionless Mean Harvested Resonant Power When $R_N \neq 1$	Zero Coupling $R \to 0$	Weak Coupling $R > 0$, $\omega \approx (K/M)^{0.5}$	Strong Coupling $R \to \infty$
Piezoelectric vibration energy harvester	$\dfrac{P_h}{\left(\frac{M \cdot A^2}{\omega \cdot \eta_M}\right)} = \frac{1}{2} \cdot R_N \cdot \alpha_N^2$ (9.16)	$\dfrac{P_h}{\left(\frac{M \cdot A^2}{\omega \cdot \eta_M}\right)} = \frac{1}{2} \cdot \dfrac{R_N \cdot \alpha_N^2}{\left[1+R_N \cdot \alpha_N^2\right]^2 + R_N^2}$ (2.23)	$\dfrac{P_h}{\left(\frac{M \cdot A^2}{\omega \cdot \eta_M}\right)} = \frac{1}{2} \cdot \dfrac{\alpha_N^2}{R_N}$ (9.31)
Electromagnetic vibration energy harvester	$\dfrac{P_h}{\left(\frac{M \cdot A^2}{\omega \cdot \eta_M}\right)} = \frac{1}{2} \cdot R_N^3 \cdot \alpha_N^2$ (9.49)	$\dfrac{P_h}{\left(\frac{M \cdot A^2}{\omega \cdot \eta_M}\right)} = \frac{1}{2} \cdot \dfrac{R_N \cdot \alpha_N^2}{\left(\frac{1}{R_N}\right)^2 + \left[1+R_N \cdot \alpha_N^2\right]^2}$ (4.25)	$\dfrac{P_h}{\left(\frac{M \cdot A^2}{\omega \cdot \eta_M}\right)} = \frac{1}{2} \cdot R_N \cdot \alpha_N^2$ (9.59)

Table 9.3 Energy Harvesting Efficiency of Piezoelectric and Electromagnetic Vibration Energy Harvesters at the Circuit Oscillation Resonance

Energy Harvesting Efficiency When $R_N = 1$	Zero Coupling $R \to 0$	Weak Coupling $R > 0$, $\omega \approx (K/M)^{0.5}$	Strong Coupling $R \to \infty$
Piezoelectric vibration energy harvester	$\eta = \alpha_N^2$ (9.21)	$\eta = \dfrac{\alpha_N^2}{2+\alpha_N^2}$ (9.7)	$\eta = \alpha_N^2$ (9.37)
Electromagnetic vibration energy harvester	$\eta = \alpha_N^2$ (9.54)	$\eta = \dfrac{\alpha_N^2}{2+\alpha_N^2}$ (9.42)	$\eta = \alpha_N^2$ (9.64)

Table 9.4 Energy Harvesting Efficiency of Piezoelectric and Electromagnetic Vibration Energy Harvesters Without the Circuit Oscillation Resonance

Energy Harvesting Efficiency When $R_N \neq 1$	Zero Coupling $R \to 0$	Weak Coupling $R > 0$, $\omega \approx (K/M)^{0.5}$	Strong Coupling $R \to \infty$
Piezoelectric vibration energy harvester	$\eta = R_N \cdot \alpha_N^2$ (9.19)	$\eta = \dfrac{R_N \alpha_N^2}{1+R_N \alpha_N^2 + R_N^2}$ (2.26)	$\eta = \dfrac{\alpha_N^2}{R_N}$ (9.35)
Electromagnetic vibration energy harvester	$\eta = R_N^3 \cdot \alpha_N^2$ (9.52)	$\eta = \dfrac{R_N \alpha_N^2}{\frac{1}{R_N^2} + \left[1+R_N \alpha_N^2\right]}$ (4.30)	$\eta = R_N \cdot \alpha_N^2$ (9.62)

efficiency are interchangeable from the zero coupling to a strong coupling when the resonant energy harvesting efficiency is divided or multiplied by $(R_N)^2$. Comparing Eqs. (9.6), (9.20), and (9.36) in Table 9.1 or comparing Eqs. (9.7), (9.21), and (9.37) in Table 9.3, it is seen that the dimensionless mean harvested resonant power is identical for the cases of the zero and strong coupling, so is the resonant energy harvesting efficiency. The dimensionless mean harvested resonant power in the cases of the zero and strong coupling is different from that in the case of the weak coupling, so is the resonant energy harvesting efficiency. This is because, for the zero coupling, the total resistance tends to be zero, and the harvesting circuit is short. For the strong coupling, the total load resistance tends to be infinity, the harvesting circuit is open. In both the cases of the zero and strong coupling, the interaction of the mechanical and electrical subsystems disappears. In the case of the weak coupling, the total load resistance is larger than zero and less than infinity, there exists the interaction of the mechanical and electrical subsystems. Similarly, analysis of a linear SDOF EMVEH system can be conducted for different electromechanical coupling in the following section.

9.3 COUPLING ANALYSIS OF A LINEAR SINGLE DEGREE OF FREEDOM ELECTROMAGNETIC VIBRATION ENERGY HARVESTING SYSTEM UNDER A HARMONIC EXCITATION

For a linear SDOF EMVEH excited by a sinusoid signal of constant vibration magnitude shown in Fig. 4.1, governing equation of the mechanical subsystem is given by

$$M \cdot \ddot{z} + c \cdot \dot{z} + K \cdot z = -M \cdot \ddot{y} - B \cdot l \cdot I \tag{4.1}$$

where y is the excitation displacement; M is the oscillator mass; c is the short circuit mechanical damping coefficient; K is the short circuit stiffness of the SDOF electromagnetic—mechanical system; z is the relative displacement of the oscillator with respect to the base; I is the current in the circuit; B is the magnetic field constant or the magnetic flux density, l is the total length of the wire that constitutes the coil. Governing equation of the electrical subsystem is given by

$$V = B \cdot l \cdot \frac{dz}{dt} - R_e \cdot I - L_e \cdot \frac{dI}{dt} \tag{4.2}$$

where V is the voltage; R_e and L_e are respectively the resistance and the self-inductance of the coil.

For a linear SDOF EMVEH connected to a single load resistor as shown in Fig. 4.1, substitution of Eq. (9.1) into Eqs. (4.1) and (4.2) gives

$$\frac{Z}{s^2 \cdot Y} = \frac{-M \cdot \left[\frac{L_e \cdot s}{R} + 1 + \frac{R_e}{R}\right]}{\left[\frac{R_e + L_e \cdot s + R}{R} \cdot M + \frac{L_e \cdot c}{R}\right] \cdot s^2 + \left[\frac{R_e \cdot c}{R} + c + \frac{K \cdot L_e}{R} + R \cdot \alpha^2\right] \cdot s + \frac{R_e}{R} \cdot K + K} \tag{9.38}$$

and

$$\frac{V}{Z} = \frac{\alpha \cdot R \cdot s}{1 + \frac{R_e + L_e \cdot s}{R}} \tag{9.39}$$

where $s = i\omega$. $\alpha = Bl/R$. The modulus of the transfer function between the base acceleration and relative oscillator displacement is then given by

$$\frac{|Z|}{A} = \frac{M \cdot \sqrt{\left(\frac{L_e \cdot \omega}{R}\right)^2 + \left(1 + \frac{R_e}{R}\right)^2}}{\sqrt{\left[\frac{R_e}{R} \cdot K + K - \frac{(R_e + R) \cdot M + L_e \cdot c}{R} \cdot \omega^2\right]^2 + \left[\frac{R_e \cdot c}{R} + c + \frac{K \cdot L_e}{R} + R \cdot \alpha^2 - \frac{L_e \cdot \omega^2 \cdot M}{R}\right]^2 \cdot \omega^2}} \tag{9.40}$$

According to Chapter 4 and Refs [26–28,30–32], the electrical load resistance and equivalent force factor are normalized as two dimensionless variables of R_N and α_N^2 which are given by:

$$\begin{cases} R_N = \dfrac{R}{L_e \cdot \omega} \\[3mm] \alpha_N^2 = \dfrac{\alpha^2 \cdot L_e \cdot \omega}{c} = \left(\dfrac{B \cdot l}{R}\right)^2 \dfrac{L_e}{M \cdot \eta_M} \end{cases} \tag{4.23}$$

where $\eta_M = \frac{c}{2 \cdot M \cdot \omega}$ is the mechanical damping loss factor. $\alpha = Bl/R$ is the equivalent force factor of the EMVEH.

When the circuit in Fig. 4.1 is switched on and R is larger than zero, it is assumed that the resistance of the coil wire is very small, compared with the external load resistance, that is, $R_e \ll R$ or $\frac{R_e}{R} \approx 0$. For example, in the experimental device in Ref. [33], the coil resistance R_e was 0.3 Ω, the coil inductance L_e was 2.4 mH, the load resistance R was 10,000 Ω. It is obvious that the coil resistance is much less than the load resistance. Consider a small damping, at the resonance, $\omega \approx (K/M)^{0.5}$ which represents a weak coupling, the dimensionless mean harvested resonant power can be derived from Chapter 4 and given by:

$$\frac{P_h}{\left(\frac{M^2 A^2}{C}\right)} = \frac{P_h}{\left(\frac{M \cdot A^2}{\omega \cdot \eta_M}\right)} = \frac{1}{2} \cdot \frac{R_N \alpha_N^2}{\left(\frac{1}{R_N}\right)^2 + \left[1 + R_N \alpha_N^2\right]^2} \tag{4.25}$$

Consider the circuit oscillation resonance, $R_N = 1$, Eq. (4.25) becomes

$$\frac{P_h}{\left(\frac{M \cdot A^2}{\omega \cdot \eta_M}\right)} = \frac{1}{2} \cdot \frac{\alpha_N^2}{2 + 2 \cdot \alpha_N^2 + \alpha_N^4} \tag{9.41}$$

The resonant energy harvesting efficiency can be derived from Chapter 4 and is given by

$$\eta = \frac{R_N \alpha_N^2}{\frac{1}{R_N^2} + \left[1 + R_N \alpha_N^2\right]} \tag{4.30}$$

Consider the circuit oscillation resonance $R_N = 1$, Eq. (4.30) becomes

$$\eta = \frac{\alpha_N^2}{2 + \alpha_N^2} \tag{9.42}$$

When the resistance R tends to be very small, which corresponds to a zero coupling, it is assumed that the resistance of the coil wire R_e is also small, that is, $\frac{R_e}{R} \rightarrow 1$, Eq. (9.38) becomes,

$$\frac{Z}{\omega^2 \cdot Y} = \frac{M \cdot \left[\frac{L_e \cdot i \cdot \omega}{R} + 2 \right]}{2 \cdot K - \left[\left(2 + \frac{L_e \cdot i \cdot \omega}{R} \right) \cdot M + \frac{L_e \cdot c}{R} \right] \cdot \omega^2 + \left[2 \cdot c + \frac{K \cdot L_e}{R} + R \cdot \alpha^2 \right] \cdot i \cdot \omega} \tag{9.43}$$

At the resonance of $\omega = \sqrt{\frac{K}{M}}$ Eq. (9.39) becomes

$$\frac{V}{Z} = \frac{\alpha \cdot R \cdot i \cdot \omega}{2 + \frac{L_e \cdot i \cdot \omega}{R}} \tag{9.44}$$

Eq. (9.43) becomes

$$\frac{|Z|}{A} = \frac{M}{c \cdot \omega} \tag{9.45}$$

From Eq. (9.43), the real part is given by

$$\mathrm{Re}\left(\frac{Z}{i\omega \cdot Y} \right) = -\frac{M}{c} \tag{9.46}$$

$$\frac{|\overline{V}|}{A} = \frac{|\overline{V}|}{|Z|} \cdot \frac{|Z|}{A} = \frac{\alpha \cdot R^2}{L_e} \cdot \frac{M}{c \cdot \omega} \tag{9.47}$$

From Eqs. (9.47) and (4.23), according to Ref. [29], the dimensionless mean harvested resonant power ratio is then given by

$$\frac{P_h}{A^2} = \frac{1}{2} \cdot \frac{\frac{|\overline{V}|^2}{A^2}}{R} = \frac{1}{2} \cdot \frac{\alpha^2 \cdot R^3}{\omega^2 \cdot L_e^2 \cdot c} \cdot \frac{M^2}{c} = \frac{1}{2} \cdot R_N^3 \cdot \alpha_N^2 \cdot \frac{M^2}{c} \tag{9.48}$$

The dimensionless mean harvested resonant power is given by

$$\frac{P_h}{\left(\frac{M^2 \cdot A^2}{c} \right)} = \frac{P_h}{\left(\frac{M \cdot A^2}{\omega \cdot \eta_M} \right)} = \frac{1}{2} \cdot R_N^3 \cdot \alpha_N^2 \tag{9.49}$$

From Eq. (9.46), according to Ref. [29], the mean input resonant power is given by

$$P_{in} = -\frac{1}{2} \cdot M \cdot A^2 \cdot \mathrm{Re}\left(\frac{Z}{i\omega \cdot Y} \right) = \frac{1}{2} \cdot \frac{M^2}{c} \cdot A^2 \tag{9.50}$$

From Eq. (9.50), the dimensionless mean input resonant power is given by

$$\frac{P_{\text{in}}}{\left(\frac{M^2 \cdot A^2}{c}\right)} = \frac{P_{\text{in}}}{\left(\frac{M \cdot A^2}{\omega \cdot \eta_M}\right)} = \frac{1}{2} \tag{9.51}$$

From Eqs. (9.49) and (9.51), the resonant energy harvesting efficiency of the linear SDOF system connected with a single load resistor is given by

$$\eta = \frac{\frac{P_h}{\left(\frac{M^2 \cdot A^2}{c}\right)}}{\frac{P_{\text{in}}}{\left(\frac{M^2 \cdot A^2}{c}\right)}} = R_N^3 \cdot \alpha_N^2 \tag{9.52}$$

Consider the circuit oscillation resonance, $R_N = 1$, Eq. (9.49) becomes

$$\frac{P_h}{\left(\frac{M \cdot |\ddot{y}|^2}{\omega \cdot \eta_M}\right)} = \frac{1}{2} \cdot \alpha_N^2 \tag{9.53}$$

Consider the circuit oscillation resonance, $R_N = 1$, Eq. (9.52) becomes

$$\eta = \alpha_N^2 \tag{9.54}$$

When the resistance R tends to be very large, which corresponds to a strong coupling, it is assumed that the resistance R_e of the coil wire is very small, compared with the external load resistance R, that is, $R_e \ll R$ or $\frac{R_e}{R} \approx 0$, at the resonance, $\omega = \sqrt{\frac{K}{M} \cdot \frac{R \cdot M}{R \cdot M + L_e \cdot c}} \approx \sqrt{\frac{K}{M}}$, Eq. (9.40) becomes,

$$\frac{|Z|}{A} = \frac{M}{c \cdot \omega} \tag{9.55}$$

From Eq. (9.38), the real part is given by

$$\text{Re}\left(\frac{Z}{i\omega \cdot Y}\right) = -\frac{M}{c} \tag{9.56}$$

From Eqs. (9.39) or (9.44) and (9.55), the modulus of the harvested voltage ratio is given by

$$\frac{V}{A} = \frac{|V|}{|Z|} \cdot \frac{|Z|}{A} = \left|\frac{\alpha \cdot R \cdot i \cdot \omega}{1 + \frac{R_e + L_e \cdot i \cdot \omega}{R}}\right| \cdot \frac{|Z|}{A} = \alpha \cdot R \cdot \frac{M}{c} \tag{9.57}$$

From Eqs. (4.23) and (9.57), according to Ref. [29], the dimensionless mean harvested resonant power ratio is then given by

$$\frac{P_h}{A^2} = \frac{1}{2} \cdot \frac{\frac{|V|^2}{A^2}}{R} = \frac{1}{2} \cdot \frac{M^2}{c} \cdot \frac{R \cdot \alpha^2}{c} = \frac{1}{2} \cdot \frac{M^2}{c} \cdot R_N \cdot \alpha_N^2 \tag{9.58}$$

The dimensionless mean harvested resonant power is given by

$$\frac{P_h}{\left(\frac{M^2 \cdot A^2}{c}\right)} = \frac{P_h}{\left(\frac{M \cdot A^2}{\omega \cdot \eta_M}\right)} = \frac{1}{2} \cdot R_N \cdot \alpha_N^2 \tag{9.59}$$

where η_M is the mechanical damping loss factor. According to Eq. (9.56) and Ref. [29], the mean input resonant power is given by

$$P_{in} = -\frac{1}{2} \cdot M \cdot A^2 \cdot \mathrm{Re}\left(\frac{Z}{i\omega \cdot Y}\right) = \frac{1}{2} \cdot \frac{M^2}{c} \cdot A^2 \tag{9.60}$$

The dimensionless mean input resonant power is given by

$$\frac{P_{in}}{\left(\frac{M^2 \cdot A^2}{c}\right)} = \frac{P_{in}}{\left(\frac{M \cdot A^2}{\omega \cdot \eta_M}\right)} = \frac{1}{2} \tag{9.61}$$

From Eqs. (9.59) and (9.61), the resonant energy harvesting efficiency of the SDOF system connected with a single load resistor gives

$$\eta = \frac{\frac{P_h}{\left(\frac{M^2 \cdot A^2}{c}\right)}}{\frac{P_{in}}{\left(\frac{M^2 \cdot A^2}{c}\right)}} = R_N \cdot \alpha_N^2 \tag{9.62}$$

Consider the circuit oscillation resonance, $R_N = 1$, Eq. (9.59) becomes

$$\frac{P_h}{\left(\frac{M \cdot |\ddot{y}|^2}{\omega \cdot \eta_M}\right)} = \frac{1}{2} \cdot \alpha_N^2 \tag{9.63}$$

Consider the circuit oscillation resonance, $R_N = 1$, Eq. (9.62) becomes

$$\eta = \alpha_N^2 \tag{9.64}$$

From Eqs. (9.49) and (9.59) in Table 9.2, the formulas of the dimensionless mean harvested resonant power are interchangeable from the zero coupling to the strong coupling when the dimensionless mean harvested resonant power is divided or multiplied by $(R_N)^2$. This feature has reflected the duality of the dimensionless mean harvested resonant power from the zero to strong coupling in term of the reciprocal $(R_N)^2$. The dimensionless mean harvested resonant power values of the zero and strong coupling not only depend on the coupling strength but also depend on how close the circuit oscillation is to the resonance, which coincided with the conclusions of Ref. [34]. Comparing Eqs. (9.41), (9.53), and (9.63) in Table 9.1, it is seen that the dimensionless mean harvested resonant power is identical for the zero and strong coupling. The dimensionless mean harvested resonant power values in the cases of the zero and strong coupling are different from those in the case of the weak coupling. This is because no interaction exists between the mechanical and electrical subsystems in the zero and strong coupling as the electrical circuit is either short or open in the two cases. The dimensionless mean harvested resonant power in the zero coupling is identical to that in the strong coupling, so is the energy harvesting efficiency. When the circuit is switched on, there develops the interaction between the mechanical and electrical subsystems, and a weak coupling develops. The interaction makes the dimensionless mean harvested resonant power or energy harvesting efficiency of a weak coupling different from that of the zero and strong coupling. From Eqs. (9.52) and (9.62) in Table 9.4, the formulas of the resonant energy harvesting efficiency are interchangeable from the zero coupling to the strong coupling when the resonant energy harvesting efficiency is divided or multiplied by $(R_N)^2$. Comparing Eqs. (9.42), (9.54), and (9.64) in Table 9.3, it is seen that the resonant energy harvesting efficiency is same for the zero and strong coupling. The resonant energy harvesting efficiency in the zero and strong coupling is different from that in the weak coupling case. The reasons have been illustrated above.

It is noticed that the electromechanical coupling has little effects on the resonant frequency of the electromagnetic harvester system, the natural frequency is always calculated by $\omega = \sqrt{\frac{K}{M}}$, which is different from that for a piezoelectric VEH as illustrated in Eq. (9.22).

The duality of the PVEH and EMVEH was shown in Chapter 5 and in Ref. [35]. It was proved that the calculation formulas of the dimensionless mean harvested resonant power or energy harvesting efficiency are interchangeable from the PVEH to EMVEH by replacing $(1/R_N)^2$ with $(R_N)^2$ where the $R_N \cdot \alpha_N^2$ term is not changed. The duality of the PVEH and EMVEH has been further verified in this chapter by comparing Eqs. (9.16) and (9.49), Eqs. (2.23) and (4.25), and Eqs. (9.31) and (9.59) shown in Table 9.2 or by comparing Eqs. (9.19) with (9.52), Eqs. (2.26) and (4.30), and Eqs. (9.35) and (9.62) shown in Table 9.4. From the pairs of Eqs. (9.20) and (9.53), Eqs. (9.6) and (9.41), Eqs. (9.36) and (9.63) shown in Table 9.1, it is seen that at the electric circuit oscillation resonance $R_N = R \cdot C_0 \cdot \omega = 1$, the dimensionless mean harvested resonant power is identical for both the PVEH and EMVEH. From the pairs of Eqs. (9.21) and (9.54), Eqs. (9.7) and (9.42), Eqs. (9.37) and (9.64) shown in Table 9.3, it is seen that at the electric circuit oscillation resonance $R_N = R \cdot C_0 \cdot \omega = 1$, the resonant energy harvesting efficiency is identical for both of PVEH and EMVEH. This is because at the electric circuit oscillation resonance and the mechanical resonance, the electrical impedances of piezoelectric and electromagnetic circuits are all equal to the mechanical impedance. The dimensionless mean harvested resonant power or energy harvesting efficiency only depends on the dimensionless force factor α_N, regardless of the VEH type of the piezoelectric or electromagnetic.

9.4 COUPLING ANALYSES OF LINEAR PIEZOELECTRIC AND ELECTROMAGNETIC VIBRATION ENERGY HARVESTERS UNDER RANDOM EXCITATIONS

The above analyses are based on a harmonic excitation. However, ambient environment excitations are often random with a finite frequency bandwidth. The frequency integrations over the bandwidth are required for the output voltage or relative displacement in the above analyses to get the harvested power and energy harvesting efficiency under random excitations. Since ambient environmental random vibration process is assumed to be a stationary and ergodic process or a Gaussian random process. The sampling time-averaged values of the process are equal to the ensemble averaged values. Therefore the PVEH and EMVEH can be studied using SEA where the mechanical and electrical subsystems are considered. This is possible because both of the SEA and the above vibration energy harvesting analysis are focused on the resonant frequencies of mechanical and electric subsystems.

From Eqs. (2.21) and (9.23), it is seen that the squared dimensionless force factor α_N^2 is equal to the electromechanical coupling factor k_e^2 divided by the mechanical damping loss factor and given by

$$\alpha_N^2 = \frac{k_e^2}{\eta_M} = \frac{\alpha^2}{K \cdot C_0 \cdot \eta_M} = \frac{\alpha^2}{\omega \cdot c \cdot C_0} \tag{9.65}$$

where

$$\eta_M = \frac{c}{\sqrt{K \cdot M}} \tag{9.66}$$

η_M is the damping loss factor of the mechanical subsystem. From Eq. (9.65), α_N^2 has been proved to be the critical coupling strength defined in Ref. [21–23]. The vibration energy harvesting system in Figs. 2.1 or 4.1 can be described by an SEA model where a mechanical subsystem is coupled to an electrical subsystem as shown in Fig. 9.1.

For a weak coupling, consider a narrow band random excitation, power transfer from the mechanical to electrical subsystem can be written as

$$P_h = \frac{\langle V^2 \rangle}{R} = \omega \cdot \eta_{ME} \cdot (E_M - E_E) \tag{9.67}$$

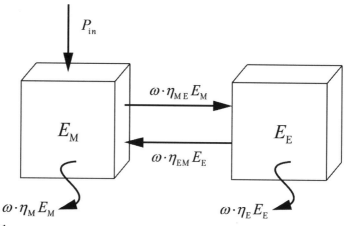

FIGURE 9.1

Statistical energy analysis model of a linear vibration energy harvesting system.

where η_{ME} is the coupling loss factor from the mechanical to electrical subsystem; $\langle V^2 \rangle$ is the mean squared voltage harvested; ω is the center frequency of the narrow excitation band which should be close to the resonant frequencies of both of the mechanical and electrical subsystems; E_M is the average energy of the mechanical subsystem; E_E is the average energy of the electrical subsystem. For the SDOF VEH system with one piezoelectric element, the modal densities of both of the mechanical and electrical subsystems are equal to one, which is to say that both of the mechanical and electrical subsystems are of SDOF. The mean energy of the mechanical subsystem is given by

$$E_M = M \cdot \langle \dot{z}^2 \rangle \tag{9.68}$$

where $\langle \dot{z}^2 \rangle$ is the mean squared relative velocity of the mechanical subsystem. The mean energy of the electrical subsystem is given by

$$E_E = \frac{\langle V^2 \rangle}{R \cdot \eta_E \cdot \omega} \tag{9.69}$$

where η_E is the damping loss factor of the electrical subsystem. The mean energy difference of the mechanical and electrical subsystems is given by

$$E_M - E_E = M \cdot \langle \dot{z}^2 \rangle - \frac{\langle V^2 \rangle}{R \cdot \eta_E \cdot \omega} \tag{9.70}$$

Substitution of Eq. (9.70) into Eq. (9.67) gives

$$\eta_{ME} = \frac{1}{R \cdot M \cdot \frac{\langle \dot{z}^2 \rangle}{\langle V^2 \rangle} \cdot \omega - \frac{1}{\eta_E}} \tag{9.71}$$

It is assumed that the energy harvesting process is a stationary and ergodic random process, the spatial and ensemble average of the squared velocity or voltage are equal to the time average of the squared velocity or voltage. For a narrow band random excitation, the time average of the squared velocity divided by the time average of the squared voltage is then equal to the squared modulus of the velocity spectrum over the voltage spectrum in the frequency domain. Eq. (9.71) becomes

$$\eta_{ME} = \frac{1}{R \cdot M \cdot \left| \frac{Z(\omega)}{V(\omega)} \right|^2 \cdot \omega^3 - \frac{1}{\eta_E}} \tag{9.72}$$

If a harmonic excitation can be looked upon as a very narrow band random excitation, from Eq. (9.3) where $s = i \cdot \omega$, it gives

$$\left| \frac{Z(\omega)}{V(\omega)} \right|^2 = \frac{(R \cdot C_0 \cdot \omega)^2 + 1}{(\alpha \cdot \omega \cdot R)^2} = \frac{C_0^2}{\alpha^2} + \frac{1}{(\alpha \cdot \omega \cdot R)^2} \tag{9.73}$$

Substitution of Eq. (9.73) into Eq. (9.72) gives

$$\eta_{ME} = \frac{1}{R \cdot M \cdot \left[\frac{C_0^2}{\alpha^2} + \frac{1}{(\alpha \cdot \omega \cdot R)^2} \right] \cdot \omega^3 - \frac{1}{\eta_E}} \tag{9.74}$$

Consider the resonance of the electrical subsystem or the circuit oscillation $R_N = R \cdot C_0 \cdot \omega = 1$, from Eqs. (2.21) and (9.65), Eq. (9.74) becomes

$$\frac{\eta_{ME}}{\eta_M} = \frac{1}{\frac{2}{\alpha_N^2} - \frac{\eta_M}{\eta_E}} \tag{9.75}$$

From Eqs. (9.65) and (9.75), it gives

$$\eta_{ME} = \frac{1}{\frac{2}{k_e^2} - \frac{1}{\eta_E}} \tag{9.76}$$

Therefore, the coupling loss factor from the mechanical to electrical subsystem η_{ME} is associated with the squared electromechanical coupling factor k_e^2 and the damping loss factor of the electrical subsystem η_E for the weak coupling. The squared electromechanical coupling factor k_e^2 is defined by Eq. (9.23) for the piezo-electric vibration energy harvesting system.

The coupling loss factor from the mechanical to electrical subsystem η_{ME} is associated with the squared force factor α_N^2 and the damping loss factors of the mechanical and electrical subsystems η_M and η_E as illustrated in Eq. (9.75).

It is assumed that the mechanical damping loss factor is equal to the electrical damping loss factor $\eta_M = \eta_E$, which is small, therefore, for a weak coupling, Eq. (9.75) becomes

$$\frac{\eta_{ME}}{\eta_M} = \frac{\alpha_N^2}{2 - \alpha_N^2} \tag{9.77}$$

Comparing Eq. (9.65) with Eq. (9.77) gives

$$\eta_{ME} = \frac{k_e^2 \cdot \eta_M}{2 \cdot \eta_M - k_e^2} \tag{9.78}$$

When $\eta_M \ll \eta_E$, for a weak coupling,

$$\frac{\eta_{ME}}{\eta_M} = \frac{\alpha_N^2}{2} \tag{9.79}$$

From Eqs. (9.65) and (9.79), it gives

$$\eta_{ME} = \frac{1}{2} \cdot k_e^2 \tag{9.80}$$

For the weak coupling, the coupling loss factor from the mechanical to electrical subsystems in the EMVEH system is given by

$$\eta_{ME} = \frac{1}{\frac{2}{k_e^2} - \frac{1}{\eta_E}} = \frac{\eta_M}{\frac{2}{\alpha_N^2} - \frac{\eta_M}{\eta_E}} = \frac{B^2 \cdot l^2 \cdot L_e \cdot \eta_E}{2 \cdot M \cdot R^2 \cdot \eta_E - B^2 \cdot l^2 \cdot L_e} \tag{9.81}$$

The relationship of the coupling loss factor η_{ME}, squared electromechanical coupling factor k_e^2, damping loss factor of the electrical subsystem η_E have not been illustrated in the past. The relationship of the coupling loss factor η_{ME}, squared dimensionless force factor α_N^2, damping loss factors of the electrical and mechanical subsystems η_E and η_M have also not been illustrated in the previous literature.

The above linear SDOF analysis should be easy to be expanded to a linear MDOF vibration energy harvesting system where the modal densities of the mechanical and electrical subsystems are larger than one. When MDOF mechanical and electrical subsystems are considered as shown in Fig. 9.1, under the assumption of a weak coupling, an excitation frequency bandwidth is selected to cover the major mechanical and electrical subsystem resonant modes [36], which leads to modal densities of the mechanical subsystem system $n_M(\omega)$ and the electrical subsystem $n_E(\omega)$ being included in Eqs. (9.75) and (9.76) which is given by:

$$\frac{\eta_{ME}}{\eta_M} = \frac{1}{\frac{n_M(\omega)}{n_E(\omega)} \cdot \frac{2}{\alpha_N^2} - \frac{\eta_M}{\eta_E}} \tag{9.82}$$

and

$$\eta_{ME} = \frac{1}{\frac{n_M(\omega)}{n_E(\omega)} \cdot \frac{2}{k_e^2} - \frac{1}{\eta_E}} \tag{9.83}$$

where k_e^2 is the mode averaged electromechanical coupling factor, α_N^2 is the mode averaged critical coupling strength. After the coupling loss factor η_{ME} and the critical coupling strength α_N^2 have been established, the SEA can be applied to predict the mean energy values and variances of the mechanical and electrical subsystems according to Eq. (3.2.13), Eqs. (4.2.13), and (4.2.14) in Ref. [36].

9.5 RELATIONSHIP BETWEEN THE VIBRATION ENERGY HARVESTING PERFORMANCE AND CRITICAL COUPLING STRENGTH

After the coupling loss factor of the vibration energy harvesting systems is derived, the relationships between the dimensionless mean harvested resonant power, energy harvesting efficiency, and critical coupling strength can be established.

From Eq. (9.65), α_N^2 has been proved to be the critical coupling strength defined in Refs [21−23]. The relationship of the critical coupling strength α_N^2 and the force factor α is presented in Eq. (2.21). The coupling quotient was defined as $v_{ME}^2 = \frac{\eta_M}{\eta_{ME}}$ in Ref. [24]. For a weak coupling, from Eq. (9.82), it gives

$$v_{ME}^2 = \frac{\eta_M}{\eta_{ME}} = \frac{n_M(\omega)}{n_E(\omega)} \cdot \frac{2}{\alpha_N^2} - \frac{\eta_M}{\eta_E} \tag{9.84}$$

The value of $v_{ME}^2(\omega)$ will determine whether the coupling between the subsystems is weak, moderate, or strong. If $v_{ME}^2 = \frac{\eta_M}{\eta_{ME}} \gg 1$, or $\frac{n_M(\omega)}{n_E(\omega)} \cdot \frac{2}{\alpha_N^2} - \frac{\eta_M}{\eta_E} \gg 1$, the coupling between the mechanical and electrical subsystems is weak, and if $v_{ME}^2 = \frac{\eta_M}{\eta_{ME}} \ll 1$ or $\frac{n_M(\omega)}{n_E(\omega)} \cdot \frac{2}{\alpha_N^2} - \frac{\eta_M}{\eta_E} \ll 1$ then the coupling is strong [24]. When the coupling loss factor from the mechanical to electrical subsystem is much larger than the damping loss factor of the mechanical subsystem, the vibration energy harvesting systems are in a strong coupling. When the coupling loss factor from the mechanical to electrical subsystem is much less than the damping loss factor of the mechanical subsystem, the vibration energy harvesting systems are in a weak coupling. When there exists $\frac{n_M(\omega)}{n_E(\omega)} \cdot \frac{2}{\alpha_N^2} - \frac{\eta_M}{\eta_E} \geq 10$, the systems are in a weak coupling. When there exists $1 \leq \frac{n_M(\omega)}{n_E(\omega)} \cdot \frac{2}{\alpha_N^2} - \frac{\eta_M}{\eta_E} \leq 10$, the systems are in a moderate coupling. When there exists $\frac{n_M(\omega)}{n_E(\omega)} \cdot \frac{2}{\alpha_N^2} - \frac{\eta_M}{\eta_E} \leq 1$, the systems are in a strong coupling. These results are more illustrative than those in Refs [21−23]. From Eqs. (2.21) and (4.23), the relationships between the center frequency of the narrow excitation band in Hz and the dimensionless force factor are, for the PVEH, $f = \frac{\alpha^2}{2\pi \cdot c \cdot C_0 \cdot \alpha_N^2}$; for the EMVEH, $f = \left(\frac{R}{B \cdot l}\right)^2 \cdot \frac{c \cdot \alpha_N^2}{L_e}$. From these relationships, the frequency ranges of the weak coupling, moderate coupling, and strong coupling of the PVEH and EMVEH systems can be determined. The relationship of the critical coupling strength α_N^2 and the coupling quotient v_{ME}^2 as shown in Eq. (9.84) and the resultant coupling strength ranges of the weak, moderate, and strong coupling have not been illustrated in the previous literature. According to Chapters 2−4 and Refs [1,3−7,26−28,30−32], the dimensionless mean harvested resonant power and energy harvesting efficiency are determined by only the dimensionless force factor α_N and dimensionless resistance R_N. The normalized dimensionless force factor α_N reflects the electromechanical coupling strength, while the dimensionless resistance R_N reflects how close the harvesting oscillation circuit is to its resonance. The physical meanings of the dimensionless variables R_N and α_N have not been clearly illustrated in previous literature.

For the linear SDOF piezoelectric VEH, consider the circuit oscillation resonance $R_N = R \cdot C_0 \cdot \omega = 1$ in Eq. (2.21), which gives

$$\alpha_N^2 = \frac{R \cdot \alpha^2}{M \cdot \eta_M \cdot \omega} \tag{9.85}$$

It is seen from Eq. (9.85) that increasing the force factor α and the external resistance will increase the critical coupling strength. Decreasing the mass and damping loss factor of mechanical subsystem or decreasing the center frequency of the excitation bandwidth will increase the critical coupling strength.

For the linear SDOF EMVEH, consider the oscillation circuit resonance $R_N = \frac{R}{L_e \cdot \omega} = 1$ in Eq. (4.23), it gives

$$\alpha_N^2 = \frac{B^2 \cdot l^2}{R \cdot M \cdot \omega} \tag{9.86}$$

It is seen from Eq. (9.86) that increasing the magnetic field intensity and the length of the coil will increase the critical coupling strength. Decreasing the external resistance, the mechanical subsystem mass, and the center frequency of the excitation band will also increase the critical coupling strength.

For a linear SDOF PVEH or EMVEH, assume $\eta_M = \eta_E$ according to the range of the critical coupling strength α_N^2, the dimensionless mean harvested resonant power for the cases of the weak and non-weak coupling are plotted over α_N^2 following the pairs of Eqs. (9.6) and (9.41) and Eqs. (9.36) and (9.63) in Table 9.1 and shown in Fig. 9.2. A line is connected between the ending and starting points of the

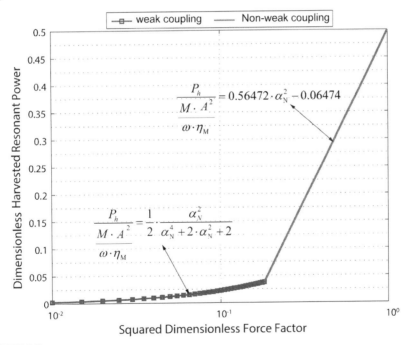

FIGURE 9.2

Dimensionless mean harvested resonant power of linear single degree of freedom piezoelectric and electromagnetic vibration energy harvesters for the cases of the weak and non-weak coupling ($\eta_M = \eta_E$).

dimensionless mean harvested resonant power curves of the weak and strong coupling. The line is used to approximate the dimensionless mean harvested resonant power of a moderate coupling as there are no such equations available for the moderate coupling. α_N^2 changes from 0 to 0.1818 for the weak coupling, from 0.1818 to 1 for the moderate coupling, and from 1 onward for the strong coupling. For the PVEH, when $f > \frac{\alpha^2}{1.142 \cdot c \cdot C_0}$, the system is in a weak coupling where the SEA is applicable; when $f < \frac{\alpha^2}{6.28 \cdot c \cdot C_0}$, the system is in a strong coupling where the SEA is not applicable. When $\frac{\alpha^2}{1.142 \cdot c \cdot C_0} < f < \frac{\alpha^2}{6.28 \cdot c \cdot C_0}$, the system is in a moderate coupling.

For the EMVEH, when $f < 0.1818 \cdot \left(\frac{R}{B \cdot l}\right)^2 \cdot \frac{c}{L_e}$, the system is in a weak coupling where the SEA is applicable; when $f > \left(\frac{R}{B \cdot l}\right)^2 \cdot \frac{c}{L_e}$, the system is in a strong coupling where the SEA is not applicable. When $\left(\frac{R}{B \cdot l}\right)^2 \cdot \frac{c}{L_e} < f < 0.1818 \cdot \left(\frac{R}{B \cdot l}\right)^2 \cdot \frac{c}{L_e}$, the system is in a moderate coupling.

It is seen from Fig. 9.2 that the dimensionless mean harvested resonant power for the cases of non-weak coupling is much larger than that for the case of the weak coupling.

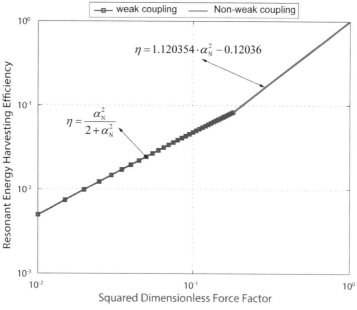

FIGURE 9.3

Resonant energy harvesting efficiency of linear single degree of freedom piezoelectric and electromagnetic vibration energy harvesters for the cases of weak and non-weak coupling ($\eta_M = \eta_E$).

According to the range of the critical coupling strength, the dimensionless energy harvesting efficiency for the cases of the weak and non-weak coupling are plotted over α_N^2 following the pairs of Eqs. (9.7) and (9.42), Eqs. (9.37) and (9.64) in Table 9.3 and shown in Fig. 9.3. A line is connected between the ending and starting points of the resonant harvesting energy efficiency curves of the weak and strong coupling. The line is used to approximate the resonant harvesting energy efficiency of a moderate coupling as there are no such equations available for the moderate coupling. α_N^2 changes from 0 to 0.1818 for the weak coupling, from 0.1818 to 1 for the moderate coupling, and from 1 onward for the strong coupling.

It is seen from Fig. 9.3 that the energy harvesting efficiency for the cases of the non-weak coupling is much larger than that for the case of the weak coupling.

It is seen from Figs. 9.2 and 9.3 that as the dimensionless force factor or the critical coupling strength α_N^2 increases, the dimensionless mean harvested resonant power and energy harvesting efficiency increase.

For a linear MDOF PVEH or EMVEH, k_e^2 should be replaced by $k_e^2 \cdot \frac{n_E(\omega)}{n_M(\omega)}$ in Eqs. (9.76), (9.78), and (9.80); α_N^2 should be replaced by $\alpha_N^2 \cdot \frac{n_E(\omega)}{n_M(\omega)}$ in Eqs. (9.75), (9.77) and (9.79) and (9.81) and in the formulas of Tables 9.1 and 9.3 and in Figs. 9.2 and 9.3.

NOMENCLATURE

B	Magnetic field constant or the magnetic flux density
l	The total length of the wire that constitutes the coil
A	The input excitation acceleration amplitude of the base
c	Short circuit mechanical damping of the single degree of freedom system
C_0	Blocking capacity of the piezoelectric material insert
E_E	Mean energy of the electrical subsystem
E_M	Mean energy of the mechanical subsystem
e	2.718281828
e_{33}	Piezoelectric constant
f_n	Natural frequency
H	Thickness of the piezoelectric material insert disk thickness
i	Square root of -1
I	Current in the circuit
K	Short circuit stiffness of the single degree of freedom system
k_e	The electromechanical coupling factor or electromechanical coupling coefficient
$\frac{k_e^2}{\xi_M}$	Critical coupling strength
ξ, ξ_M	The mechanical damping ratio
L_e	Self-inductance of the coil
R_e	The resistance of the coil
M	Oscillator mass of the single degree of freedom system
P_h	Mean harvested resonant power
P_{in}	Mean input resonant power

$\frac{P_{\mathrm{h}}}{\left(\frac{M^2 A^2}{c}\right)}$	Dimensionless resonant mean harvested power
A	The input excitation acceleration amplitude
s	Laplace variable
t	Time instances
V	Output voltage of the single degree of freedom system
$\dot{V}$	The change rate of the output voltage
$\overline{V}, \overline{V}(\omega)$	The Fourier transform of the output voltage
$|\overline{V}|$	Modulus or amplitude of the output voltage
y	Base excitation displacement
$\dot{y}$	Base excitation velocity
$\ddot{y}$	Base excitation acceleration
$Y, Y(\omega)$	The Fourier transform of y
z	Relative displacement of the mass with respect to the base
$\dot{z}$	Relative velocity of the mass with respect to the base
$\ddot{z}$	Relative acceleration of the mass with respect to the base
$Z, Z(\omega)$	The Fourier transform of z
$|Z|$	Modulus or amplitude of the relative displacement
α	Force factor of the piezoelectric material insert or equivalent force factor of the electromagnetic device/transducer
α_{N}	Dimensionless force factor
R_{N}	Dimensionless resistance
ν_{ME}^2	The coupling quotient from the mechanical to the electrical subsystems
η	Resonant energy harvesting efficiency
η_{M}	Damping loss factor of mechanical subsystem or mechanical damping loss factor
η_{E}	Damping loss factor of electrical subsystem or electrical damping loss factor
η_{ME}	Coupling loss factor from the mechanical to the electrical subsystems
$n_{\mathrm{M}}(\omega)$	Modal densities of the mechanical subsystem
$n_{\mathrm{E}}(\omega)$	Modal densities of the electrical subsystem
ω	The center radial frequency of the narrow excitation band close to the resonant frequency in rad/s
f	The center frequency of the narrow excitation band close to the resonant frequency in Hz
ω_{n}	Natural frequency in rad/s
π	3.1415927
Ω_{sc}	The frequency ratio of the short circuit
Ω_{oc}	The frequency ratio of the open circuit
$\frac{\mathrm{d}}{\mathrm{d}t}$	The first differential

Subscripts

0	Blocking capacity of the piezoelectric material insert
33	Piezoelectric working mode having the same direction of loading and electric poles
M	Mechanical subsystem
E	Electrical subsystem
h	Harvested energy
in	Input
N	Normalized

Superscripts

−1   Inverse
*   Complex conjugate
__   Time average
.   The first differential
..   The second differential

Special Function

$<>$ Time and spatial averaged
$|\,|$ Modulus or absolute value
Re[ ]   The real part of a complex variables

Abbreviations

SDOF   Single degree of freedom
SEA   Statistical energy analysis

REFERENCES

[1]   Wang X, Lin LW. Dimensionless optimization of piezoelectric vibration energy harvesters with different interface circuits. Smart Mater Struct 2013;22:20, 085011.

[2]   Arroyoa E, Badela A, Formosaa F, Wu Y, Qiu J. Comparison of electromagnetic and piezoelectric vibration energy harvesters: model and experiments. Sens Actuators A 2012;183:148−56.

[3]   Xiao H, Wang X, John S. A dimensionless analysis of a 2DOF piezoelectric vibration energy harvester. Mech Syst Signal Process 2015;58−59:355−75.

[4]   Xiao H, Wang X, John S. A multi degree of freedom piezoelectric vibration energy harvester with piezoelectric elements inserted between two nearby oscillators. Mech Syst Signal Process February 2016;68−69:138−54.

[5]   Xiao H, Wang X. A review of piezoelectric vibration energy harvesting techniques. Int Rev Mech Eng 2014;8(3). ISSN: 1970-8742:609−20.

[6]   Wang X, Xiao H. Dimensionless analysis and optimization of piezoelectric vibration energy harvester. Int Rev Mech Eng 2013;7(4). ISSN: 1970-8742:607−24.

[7]   Cojocariu B, Hill A, Escudero A, Xiao H, Wang X. Piezoelectric vibration energy harvester − design and prototype. In: Proceedings of ASME 2012 International Mechanical Engineering Congress & Exposition, ISBN 978-0-7918-4517-2.

[8]   Badel A, Guyomar D, Lefeuvre E, Richard C. Efficiency enhancement of a piezoelectric energy harvesting device in pulsed operation by synchronous charge inversion. J Intell Mater Syst Struct 2005;16:889−901.

[9]   Dutoit NE, Wardle BL. Performance of microfabricated piezoelectric vibration energy harvesters. Integr Ferroelectr 2006;83(1):13−32.

[10] Guyomar D, et al. Toward energy harvesting using active materials and conversion improvement by nonlinear processing. Ultrasonics, Ferroelectr Freq Control, IEEE Trans 2005;52(4):584−95.

[11] Guyomar D, et al. Energy harvester of 1.5 cm^3 giving output power of 2.6 mW with only 1 G acceleration. J Intell Mater Syst Struct 2010;22(5):415−20.

[12] Lefeuvre E, Badel A, Richard C, Petit L, Guyomar D. A comparison between several vibration-powered piezoelectric generators for standalone systems. Sens Actuators A 2006;126:405−16.

[13] Lefeuvre E. Piezoelectric energy harvesting device optimization by synchronous electric charge extraction. J Intell Mater Syst Struct 2005;16(10):865−76.

[14] Liao Y, Sodano HA. Optimal parameters and power characteristics of piezoelectric energy harvesters with an RC circuit. Smart Mater Struct 2009;18(4):045011.

[15] Lien IC, et al. Revisit of series-SSHI with comparisons to other interfacing circuits in piezoelectric energy harvesting. Smart Mater Struct 2010;19(12):125009.

[16] Shafer MW, et al. Erratum: designing maximum power output into piezoelectric energy harvesters. Smart Mater Struct 2012;21(10):109601.

[17] Tang LH, Yang YW. Analysis of synchronized charge extraction for piezoelectric energy harvesting. Smart Mater Struct 2011;20:15, 085022.

[18] Wickenheiser AM, Garcia E. Design of energy harvesting systems for harnessing vibrational motion from human and vehicular motion. In: Proceedings of SPIE, vol. 7643; 2010. 76431B-76431B-76413.

[19] Wickenheiser AM, et al. Modeling the effects of electro-mechanical coupling on energy storage through piezoelectric energy harvesting. Mechatronics. IEEE/ASME Trans 2010;15(3):400−11.

[20] Wickenheiser AM. Design optimization of linear and non-linear cantilevered energy harvesters for broadband vibrations. J Intell Mater Syst Struct 2011;22(11):1213−25.

[21] Shu YC, Lien IC. Analysis of power output for piezoelectric energy harvesting systems. Smart Mater Struct 2006;15:1499−512.

[22] Shu YC, Lien IC. Efficiency of energy conversion for a piezoelectric power harvesting system. J Micromech Microeng 2006;16:2429−38.

[23] Shu YC, Lien IC, Wu W. An improved analysis of the SSHI interface in piezoelectric energy harvesting. Smart Mater Struct 2007;16:2253−64.

[24] Maidanik G, Dickey J. On the external input power into coupled structures. Solid Mech Its Appl 1999;67:197−208.

[25] Deleted in review.

[26] Wang X, Liang XY, Hao ZY, Du HP, Zhang N, Qian M. Comparison of electromagnetic and piezoelectric vibration energy harvesters with different interface circuits. Mech Syst Signal Process 2016 [accepted by October 29, 2015].

[27] Wang X. Coupling loss factor of linear vibration energy harvesting systems in a framework of statistical energy analysis. J Sound Vib 2016;362:125−41.

[28] Wang X, Liang XY, Shu GQ, Watkins S. Coupling analysis of linear vibration energy harvesting systems. Mech Syst Signal Process March 2016;70−71:428−44.

[29] Wang X, Koss LL. Frequency response function for power, power transfer ratio and coherence. J Sound Vib 1992;155(1):55−73.

[30] Wang X, et al. Similarity and duality of electromagnetic and piezoelectric vibration energy harvesters. Mech Syst Signal Process 2014;52−53.

[31] Wang X, Liang XY, Wei HQ. A study of electromagnetic vibration energy harvesters with different interface circuits. Mech Syst Signal Process 2015;58−59:376−98.

[32] Bawahab M, Xiao H, Wang X. A study of linear regenerative electromagnetic shock absorber system. 2015. SAE 2015-01-0045.

[33] Gatti RR. Spatially varying multi degree of freedom electromagnetic energy harvesting [Ph.D. thesis]. Australia: Curtin University; 2013.

[34] Stephen NG. On energy harvesting from ambient vibration. J Sound Vib 2006; 293(1−2):409−25.

[35] Poulin G, Sarraute E, Costa F. Generation of electrical energy for portable devices. Sens Actuators A Phys 2004;116(3):461−71.

[36] Lyon RH, DeJong RG. Theory and application of statistical energy analysis. Boston: Butterworth-Heinemann; 1995.

Correlation and Frequency Response Analyses of Input and Harvested Power Under White Noise, Finite Bandwidth Random and Harmonic Excitations

10

CHAPTER OUTLINE

10.1 INTRODUCTION

Method for analysis and measurement of input and harvested power and energy harvesting efficiency is important for performance evaluation and design optimization of vibration energy harvesters. In the previous chapters, dynamic differential equations of piezoelectric vibration energy harvester (PVEH) and electromagnetic vibration energy harvester (EMVEH) are established, from which the frequency response analysis method based on the Fourier and Laplace transforms, the time domain integration, and state space analysis methods are used to calculate the response displacement, output voltage, input and harvested power, and energy harvesting efficiency [1−8]. The frequency response analysis and state space analysis methods are only applicable to linear vibration energy harvesting systems, while the integration method is applicable to both the linear and nonlinear vibration energy harvesting systems. The three methods have been validated by one another and by experiments [9,10,16]. The three methods in the previous chapters are only developed under the harmonic excitations. The linear vibration energy harvesting systems under random excitations have not been investigated in the previous chapters. As most of vibration signals from ambient environment are random, analysis of the linear vibration energy harvesting systems under random excitations are essential. Therefore, motivation of this chapter is to introduce a new method for analysis of the linear vibration energy harvesting systems under both harmonic and random excitations. The analysis results using the new method should be versified by those using the three methods in the previous chapters. The new method should enable direct measurement and analysis of the input and harvested power and energy harvesting efficiency for the linear vibration energy harvesting systems under both harmonic and random excitations. The new method will involve correlation and frequency response analysis of power variables.

The concept of correlation of power variables was introduced by Koss in 1987 and 1988 [11−14] where the fluctuations of power variables were defined and illustrated at different locations of a system. Several questions were answered regarding coherence of power fluctuation quantities and transfer of power from an input point to another point in a vibratory system. For example, under what conditions of the force input would the coherence between power quantities be high, or near or equal to unity? Which part of the frequency response function represents the mean power transfer from a driving point to another point on the system? Which part of the frequency response function would represent the fluctuating power transfer from a driving point to another point in the system? The previous research studies were only focused on linear structures, mechanical, or machinery systems. Vibration energy harvesting systems of electromechanical couplings were not involved.

This chapter will demonstrate a new method of frequency response analysis of power variables in linear PVEH and EMVEH systems. Two types of frequency response functions will be used for the work presented here. The first type is the frequency response functions for linear field variables such as the frequency response function between the input excitation force and output response velocity which is also called mobility function. If a single degree of freedom (SDOF) spring-mass-dashpot system is inserted with a piezoelectric material or an electromagnetic device, it becomes a vibration energy harvesting system. The frequency response function could also be the frequency response function between the excitation force and the output voltage of the vibration energy harvesting system. The second type of frequency response function is that which relates power variables such as input power and output voltage squared or output power. The frequency response function of power variables is given the acronym FRFP to distinguish it from the frequency response function of linear field variables (FRF). FRFP is a function of the power fluctuation frequency while FRF is a function of the excitation force frequency. Power at the zero power fluctuation frequency (dc value) represents the mean value of the power variable whereas the excitation force frequency is not equal to zero. Power fluctuation frequency ω is two times of the force frequency ω_0, when it is not equal to zero. DC values of power variables correspond to the time-averaged power values, while the non-DC values of power variables represent the fluctuating values of the power variables around the mean.

In the previous chapters, the input and harvested power and efficiency have been analyzed and normalized as functions of dimensionless resistance and force factor for the PVEH and EMVEH [4,5]. This chapter aims to establish the relationship between the DC values of the input and harvested power variables and the mean input and harvested power and the relationship between the DC value of the frequency response function for the power variables and the energy harvesting efficiency. Novelty of this chapter is that from frequency response measurement of the product of the excitation input force and the driving point velocity and the product of the harvested voltage and current, the mean input and harvested power and energy harvesting efficiency can be determined; therefore the harvested power and energy harvesting efficiency can be directly measured from experiments.

10.2 CORRELATION AND FREQUENCY RESPONSE ANALYSIS OF POWER VARIABLES

The expected value of the product of four Gaussian random variables x_1, x_2, x_3, and x_4, $E[x_1 \cdot x_2 \cdot x_3 \cdot x_4]$ can be reduced to a sum of the expected values of the products of any two variables [15] (see p. 72), as follows:

$$E[x_1 \cdot x_2 \cdot x_3 \cdot x_4] = E[x_1 \cdot x_2] \cdot E[x_3 \cdot x_4] + E[x_1 \cdot x_3] \cdot E[x_2 \cdot x_4] + E[x_1 \cdot x_4] \cdot E[x_2 \cdot x_3]$$
$$- 2 \cdot \mu_{x_1} \cdot \mu_{x_2} \cdot \mu_{x_3} \cdot \mu_{x_4}$$

$$(10.1)$$

where μ_{x_i} is the average value of the x_i process $i = 1, 2, 3, 4$. With $x_1(t) = F(t)$ and $x_2(t) = u(t)$ the force and velocity at a driving point $x_3(t) = V(t)$ is the output voltage and $x_4(t) = I(t)$ is the output current, the input mechanical power is given by

$$P_M(t) = F(t) \cdot u(t) \qquad (10.2)$$

The output electric power $P_E(t)$ is given by

$$P_E(t) = I(t) \cdot V(t) \qquad (10.3)$$

For a Gaussian random force input into a structure, the cross-correlation function between input mechanical power and harvested electric power can be calculated from Eq. (10.1) and given by

$$R_{P_M P_E}(\tau) = \overline{P_M} \cdot \overline{P_E} + R_{FV}(\tau) \cdot R_{uI}(\tau) + R_{FI}(\tau) \cdot R_{uV}(\tau) \qquad (10.4)$$

where τ is the time delay; $R_{FV}(\tau)$ is the cross-correlation function between the input force $F(t)$ and output voltage $V(t)$; $R_{uI}(\tau)$ is the cross-correlation function between the driving point velocity $u(t)$ and output current $I(t)$; $R_{FI}(\tau)$ is the cross-correlation function between the input force $F(t)$ and output current $I(t)$; $R_{uV}(\tau)$ is the cross-correlation function between the driving point velocity $u(t)$ and output voltage $V(t)$; $\overline{P_M}$ and $\overline{P_E}$ are the mean input mechanical power and the mean output electric power. It is assumed that the average values of the excitation force $F(t)$, the driving point velocity $u(t)$, the output voltage $V(t)$, and the output current $I(t)$ are all equal to zero. The electricity is generated in alternative current (AC). If the constant term $\overline{P_M} \cdot \overline{P_E}$ in Eq. (10.4) is subtracted out, then by use of the frequency convolution theorem, the two-sided cross-spectral density function of $S_{P_M P_E}(\omega)$ is found to be

$$S_{P_M P_E}(\omega) = \frac{1}{2 \cdot \pi} \left[\int_{-\infty}^{\infty} S_{FV}(\eta) \cdot S_{uI}(\omega - \eta) d\eta + \int_{-\infty}^{\infty} S_{FI}(\eta) \cdot S_{uV}(\omega - \eta) d\eta \right]$$

(10.5)

where $S_{FV}(\eta)$ is the two-sided cross-spectral density function between the force $F(t)$ and output voltage $V(t)$; $S_{uI}(\eta)$ is the two-sided cross-spectral density function between the driving point velocity $u(t)$ and output current $I(t)$; $S_{FI}(\eta)$ is the two-sided cross-spectral density function between the force $F(t)$ and output current $I(t)$; $S_{uV}(\eta)$ is the two-sided cross-spectral density function between the driving point velocity $u(t)$ and output voltage $V(t)$; ω is the power fluctuation frequency, η is a dummy variable. Eq. (10.5) gives the cross-spectral density function between input mechanical power fluctuation $P_M(t)$ and the output electric power fluctuation $P_E(t)$. In the same way, autospectral densities of input mechanical power $P_M(t)$ and the output electric power $P_E(t)$ are given by

$$S_{P_M P_M}(\omega) = \frac{1}{2 \cdot \pi} \left[\int_{-\infty}^{\infty} S_{Fu}(\eta) \cdot S_{uF}(\omega - \eta) d\eta + \int_{-\infty}^{\infty} S_{FF}(\eta) \cdot S_{uu}(\omega - \eta) d\eta \right]$$

(10.6)

and

$$S_{P_E P_E}(\omega) = \frac{1}{2 \cdot \pi} \left[\int_{-\infty}^{\infty} S_{VI}(\eta) \cdot S_{IV}(\omega - \eta) d\eta + \int_{-\infty}^{\infty} S_{VV}(\eta) \cdot S_{II}(\omega - \eta) d\eta \right]$$

(10.7)

where $S_{FF}(\omega)$ and $S_{uu}(\eta)$ are the two-sided autospectral density functions of the input force $F(t)$ and the driving point velocity $u(t)$. Also $S_{Fu}(\eta)$ and $S_{uF}(\eta)$ are two-sided cross-spectral density functions between the input force $F(t)$ and driving point velocity $u(t)$; $S_{VV}(\eta)$ and $S_{II}(\eta)$ are the two-sided autospectral density functions of the output voltage and current; $S_{VI}(\eta)$ and $S_{IV}(\eta)$ are two-sided cross-spectral density functions between the output voltage and current. From Eqs. (10.5)–(10.7), the frequency response function for the least square fitting linear system $H_{P_M P_E}(\omega)$ which relates input and output and the coherence function $\gamma^2_{P_M P_E}(\omega)$ is determined and given by

$$H_{P_M P_E}(\omega) = \frac{\int_{-\infty}^{\infty} S_{FV}(\eta) \cdot S_{uI}(\omega - \eta)\mathrm{d}\eta + \int_{-\infty}^{\infty} S_{FI}(\eta) \cdot S_{uV}(\omega - \eta)\mathrm{d}\eta}{\int_{-\infty}^{\infty} S_{Fu}(\eta) \cdot S_{uF}(\omega - \eta)\mathrm{d}\eta + \int_{-\infty}^{\infty} S_{FF}(\eta) \cdot S_{uu}(\omega - \eta)\mathrm{d}\eta} \quad (10.8)$$

$$\gamma^2_{P_M P_E}(\omega) = \frac{|S_{P_M P_E}(\omega)|^2}{S_{P_M P_M}(\omega) \cdot S_{P_E P_E}(\omega)} \quad (10.9)$$

The autospectral density terms and cross-spectral density terms can be obtained from the frequency response functions of field variables and autospectral density terms of the input force. Thus, Eqs. (10.8) and (10.9) can be rewritten as

$$H_{P_M P_E}(\omega) = \frac{\int_{-\infty}^{\infty} S_{FF}(\eta) \cdot S_{FF}(\omega - \eta) \cdot |H_{Fu}(\omega - \eta)|^2 \cdot [H_{FV}(\eta) \cdot H_{uI}(\omega - \eta) + H_{FI}(\eta) \cdot H_{uV}(\omega - \eta)]\mathrm{d}\eta}{\int_{-\infty}^{\infty} S_{FF}(\eta) \cdot S_{FF}(\omega - \eta) \cdot \left[H^*_{Fu}(\omega - \eta) \cdot H_{Fu}(\eta) + |H_{Fu}(\omega - \eta)|^2 \right]\mathrm{d}\eta}$$

$$(10.10)$$

$$\gamma^2_{P_M P_E}(\omega) = \left| \int_{-\infty}^{\infty} S_{FF}(\eta) \cdot S_{FF}(\omega - \eta) \cdot |H_{Fu}(\omega - \eta)|^2 \cdot [H_{FV}(\eta) \cdot H_{uI}(\omega - \eta) + H_{FI}(\eta) \cdot H_{uV}(\omega - \eta)]\mathrm{d}\eta \right|$$

$$\Big/ \left\{ \int_{-\infty}^{\infty} S_{FF}(\eta) \cdot S_{FF}(\omega - \eta) \cdot \left[H^*_{Fu}(\omega - \eta) \cdot H_{Fu}(\eta) + |H_{Fu}(\omega - \eta)|^2 \right]\mathrm{d}\eta \cdot \int_{-\infty}^{\infty} S_{FF}(\eta) \cdot S_{FF}(\omega - \eta) \cdot \right.$$

$$\left. \left[H_{VI}(\eta) \cdot |H_{FV}(\eta)|^2 \cdot H^*_{VI}(\omega - \eta) \cdot |H_{FV}(\omega - \eta)|^2 + |H_{FV}(\eta)|^2 \cdot |H_{FV}(\omega - \eta)|^2 \right]\mathrm{d}\eta \right\}$$

$$(10.11)$$

where $H_{Fu}(\eta)$, $H_{FV}(\eta)$, and $H_{FI}(\eta)$ are frequency response functions between the excitation force F and the driving point velocity u, between the excitation force F and output voltage V and between the excitation force F and output current I. $H_{uI}(\eta)$ and $H_{uV}(\eta)$ are frequency response functions between the driving point velocity u and the output current I and between the driving point velocity u and output voltage V. The asterisk (*) represents complex conjugation; $H_{VI}(\eta)$ and $H^*_{VI}(\omega - \eta)$ are equal to resistance $1/R$ for the system in Figs. 2.1 or 4.1. Eq. (10.10) reflects the frequency response function of the fluctuating input mechanical and output electric power.

10.3 HARVESTED RESONANT POWER AND ENERGY HARVESTING EFFICIENCY UNDER WHITE NOISE RANDOM EXCITATION

For linear vibration energy harvesting systems under white noise random excitation as shown in Fig. 10.1, if only the DC term in Eq. (10.4) is considered, then, the DC value of the frequency response function for the input and output power is given by

Wide-band record

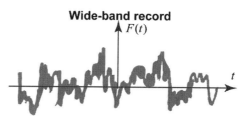

Autocorrelation

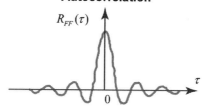

Autospectrum

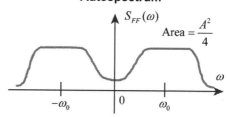

FIGURE 10.1

Time record, autocorrelation, and autospectrum functions of broadband random or white noise excitation input of the inertia force.

$$H_{P_M P_E}(0) = \eta_E = \frac{\overline{P}_E}{\overline{P}_M} \tag{10.12}$$

which is the energy harvesting efficiency; $\overline{P}_E$ is given by

$$\overline{P}_E = R_{VI}(0) = \int_{-\infty}^{\infty} S_{VI}(\eta)\mathrm{d}\eta = \int_{-\infty}^{\infty} S_{FF}(\eta)\cdot|H_{FV}(\eta)|^2\cdot H_{VI}(\eta)\mathrm{d}\eta \tag{10.13}$$

And, similarly, $\overline{P}_M$ is given by

$$\overline{P}_M = R_{Fu}(0) = \int_{-\infty}^{\infty} S_{Fu}(\eta)\mathrm{d}\eta = \int_{-\infty}^{\infty} S_{FF}(\eta)\cdot H_{Fu}(\eta)\mathrm{d}\eta \tag{10.14}$$

Eq. (10.12) is thus rewritten as

$$H_{P_M P_E}(0) = \eta_E = \frac{\int_{-\infty}^{\infty} S_{FF}(\eta)\cdot|H_{FV}(\eta)|^2\mathrm{d}\eta}{R\cdot \int_{-\infty}^{\infty} S_{FF}(\eta)\cdot H_{Fu}(\eta)\mathrm{d}\eta} \tag{10.15}$$

If the input inertia force is $-M \cdot \ddot{y}$, where M is the oscillator mass of a structure or machine and $\ddot{y}$ is the acceleration of an excitation base, and if $\dot{y}$ is the velocity of the base, $\dot{z}$ is the relative velocity of the driving point with respect to the base, then the driving point absolute velocity is $\dot{z} + \dot{y}$, and $S_{FF}(\eta)$ is then given by

$$S_{FF}(\eta) = M^2 \cdot S_{\ddot{y}\ddot{y}}(\eta) \tag{10.16}$$

where $S_{\ddot{y}\ddot{y}}(\eta)$ is the two-sided autospectral density function of the base excitation acceleration $\ddot{y}$. The frequency response function $H_{Fu}(\eta)$ is then given by

$$H_{Fu}(\eta) = \frac{-i \cdot \eta}{M} \cdot \left[H_{\ddot{y}z}(\eta) - 1/\eta^2 \right] \tag{10.17}$$

The frequency response function $H_{FV}(\eta)$ is then given by

$$H_{FV}(\eta) = \frac{-1}{M} \cdot H_{\ddot{y}V}(\eta) \tag{10.18}$$

Substitution of Eqs. (10.16)–(10.18) into Eqs. (10.13)–(10.15) gives

$$\overline{P}_{\mathrm{E}} = R_{VI}(0) = \frac{1}{R} \cdot \int_{-\infty}^{\infty} S_{\ddot{y}\ddot{y}}(\eta) \cdot \left| H_{\ddot{y}V}(\eta) \right|^2 \mathrm{d}\eta \tag{10.19}$$

$$\overline{P}_{\mathrm{M}} = M \cdot \int_{-\infty}^{\infty} S_{\ddot{y}\ddot{y}}(\eta) \cdot \left[i/\eta - i \cdot \eta \cdot H_{\ddot{y}z}(\eta) \right] \mathrm{d}\eta \tag{10.20}$$

$$H_{P_{\mathrm{M}}P_{\mathrm{E}}}(0) = \eta_{\mathrm{E}} = \frac{\int_{-\infty}^{\infty} S_{\ddot{y}\ddot{y}}(\eta) \cdot \left| H_{\ddot{y}V}(\eta) \right|^2 \mathrm{d}\eta}{R \cdot M \cdot \int_{-\infty}^{\infty} S_{\ddot{y}\ddot{y}}(\eta) \cdot \left[i/\eta - i \cdot \eta \cdot H_{\ddot{y}z}(\eta) \right] \mathrm{d}\eta} \tag{10.21}$$

If $S_{\ddot{y}\ddot{y}}(\eta)$ has a constant amplitude of S_0 and $H_{\ddot{y}V}(\eta) = \frac{V(\eta)}{(i \cdot \eta)^2 \cdot Y(\eta)}$ and $H_{\ddot{y}z}(\eta) = \frac{Z(\eta)}{(i \cdot \eta)^2 \cdot Y(\eta)}$ of a vibration energy harvester are given, harvested power and efficiency can be calculated from Eqs. (10.19) and (10.21).

For example, if $H_{\ddot{y}V}(\eta) = \frac{V(\eta)}{(i \cdot \eta)^2 \cdot Y(\eta)}$ and $H_{\ddot{y}z}(\eta) = \frac{Z(\eta)}{(i \cdot \eta)^2 \cdot Y(\eta)}$ of SDOF, PVEH and EMVEH can be calculated from Chapters 2 and 4, the harvested power and efficiency can be calculated from Eqs. (10.19) and (10.21) for a white noise random excitation.

10.4 HARVESTED RESONANT POWER AND ENERGY HARVESTING EFFICIENCY UNDER FINITE BANDWIDTH RANDOM EXCITATION

For a finite bandwidth random excitation in the frequency range from ω_1 to ω_2, as shown in Fig. 10.2, it is assumed that the double-sided autospectrum function of the excitation acceleration becomes a pair of symmetric narrow rectangular band with a bandwidth of $\Delta\omega = \omega_2 - \omega_1$ centered at $-\omega_0$ and ω_0. Eqs. (10.19)–(10.21) becomes

Narrow-band record

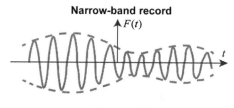

Autocorrelation

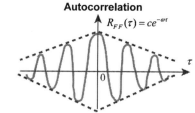

Autospectrum

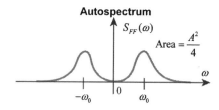

FIGURE 10.2

Time record, autocorrelation, and autospectrum functions of finite narrow bandwidth random excitation input of the inertia force.

$$\overline{P}_E = \frac{2}{R} \cdot \int_{\omega_1}^{\omega_2} S_{\ddot{y}\ddot{y}}(\eta) \cdot \left|H_{\ddot{y}V}(\eta)\right|^2 d\eta \tag{10.22}$$

$$\overline{P}_M = -2 \cdot M \cdot \int_{\omega_1}^{\omega_2} S_{\ddot{y}\ddot{y}}(\eta) \cdot \text{Re}\left[i \cdot \eta \cdot H_{\ddot{y}z}(\eta)\right] d\eta \tag{10.23}$$

$$H_{P_M P_E}(0) = \eta_E = \frac{\int_{\omega_1}^{\omega_2} S_{\ddot{y}\ddot{y}}(\eta) \cdot \left|H_{\ddot{y}V}(\eta)\right|^2 d\eta}{-R \cdot M \cdot \int_{\omega_1}^{\omega_2} S_{\ddot{y}\ddot{y}}(\eta) \cdot \text{Re}\left[i \cdot \eta \cdot H_{\ddot{y}z}(\eta)\right] d\eta} \tag{10.24}$$

If $S_{\ddot{y}\ddot{y}}(\eta)$ has a constant amplitude of S_0 and $H_{\ddot{y}V}(\eta) = \frac{V(\eta)}{(i \cdot \eta)^2 \cdot Y(\eta)}$ and $H_{\ddot{y}z}(\eta) = \frac{Z(\eta)}{(i \cdot \eta)^2 \cdot Y(\eta)}$ of a vibration energy harvester are given, harvested power and efficiency can be calculated from Eqs. (10.22) and (10.24).

For example, if $H_{\ddot{y}V}(\eta) = \frac{V(\eta)}{(i \cdot \eta)^2 \cdot Y(\eta)}$ and $H_{\ddot{y}z}(\eta) = \frac{Z(\eta)}{(i \cdot \eta)^2 \cdot Y(\eta)}$ of SDOF, PVEH and EMVEH can be calculated from Chapters 2 and 4, the harvested power and efficiency can be calculated from Eqs. (10.22) and (10.24) for a finite bandwidth random excitation.

For the PVEH system in Fig. 2.1, from the Laplace transform transfer functions of Eqs. (2.8) and (2.10), $H_{\ddot{y}z}(\eta)$ and $H_{\ddot{y}V}(\eta)$ are given by

$$H_{\ddot{y}z}(\eta) = \frac{Z(\eta)}{(i \cdot \eta)^2 \cdot Y(\eta)}$$

$$= \frac{-M \cdot R \cdot C_0 \cdot \eta \cdot i - M}{\left(-R \cdot C_0 \cdot M \cdot \eta^3 \cdot i - (R \cdot C_0 \cdot c + M) \cdot \eta^2 + (c + R \cdot C_0 \cdot K + \alpha^2 \cdot R) \cdot \eta \cdot i + K\right)}$$

(10.25)

$$H_{\ddot{y}V}(\eta) = \frac{V(\eta)}{(i \cdot \eta)^2 \cdot Y(\eta)}$$

$$= \frac{-M \cdot \alpha \cdot R \cdot \eta \cdot i}{\left(-R \cdot C_0 \cdot M \cdot \eta^3 \cdot i - (R \cdot C_0 \cdot c + M) \cdot \eta^2 + (c + R \cdot C_0 \cdot K + \alpha^2 \cdot R) \cdot \eta \cdot i + K\right)}$$

(10.26)

where $s = i\eta$.

Substitution of Eqs. (10.25) and (10.26) into Eqs. (10.19)–(10.21) or Eqs. (10.22)–(10.24) calculates the mean input mechanical and output electric power and energy harvesting efficiency of the PVEH system for a white noise or finite bandwidth random excitation.

10.5 HARVESTED RESONANT POWER AND ENERGY HARVESTING EFFICIENCY UNDER A HARMONIC EXCITATION

For a harmonic excitation of frequency ω_0 and amplitude A, the double-sided auto-spectrum function of the excitation acceleration $S_{\ddot{y}\ddot{y}}(\eta)$ becomes a pair of impulse delta function with amplitude of $\frac{A^2}{4}$ centered at $-\omega_0$ and ω_0, as shown in Fig. 10.3. From the integration property of the impulse delta function, Eqs. (10.19)–(10.21) become

$$\overline{P}_E = \frac{2}{R} \cdot \left| H_{\ddot{y}V}(\omega_0) \right|^2 \cdot A^2$$

(10.27)

$$\overline{P}_M = -2 \cdot M \cdot \mathrm{Re}\left[i \cdot \omega_0 \cdot H_{\ddot{y}z}(\omega_0)\right] \cdot A^2$$

(10.28)

$$H_{P_M P_E}(0) = \eta_E = \frac{\left| H_{\ddot{y}V}(\omega_0) \right|^2}{-R \cdot M \cdot \mathrm{Re}\left[i \cdot \omega_0 \cdot H_{\ddot{y}z}(\omega_0)\right]}$$

(10.29)

Substitution of Eqs. (10.25) and (10.26) into Eqs. (10.27) and (10.29) gives

$$\overline{P}_E = P_h = \frac{1}{2} \cdot \frac{M^2 \cdot A^2}{c} \cdot \frac{R \cdot \alpha^2 \cdot c}{\alpha^4 \cdot R^2 + 2 \cdot R \cdot \alpha^2 \cdot c + \left(1 + R^2 \cdot C_0^2 \cdot \omega^2\right) \cdot c^2}$$

(10.30)

and

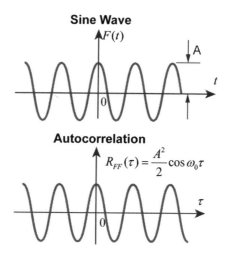

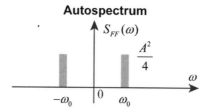

FIGURE 10.3

Time record, autocorrelation, and autospectrum functions of harmonic excitation input of the inertia force.

$$H_{P_M P_E}(0) = \eta_E = \frac{\alpha^2 \cdot R}{\left[(R \cdot C_0)^2 \cdot c \cdot \omega_0^2 + c + \alpha^2 \cdot R \right]} \tag{10.31}$$

where $\overline{P}_E = P_h$ and $A^2 = \left| (i\omega_0)^2 \, Y(\omega_0) \right|^2$.

From Eq. (2.21),

$$\begin{cases} R_N = R \cdot C_0 \cdot \omega \\ \alpha_N = \sqrt{\dfrac{\alpha^2}{c \cdot C_0 \cdot \omega}} \end{cases} \tag{2.21}$$

Eqs. (10.30) and (10.31) become

$$\frac{\overline{P}_E}{\left(\frac{M^2 \cdot A^2}{c} \right)} = \frac{P_h}{\left(\frac{M^2 \cdot \left| (i\omega_0)^2 Y(\omega_0) \right|^2}{c} \right)} = \frac{1}{2} \cdot \frac{R_N \cdot \alpha_N^2}{\alpha_N^4 \cdot R_N^2 + 2 \cdot R_N \cdot \alpha_N^2 + \left(1 + R_N^2 \right)} \tag{10.32}$$

and

$$H_{P_M P_L}(0) = \eta_E = \frac{\alpha_N^2 \cdot R_N}{R_N^2 + 1 + \alpha_N^2 \cdot R_N} \tag{10.33}$$

Eqs. (10.32) and (10.33) are same as Eqs. (2.23) and (2.26). From Eq. (2.21), it is seen that when the mechanical damping c tends to be very small, close to zero, α_N tends to be very large. If R_N is kept constant, from Eqs. (10.32) and (10.33), when α_N tends to be very large, the harvested power tends to be very small or close to zero; however, the energy harvesting efficiency tends to be 100%. When the mechanical damping c tends to be very large, α_N tends to be very small or close to zero. If R_N is kept constant, from Eqs. (10.32) and (10.33), when α_N tends to be very small or close to zero, the harvested power tends to be very large; however, the energy harvesting efficiency tends to be very small or close to zero.

For the EMVEH system in Fig. 4.1, from Eqs. (4.8) and (4.9), $H_{\ddot{y}z}(\eta)$ and $H_{\ddot{y}V}(\eta)$ are given by

$$H_{\ddot{y}z}(\eta) = \frac{Z(\eta)}{(i \cdot \eta)^2 \cdot Y(\eta)}$$

$$= \frac{-M \cdot \left[\dfrac{L_e \cdot \eta \cdot i}{R} + 1 + \dfrac{R_e}{R}\right]}{-\left[\dfrac{R_e + L_e \cdot \eta \cdot i + R}{R} \cdot M + \dfrac{L_e \cdot c}{R}\right] \cdot \eta^2 + \left[\dfrac{R_e \cdot c}{R} + c + \dfrac{K \cdot L_e}{R} + R \cdot \alpha^2\right] \cdot \eta \cdot i + \dfrac{R_e}{R} \cdot K + K} \tag{10.34}$$

$$H_{\ddot{y}V}(\eta) = \frac{V(\eta)}{(i \cdot \eta)^2 \cdot Y(\eta)}$$

$$= \frac{-M \cdot \alpha \cdot R^2 \cdot \eta \cdot i}{-L_e \cdot M \cdot \eta^3 \cdot i - (R_e \cdot M + R \cdot M + L_e \cdot c) \cdot \eta^2 + (R_e \cdot c + K \cdot L_e + R \cdot c + R^2 \cdot \alpha^2) \cdot \eta \cdot i + R_e \cdot K + R \cdot K} \tag{10.35}$$

where $s = i\eta$; substitution of Eqs. (10.34) and (10.35) into Eqs. (10.27)–(10.29) gives the mean input mechanical and output electric power as well as energy harvesting efficiency of the EMVEH system. It is assumed that the internal resistance of the coil wire is very small, comparing with the external load resistance, that is, $R_e \ll R$ or $\frac{R_e}{R} \approx 0$, consider Eq. (4.23),

$$\begin{cases} R_N = \dfrac{R}{L_e \cdot \omega} \\ \\ \alpha_N = \sqrt{\dfrac{\alpha^2 \cdot L_e \cdot \omega}{c}} = \dfrac{B \cdot l}{R} \cdot \sqrt{\dfrac{L_e \cdot \omega}{c}} \end{cases} \tag{4.23}$$

Eq. (10.29) becomes

$$\frac{\overline{P_E}}{\left(\frac{M^2 \cdot A^2}{c}\right)} = \frac{P_h}{\left(\frac{M^2 \cdot \left|(i\omega_0)^2 Y(\omega_0)\right|^2}{c}\right)} \approx \frac{1}{2} \cdot \frac{R_N \cdot \alpha_N^2}{\left(\frac{1}{R_N}\right)^2 + \left[1 + R_N \cdot \alpha_N^2\right]^2} \tag{10.36}$$

Eq. (10.31) becomes

$$H_{P_M P_E}(0) = \eta_E = \frac{R_N \cdot \alpha_N^2}{\frac{1}{R_N^2} + 1 + R_N \cdot \alpha_N^2} \tag{10.37}$$

Eq. (10.36) and (10.37) are same as Eqs. (4.25) and (4.30). From Eq. (4.23), it is seen that when the mechanical damping c tends to be very small, close to zero, α_N tends to be very large. If R_N is kept constant, from Eq. (10.36), when α_N tends to be very large, the harvested power tends to be very small or close to zero; however, from Eq. (10.37), the energy harvesting efficiency tends to be 100%. When the mechanical damping c tends to be very large, α_N tends to be very small or close to zero. If R_N is kept constant, from Eq. (10.36), when α_N tends to be very small or close to zero, the harvested power tends to be small or close to zero; however, from Eq. (10.37), the energy harvesting efficiency tends to be very small or close to zero.

In general, the frequency response analysis of power variables has been proved to be able to calculate the mean input and harvested power and energy harvesting efficiency of PVEH and EMVEH of single and multiple degrees of freedom for white noise, finite bandwidth random and harmonic excitations. In addition, it provides a direct method to measure the mean input and harvested power and energy harvesting efficiency of PVEH and EMVEH as shown in Fig. 10.4.

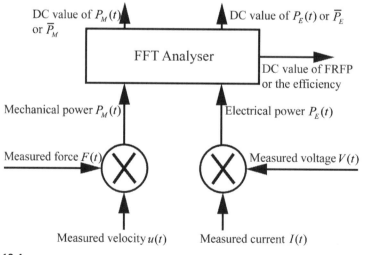

FIGURE 10.4

Schematic for the direct methods for measurement of the mean input and harvested power and energy harvesting efficiency of a vibration energy harvester. *DC*, direct current; *FFT*, fast Fourier transform; *FRFP*, frequency response function of power fluctuation.

It is seen from Fig. 10.4 that the measured force and velocity signals are passed into a multiplier box which outputs mechanical power. The measured voltage and current signals are passed into a multiplier box which outputs electric power. The mechanical and electrical power signals are fed into a two-channel FFT analyzer to measure the root mean square (RMS) values of the mechanical and electrical power signals which produces the mean input and harvested power $\left(\overline{P}_M \text{ and } \overline{P}_E\right)$ and to measure the DC value of FRFP which produces the energy harvesting efficiency $H_{P_M P_E}(0) = \eta = \frac{\overline{P}_E}{\overline{P}_M}$. The DC value of the electrical power $\overline{P}_E$ is the RMS mean harvested power. The DC value of the cross correlation function of the measurement force and measurement velocity gives the mean input power; the DC value of the cross correlation function of the measurement voltage and measurement current gives the mean harvested power. Fig. 10.4 has utilized two multiplier boxes and one FFT analyzer to directly measure the mean input and harvested power and energy harvesting efficiency of either PVEH or EMVEH, from the measured four field signals of force, velocity, voltage, and current. Alternatively, the four field signals can be measured and taken as digital data files which will be used as inputs of a Matlab program code for calculation of the DC value of FRFP and the RMS mean input and harvested power. In this case, the FFT analyzer and two-multiplier box are not required.

NOMENCLATURE

$V, V(t)$	The output voltage
$V(\eta)$	The Laplace or Fourier transform of V or $V(t)$
$I, I(t)$	The output current
$P_M, P_M(t)$	The input mechanical power
$P_E, P_E(t)$	The harvested or output electric power
η_E	The energy harvesting efficiency
$F, F(t)$	The input force
i	The square root of -1
t, t_0	Time instances
τ	The time delay
η	The integral variable or the variable of the Laplace or Fourier transform
ω	The power fluctuation frequency
$R_{FV}(\tau)$	The cross-correlation function between the input force (F) and output voltage (V)
$u, u(t)$	The driving point velocity
$R_{uI}(\tau)$	The cross-correlation function between the driving point velocity (u) and output current (I)
$R_{FI}(\tau)$	The cross-correlation function between the input force (F) and output current (I)
$R_{uV}(\tau)$	The cross-correlation function between the driving point velocity (u) and output voltage (V)
$\overline{P}_M$	The mean input mechanical power
$\overline{P}_E$	The mean output electric power
P_h	The harvested resonant power

$R_{P_M P_E}(\tau)$	The cross-correlation function between the input mechanical power (P_M) and harvested electric power P_E
R_{VI}	The cross-correlation function between the output voltage (V) and output current (I)
R_{Fu}	The cross-correlation function between the input force (F) and the driving point velocity (u)
$S_{FV}(\eta)$	The two-sided cross-spectral density function between the input force (F) and output voltage (V)
$S_{uI}(\eta)$	The two-sided cross-spectral density function between the driving point velocity (u) and output current (I)
$S_{FI}(\eta)$	The two-sided cross-spectral density function between the input force (F) and output current (I)
$S_{uV}(\eta)$	The two-sided cross-spectral density function between the driving point velocity (u) and output voltage (V)
$S_{P_M P_E}(\omega)$	The two-sided cross-spectral density function between the input mechanical power and harvested electric power
$S_{P_M P_M}(\omega)$	The autospectral density of the input mechanical power
$S_{P_E P_E}(\omega)$	The autospectral density of the harvested electric power
$S_{FF}(\eta)$	The two-sided autospectral density function of the input force
$S_{uu}(\eta)$	The two-sided autospectral density function of the driving point velocity
$S_{Fu}(\eta), S_{uF}(\eta)$	The two-sided cross-spectral density function between the input force and driving point velocity
$S_{VV}(\eta)$	The two-sided autospectral density function of the output voltage
$S_{II}(\eta)$	The two-sided autospectral density function of the output current
$S_{VI}(\eta), S_{IV}(\eta)$	The two-sided cross-spectral density function between the output voltage and current
$S_{\ddot{y}\ddot{y}}(\eta)$	The two-sided autospectral density function of the base excitation acceleration $\ddot{y}$
$H_{P_M P_E}(\omega)$	The frequency response function between the input mechanical power and the harvested electric power
$H_{P_M P_E}(0)$	The DC value of the frequency response function between the input mechanical power and the harvested electric power
$\gamma^2_{P_M P_E}(\omega)$	The coherence function
$H_{Fu}(\eta)$	The frequency response function between the input force (F) and the driving point velocity (u)
$H_{FV}(\eta)$	The frequency response function between the input force (F) and output voltage (V)
$H_{FI}(\eta)$	The frequency response function between the input force (F) and output current (I)
$H_{VI}(\eta)$	The frequency response function between the output voltage (V) and the output current (I)
$H_{uI}(\eta)$	The frequency response function between the driving point velocity (u) and the output current (I)
$H_{uV}(\eta)$	The frequency response function between the driving point velocity (u) and the output voltage (V)
$H_{\ddot{y}V}(\eta)$	The frequency response function between the excitation acceleration ($\ddot{y}$) and the output voltage (V)
$H_{\ddot{y}z}(\eta)$	The frequency response function between the excitation acceleration ($\ddot{y}$) and the relative displacement (z)

x_1, x_2, x_3, x_4	The Gaussian random variable
$E[x_1 \cdot x_2 \cdot x_3 \cdot x_4]$	The expected value of the product of four Gaussian random variables x_1, x_2, x_3, x_4
$E[x_1 \cdot x_2]$	The expected value of the product of x_1, x_2
$E[x_1 \cdot x_3]$	The expected value of the product of x_1, x_3
$E[x_1 \cdot x_4]$	The expected value of the product of x_1, x_4
$E[x_2 \cdot x_3]$	The expected value of the product of x_2, x_3
$E[x_2 \cdot x_4]$	The expected value of the product of x_2, x_4
$E[x_3 \cdot x_4]$	The expected value of the product of x_3, x_4
μ_{x_i}	The average value of the x_i, $i = 1, 2, 3, 4$
π	3.1415927
A	The amplitude of a harmonic excitation frequency
ω_0	The radial frequency of a harmonic excitation acceleration in rad/s
S_0	The auto power spectrum amplitude of the input excitation acceleration
M	Oscillator mass of a structure or a machine or of a single degree of freedom spring-mass-dashpot system
$\ddot{y}$	The acceleration of an excitation base
$\dot{y}$	The velocity of an excitation base
y	The displacement of an excitation base
$Y, Y(\eta), Y(\omega_0)$	The Laplace or Fourier transform of y
$\dot{z}$	The relative velocity of the driving point with respect to the base
z	The relative displacement of the driving point with respect to the base
$Z, Z(\eta)$	The Laplace or Fourier transform of z
ω_1	The lower limit of the narrow frequency band of the input excitation acceleration
ω_2	The upper limit of the narrow frequency band of the input excitation acceleration
$\Delta\omega$	The bandwidth of the narrow frequency band of the input excitation acceleration
ω_0	The center frequency of the narrow frequency band of the input excitation acceleration
R_N	The dimensionless resistance
α_N	The dimensionless force factor
R	The sum of the external load resistance
R_c	The resistance of the coil
L_c	Self-inductance of the coil
c	The short circuit mechanical damping of the single degree of freedom system
C_0	Blocking capacity of the piezoelectric material insert
K	The short circuit stiffness of the single degree of freedom system
∞	Infinity
$\int_{-\infty}^{\infty} d\eta$	The integration from $-\infty$ to ∞
$\int_{\omega_1}^{\omega_2} d\eta$	The integration from ω_1 to ω_2

Superscripts

$\cdot$	The first differential
$\cdot\cdot$	The second differential
$*$	Complex conjugate
$-$	Time average

Special Functions

$\|\ \|$	Modulus or absolute value
$\langle\ \rangle$	Time averaged
$Re[\ ]$	The real part of a complex variable

Abbreviations

FRF	Frequency response function.
FRFP	Frequency response function of power fluctuation.
DC	Direct current.
AC	Alternative current.
RMS	Root mean square.
FFT	Fast Fourier transform.
SDOF	Single degree of freedom.

REFERENCES

[1] Xiao H, Wang X. A review of piezoelectric vibration energy harvesting techniques. Int Rev Mech Eng 2014;8(3):609−20. ISSN: 1970-8742.

[2] Wang X. Coupling loss factor of linear vibration energy harvesting systems in a framework of statistical energy analysis. J Sound Vib 2015;362:125−41.

[3] Wang X, Liang XY, Shu GQ, Watkins S. Coupling analysis of linear vibration energy harvesting systems. Mech Syst Signal Process 2016;70−71:428−44.

[4] Wang X, Liang XY, Hao ZY, Du HP, Zhang N, Mao Q. Comparison of electromagnetic and piezoelectric vibration energy harvesters with different interface circuits. Mech Syst Signal Process 2016;72−73:906−24.

[5] Wang X, John S, Watkins S, Yu XH, Xiao H, Liang XY, et al. Similarity and duality of electromagnetic and piezoelectric vibration energy harvesters. Mech Syst Signal Process 2015;52−53:672−84.

[6] Wang X, Liang XY, Wei HQ. A study of electromagnetic vibration energy harvesters with different interface circuits. Mech Syst Signal Process 2015;58−59:376−98.

[7] Wang X, Xiao H. Dimensionless analysis and optimization of piezoelectric vibration energy harvester. Int Rev Mech Eng 2013;7(4):607−24. ISSN: 1970-8742.

[8] Wang X, Lin LW. Dimensionless optimization of piezoelectric vibration energy harvesters with different interface circuits. Smart Mater Struct 2013;22:1−20. ISSN: 0964-1726.

[9] Xiao H, Wang X, John S. A multi degree of freedom piezoelectric vibration energy harvester with piezoelectric elements inserted between two nearby oscillators. Mech Syst Signal Process 2016;68−69:138−54.

[10] Xiao H, Wang X, John S. A dimensionless analysis of a 2DOF piezoelectric vibration energy harvester. Mech Syst Signal Process 2015;58−59:355−75.

[11] Koss LL. Correlation of power variables. J Sound Vib 1987;115(2):275−87.

[12] Koss LL. Correlation of power variables—frequency domain. J Sound Vib 1988;125(3): 511−22.

[13] Wang X, Koss LL. Frequency response functions for structural intensity. J Sound Vib 1995;185(2):299−334. ISSN: 0022-460X.

[14] Wang X, Koss LL. Frequency response function for power, power transfer ratio and coherence. J Sound Vib 1992;155(1):55−73. ISSN: 0022-460X.
[15] Bendat JS, Piersol AG. Random data: analysis and measurement procedures. 1986.
[16] Bawahab M, Xiao H, Wang X. A study of linear regenerative electromagnetic shock absorber system. 2015. SAE 2015-01-0045.

Ocean Wave Energy Conversion Analysis

11

CHAPTER OUTLINE

11.1 INTRODUCTION

The quest for new technologies to solve growing environmental problems, including the imminent energy shortage, and the high cost of energy and new power plants has been a scientific concern over the last three decades. Recently, international interest in utilizing sustainable and renewable energy resources such as wind, solar, biomass, and hydro has increased sharply. The main factors stimulating research and development activities in this area include concerns about fossil fuels, energy independence, and environmental protection. In this respect, ocean wave energy is considered as one of the renewable energy sources with a huge potential all around the world. Ocean wave carries both kinetic and gravitational potential energy. The total energy of an ocean wave depends roughly on two factors: its height (H) and its period (T). The power carried by the wave is proportional to the height square H^2 and to the period T, and is usually given in watt per meter of incident wave front. For example, the coastline of Western Europe is "blessed" with an average wave climate of about 50 kW of power for each meter width of wave front. The overall resource is of the same order of magnitude as the world's electricity consumption. A conservative estimate is that it is possible to extract 10–25% of this, suggesting that wave power could make a significant contribution to the energy mix.

A lot of kinds of apparatuses have been designed to scavenge wave energy by different ways, ie, underwater turbines, floats, buoys, pitching devices, oscillating water column (OWC), point absorber (PA), tidal barrage or dam, attenuators, tapered channel or overtopping device, pendulum, rubber hose (Anaconda, snake), salter duck, tidal bridges, and fences. Wave energy converters are commonly adopted in the forms of hydraulic actuation type of Pelamis [1,2], the buoy-type PA of linear electromagnetic generator, Archimedes Wave Swing (AWS) [3,4], rotational turbine type such as OWCs, wave dragon (WD) [5] as shown in Figs. 11.1–11.3 separately. OWC and WD convert the wave motion into air flow or fluid flow which drives a rotational electromagnetic generator through turbine. Pelamis consists of large, articulated cylinders that are partially submerged and joined by hinges. The wave motion creates movement between these sections, activating a hydraulic system connected to a rotational electromagnetic generator. PA or AWS is a floating structure capable of absorbing wave energy from any direction. The device moves up and down with the movement of the waves, and this movement is then converted into electric energy through a linear electromagnetic generator. The idea of the direct drive through linear electromagnetic generator is appealing since it enables simple systems with few intermediate conversion steps and reduces mechanical complexity, but the requirements on the generator increase. Those requirements are to a great extent determined by the nature of ocean waves. The generator output also needs to be connected to the electric grid, which must be taken into account in the design specification. Linear electromagnetic generator has appeared as one of the most commonly used power take-off devices in PA or AWS wave energy converters [8].

The ocean wave energy converters have been investigated for decades; they can provide power scales on the order of 100 kW and beyond with the frequency range

FIGURE 11.1

Hydraulic actuation type of Pelamis [2].

Courtesy of R. Barros.

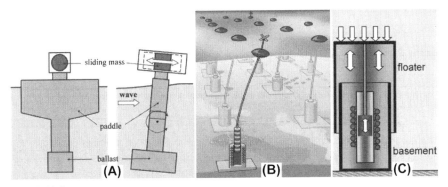

FIGURE 11.2

Wave energy converter types: (A) the PS Frog point absorber; (B) heaving buoy point absorber; (C) archimedes wave swing [2,6].

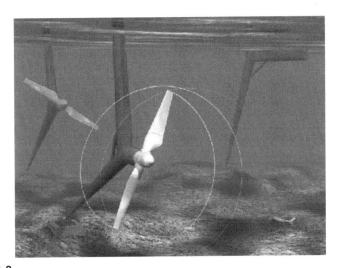

FIGURE 11.3

Rotational turbine type [7].

Tidal Stream. http://www.tidalstream.co.uk/Development/Background/background.html

from 0.075 to 0.2 Hz. The power characteristics of linear electromagnetic generators are very attractive, as the generators have low output impedance, adjustable but high current. However these advantages are offset by the drawback of very low voltage [9].

The requirements for an energy harvester are high power output, efficiency, and high durability. Most of the common linear mass-spring-damper systems are poorly suitable for energy harvesting from ocean wave motions because the output power of a linear electromagnetic harvester drops dramatically under off-resonance conditions [10]. Elliott [11] illustrated that a large size of single degree of freedom (SDOF) linear electromagnetic dynamic energy harvester will achieve a large energy harvesting efficiency. Foisal et al. [12] presented an array of harvesters with magnetic spring and demonstrated the possibility of harvesting energy from vibration of multiple frequencies. A similar electromagnetic vibration energy harvester using multi-pole magnet was studied by the same group [13]. To be more accurate, the spatial variation of the magnetic field with respect to the spatial dimension along the motion direction of the magnetic translator should be considered rather than the approximate approach of treating the magnetic field of a constant value. Cepnik [14] used the spatial variation of the magnetic field for calculating the voltage. Halbach [15] proposed a special arrangement of polarities of magnets which gives very strong magnetic field intensity on one side, but very weak magnetic field intensity on the other side.

Shahosseini and Najafi [16] applied a cylindrical Halbach magnet array to an electromagnetic vibration energy harvester. Zhu et al. [17] applied a Halbach array to a planar electromagnetic vibration energy harvester. Cheung [18] proposed to apply ferrofluid bearings between the axial oriented cylindrical magnets and tube surface of nonmagnetic material to reduce the sliding friction. This design had developed frictionless oscillators and enhanced vibration energy harvesting performance. Cheung [18] proposed to insert a washer between two magnets of opposite oriented polarities to increase local magnetic flux change rate. It was suggested that two coils should be placed in a distance of the oscillator length to ensure one pass in a single sweep from the area of highest magnetic field strength to the area of lowest field strength.

A number of intelligent strategies have been adopted in the recent years to improve the performance of the buoy-type wave energy harvester with linear electromagnetic generators. While some target the increase in voltage, others contribute in broadening the operating frequency range and maximizing the power output. In regular waves, energy is captured most efficiently in a PA-type wave energy converter (WEC) with an SDOF linear oscillator when the undamped natural frequency of the device is close to the dominant frequency of the incident wave [19]. At resonance, the velocity of the oscillator is in phase with the dynamic pressure (and hence force) of the incoming wave, resulting in a substantial transfer of energy from the wave to the oscillator [20]. However, the ocean wave period is varied in a range, and does not always coincide with the natural period of the oscillator of the PA-type WEC. When the wave period is longer than the natural period of the oscillator, using half the difference between the wave period and the natural period of the oscillator as the latching duration would be a good approximation of the latching delay [20]. In latching control, the phase of the oscillator displacement can be continuously controlled by a hold-and-release mechanism to achieve zero phase discrepancy between velocity of the oscillator and the wave force. Latching control can improve wave energy conversion performance and frequency bandwidth of a linear oscillator, although it is discrete, highly nonlinear, and by its nature suboptimal. The challenge with the strategy is determining the optimum time to release the buoy from the latched phase, as this is the control variable. Reactive control can also be used in the PA-type WEC to tune the natural frequency of the oscillator toward the variable wave frequency to increase wave energy conversion efficiency and frequency bandwidth [21]. By choosing power take-off parameters such as power-absorbing damping coefficient, spring coefficient, and inertia coefficient at each frequency, one can improve energy absorption at that frequency. Both the latching and reactive controls have limited improvement for the performance of the buoy-type PA WEC.

Widening the operating frequency range of energy harvesters is another way of maximizing the power. This is because the ocean wave source consists of random wave motions characterized by a certain range of frequencies. A two degrees of freedom linear electromagnetic dynamic energy harvester was studied by Tang [22] where a dual mass system—based linear electromagnetic vibration energy harvesting system was demonstrated to be better than an SDOF system. Gatti [23] concluded that the harvested vibration power and frequency bandwidth of multiple degrees of freedom (MDOF) oscillators are larger than those of SDOF oscillator for an electromagnetic vibration energy harvester.

A number of nonlinearization methods have been reported recently for achieving broadband resonances of the harvester oscillator. The nonlinear methods to design energy harvesting device can be divided into two catalogs:

1. To replace a linear resonator with a nonlinear resonator, this approach hardens the frequency response to the larger frequency range under large amplitude excitation.
2. To replace a mono-stable system with a bistable system which is pursued by designing a system to have two wells between which their potentials can switch subject to periodic or stochastic ambient excitations. The main advantages of a nonlinear oscillator for the wave energy conversion could be summarized as:
   a. Increased harvesting capability for the low-frequency ocean wave energy source.
   b. The capability of handling high-level periodic wave forces.
   c. The capability of automatically adapting the ocean wave energy source frequency after initial tuning.
   d. The capability of harvesting energy from stochastic wave excitation.
   e. The capability of tuning the resonant frequency of a wave energy conversion system without additional energy input.

Harne and Wang [24] presented a review which covered recent research efforts on bistable systems, for which the bistable energy harvesting technique was introduced. Beeby et al. [25] compared linear and nonlinear vibration energy harvesting technology based on real vibration data. They highlighted the importance of designing or selecting the most suitable vibration energy harvester according to the characteristics of ambient vibration source. It was found that a linear vibration energy harvester has the highest power output in most cases, while a nonlinear energy harvester has a wider harvesting frequency bandwidth, and the bistable technology can extract more electrical energy from white noise (random) vibration. Owens [26] suggested that nonlinear coupling was better than linear coupling for broadening the frequency response bandwidth. A theoretical analysis of the nonlinear Duffing oscillator using magnetic levitation was investigated by Mann and Sims [27]. Dallago et al. [28] took into account the nonlinear effect and built an analytical model of the magnetic spring–based energy harvester. Experimental results have validated the model which is able to predict the voltage performance of the harvester. Barton [29] validated the use of highly permeable NdFeB magnets characterized by nonlinear magnetic properties. Using nonlinear springs is one of the options for tuning of the wideband energy harvesting.

Three major research challenges for the buoy-type WEC with a linear electromagnetic generator are summarized from the above literature study:

1. Low-frequency problem—A significant challenge is the conversion of the slow (~0.2 Hz), random, and high-force oscillatory motion into useful electric energy through a generator with output quality acceptable to the utility network. To produce smooth electrical output, some type of energy conversion and conditioning circuits, or other means of compensation such as an array of devices are essential [32].
2. Efficiency problem—The converted resonant power and efficiency of the buoy-type WEC with light damping MDOF oscillators are expected to be larger than those with a conventional SDOF oscillator. The magnet and coil stacking patterns should be designed to maximize magnetic field intensity and its change rate around the coils.
3. Bandwidth problem—To widen the frequency bandwidth of the buoy-type WEC. The adoption of light damping MDOF nonlinear oscillators is expected to have larger operational frequency bandwidth than that of a conventional SDOF linear oscillator.

This chapter first proposes a concept of a cylindrical tube generator where a cylindrical magnet oscillator slides inside the tube passing the coils wired outside the tube. This chapter will study the SDOF nonlinear oscillator in a cylindrical tube generator for a feasible design methodology of the cylindrical tube generator.

This chapter will address the above three research challenges by developing theoretical analysis and simulation methods. A continuous collection and transfer of electric energy converted from ocean wave energy to the electric grid for a long period would lead to a sustainable and renewable energy supply source.

11.2 ANALYSIS OF A SINGLE DEGREE OF FREEDOM NONLINEAR OSCILLATOR IN A CYLINDRICAL TUBE GENERATOR USING THE TIME DOMAIN INTEGRATION METHOD

The equation that describes the dynamics of a general nonlinear oscillator can be written as:

$$m \cdot \ddot{z} + c \cdot \dot{z} + \frac{\mathrm{d}U(z)}{\mathrm{d}z} + \alpha \cdot v = -m \cdot \ddot{y} \qquad (11.1)$$

where m is the mass of the oscillator; c is the damping coefficient; α is the equivalent force factor, z is the relative displacement of the oscillator with respect to the base; y is the base displacement; $U(z)$ is the potential energy of the spring element. The nonlinearity of Eq. (11.1) could be caused by the following three types of nonlinearity: (1) spring force nonlinearity, (2) damping force nonlinearity, and (3) electromagnetic force nonlinearity. If the nonlinearity of Eq. (11.1) is caused by the spring force nonlinearity, there is one condition with a nonlinear oscillator that is different from a linear one, that is

$$U(z) \neq \frac{1}{2} \cdot k \cdot z^2 \tag{11.2}$$

where k is the spring stiffness coefficient of the linear displacement term of the oscillator. This means that the potential energy of a nonlinear oscillator is not proportional to a quadratic of the displacement. For the potential energy function $U(x)$, there are some expressions reported in the literature [30,31], one of which is given by

$$U(z) = \sum_{n=1}^{n} k_{2n-1} \cdot z^{2n} \tag{11.3}$$

where k_{2n-1} is the spring stiffness coefficient of linear or nonlinear displacement term or the potential energy coefficient of the nonlinear oscillator. For a duffing-type oscillator, the potential energy function can be defined as:

$$U(z) = \frac{1}{2} \cdot k_1 \cdot z^2 + \frac{1}{4} \cdot k_3 \cdot z^4 \tag{11.4}$$

where k_1 is the spring stiffness coefficient of the linear displacement term; k_3 is the spring stiffness coefficient of nonlinear displacement term. k_1 and k_3 are the potential energy coefficients of the nonlinear oscillator.

A typical cylindrical tube generator of an SDOF oscillator is suspended by two long strings and excited by a shaker or a sway push and release as shown in Fig. 11.4 where a cylindrical oscillator of properly stacked magnets slides freely inside the tube, and two coils are wired on the outside surface of the tube and separated by a distance of the oscillator axial length. There are two fixed magnets in the two end caps of the tube which have opposite polarities to those of the oscillator magnets, the magnetic fields between the magnets in the end caps and the oscillator magnets act as nonlinear magnetic springs for the nonlinear oscillator.

The displacement of the tube is assumed to be y, which is the displacement amplitude of the shaker excitation. The displacement of the oscillator is assumed to be x, the relative displacement of the oscillator with respect to the tube is then equal to $x-y$. For the study object of the oscillator, it is subjected to the elastic restoring force of the magnetic spring $F_k = -k_1 \cdot (x-y) - k_3 \cdot (x-y)^3$, the damping force of the magnetic spring is $F_c = -c \cdot (\dot{x} - \dot{y})$, and the electromagnetic force from the coils wired on the outside surface of the tube carrying current is $F_e = -Bl \cdot I$.

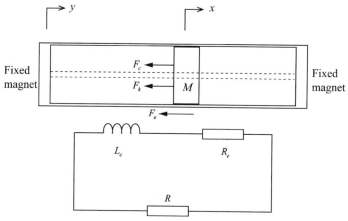

FIGURE 11.4

A single degree of freedom nonlinear oscillator in a cylindrical tube generator.

From the Newton's second law, the dynamic equation of the duffing-type oscillator is given by

$$m \cdot \ddot{x} + k_1 \cdot (x - y) + k_3 \cdot (x - y)^3 + c \cdot (\dot{x} - \dot{y}) + Bl \cdot I = 0 \qquad (11.5)$$

where m is the mass of the oscillator, k_1 is the linear spring constant, k_3 is the nonlinear spring constant, c is the linear damping coefficient, B is the magnetic flux density, I is the current in the coils, l is the total length of the coils where $l = \pi \cdot D_0 \cdot N$. N is the wiring number of each coil; D_0 is the outer diameter of the tube. The coils are connected in parallel to an external resistance R. The parallel connected coils have an internal resistance of R_e and an inductance of L_e. The dynamic equation of the circuit of the coils is given by

$$V + L_e \cdot \frac{dI}{dt} - Bl \cdot (\dot{x} - \dot{y}) = 0 \qquad (11.6)$$

where $V = IR$ is the voltage across the two output terminals of the circuit. Eqs. (11.5) and (11.6) can be solved by using the integration method introduced in Section 1.2 of Chapter 1. Eqs. (11.5) and (11.6) can be written as

$$\begin{cases} (\ddot{x} - \ddot{y}) = -\dfrac{c}{m} \cdot (\dot{x} - \dot{y}) - \dfrac{k_1}{m} \cdot (x - y) - \dfrac{k_3}{m} \cdot (x - y)^3 - \ddot{y} - \dfrac{Bl}{m \cdot (R + R_e)} \cdot V \\[3mm] \dfrac{dV}{dt} = -\dfrac{R + R_e}{L_e} \cdot V + \dfrac{Bl \cdot (R + R_e)}{L_e} \cdot (\dot{x} - \dot{y}) \end{cases}$$

$$(11.7)$$

Eq. (11.7) can be wired and programmed as a code in Matlab Simulink as shown in Fig. 11.5. For a given excitation acceleration of the tube in a sine wave with a frequency and amplitude, the output time response of the relative displacement of the

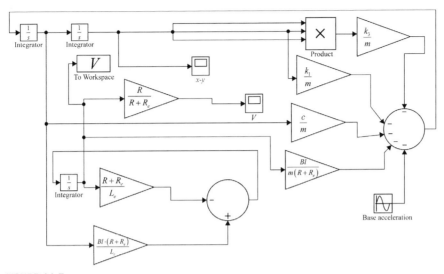

FIGURE 11.5

Simulation schematic for Eq. (11.7) for prediction of the oscillator relative displacement response $(x\text{-}y)$ and output voltage v from a sine wave base excitation acceleration input $\ddot{y}$ at a frequency using Matlab Simulink time domain integration method.

oscillator with respect to the tube and the voltage of the two terminals of the circuit can be predicted. The parameters of the tube system such as m, k_1, k_3, c, B, l, and R can be identified and measured from experiments.

It is seen from Fig. 11.5 that the base acceleration $\ddot{y}$ is fed as a sine wave into the system, the time trace outputs of the relative displacement response $(x\text{-}y)$ and output voltage v can be solved and scoped. With inputs of different excitation frequencies and amplitude of the tube acceleration $\ddot{y}$, the output voltage and power can be calculated. In the above calculations, the inductance L_e can be calculated by

$$L_e = \pi \cdot \mu_0 \cdot \mu_r \cdot \frac{N^2 \cdot D_0^2}{4 \cdot h_c} \tag{11.8}$$

where $\mu_0 = 4\pi \times 10^{-7}$ N/m^2 is the permeability of the coil with air core; μ_r is the permeability coefficient of the coil, for the iron core, $\mu_r = 1450$, for the air core, $\mu_r = 1$; N is the wiring number per coil; D_0 is the diameter of the tube; h_c is the height of each coil. The magnetic flux density can be either measured by experiments using Gaussian meter or calculated by simulation using the ANSYS Maxwell & Simplorer and Electromagnetics, which is based on the average magnetic flux density of the multiple points in the magnetic field around the coil. With inputs of different external resistance R, inductance L_e, the magnetic flux density B, and total length of the coils l, the output voltage and power can be calculated and optimized.

Since Eq. (11.5) is a nonlinear equation, Eqs. (11.5) and (11.6) are not able to be solved using the transfer function and state space analysis methods introduced in Chapter 1. Eqs. (11.5) and (11.6) can be solved using the time domain integration method introduced in Chapter 1 and harmonic balance method which is illustrated below.

11.3 ANALYSIS OF A SINGLE DEGREE OF FREEDOM NONLINEAR OSCILLATOR IN A CYLINDRICAL TUBE GENERATOR USING THE HARMONIC BALANCE METHOD

To solve Eqs. (11.5) and (11.6), it is assumed that

$$z = z_0 \cdot \cos(\omega t) \quad V = V_0 \cdot \cos(\omega t + \phi_1) \quad y = y_0 \cdot \cos(\omega t + \phi_2) \tag{11.9}$$

where $z = x - y$ is the relative displacement of the magnet oscillator with respect to the tube, z_0 is the amplitude of z, ω is the excitation frequency of the tube, t is the time variable; V is the output voltage of the coils connected in parallel to an external resistance R, V_0 is the amplitude of V, φ_1 is the phase difference between the output voltage V and the relative displacement z; y is the excitation displacement of the base (tube), y_0 is the amplitude of y, φ_2 is the phase difference between the tube displacement y and the relative displacement of the oscillator z. Substituting Eq. (11.9) into Eq. (11.6) gives

$$V_0 \cdot \cos(\omega t + \varphi_1) - \frac{L_e}{R} \cdot V_0 \cdot \omega \cdot \sin(\omega t + \varphi_1) + \omega \cdot Bl \cdot z_0 \cdot \sin \omega t = 0 \tag{11.10}$$

Let

$$\begin{cases} \cos \theta = \cos \varphi_1 = \dfrac{\dfrac{L_e}{R} \cdot \omega}{\sqrt{\left(\dfrac{L_e}{R} \cdot \omega\right)^2 + 1}} \\[4mm] \sin \theta = \sin \varphi_1 = \dfrac{1}{\sqrt{\left(\dfrac{L_e}{R} \cdot \omega\right)^2 + 1}} \end{cases} \tag{11.11}$$

Eq. (11.10) becomes

$$V_0 \cdot \sqrt{\left(\frac{L_e}{R} \cdot \omega\right)^2 + 1} \cdot \sin(\omega t + \varphi_1 - \theta) = Bl \cdot \omega \cdot z_0 \cdot \sin \omega t \tag{11.12}$$

where

$$V_0 = \frac{Bl \cdot \omega \cdot z_0}{\sqrt{\left(\frac{L_e}{R} \cdot \omega\right)^2 + 1}} \tag{11.13}$$

It is assumed that

$$A = \omega^2 \cdot y_0 \quad \alpha = \frac{Bl}{R} \tag{11.14}$$

Substituting Eq. (11.9) into Eq. (11.5) gives

$$-m \cdot z_0 \cdot \omega^2 \cdot \cos \omega t - c \cdot \omega \cdot z_0 \cdot \sin \omega t + k_1 \cdot z_0 \cdot \cos \omega t + k_3 \cdot z_0^3 \cdot \cos^3 \omega t$$
$$= -\alpha \cdot V_0 \cdot \cos(\omega t + \varphi_1) + m \cdot A \cdot \cos(\omega t + \varphi_2)$$

which is arranged as

$$\left(-m \cdot z_0 \cdot \omega^2 + k_1 \cdot z_0 + \frac{3}{4} \cdot k_3 \cdot z_0^3 + \frac{\alpha \cdot V_0 \cdot \frac{L_e}{R} \cdot \omega}{\sqrt{\left(\frac{L_e}{R} \cdot \omega\right)^2 + 1}} \right)$$

$$\cdot \cos \omega t - \left(\frac{\alpha \cdot V_0}{\sqrt{\left(\frac{L_e}{R} \cdot \omega\right)^2 + 1}} + c \cdot \omega \cdot z_0 \right) \cdot \sin \omega t \tag{11.15}$$

$$= m \cdot A \cdot \cos \omega t \cdot \cos \varphi_2 - m \cdot A \cdot \sin \omega t \cdot \sin \varphi_2$$

From Eq. (11.15), it gives

$$\left(-m \cdot z_0 \cdot \omega^2 + k_1 \cdot z_0 + \frac{3}{4} \cdot k_3 \cdot z_0^3 + \frac{\alpha \cdot V_0 \cdot \frac{L_e}{R} \cdot \omega}{\sqrt{\left(\frac{L_e}{R} \cdot \omega\right)^2 + 1}} \right)^2$$

$$+ \left(\frac{\alpha \cdot V_0}{\sqrt{\left(\frac{L_e}{R} \cdot \omega\right)^2 + 1}} + c \cdot \omega \cdot z_0 \right)^2 = m^2 \cdot A^2 \tag{11.16}$$

$$\begin{cases} \cos \varphi_2 = \dfrac{1}{m \cdot A} \left(-m \cdot z_0 \cdot \omega^2 + k_1 \cdot z_0 + \dfrac{3}{4} \cdot k_3 \cdot z_0^3 + \dfrac{\alpha \cdot V_0 \cdot \dfrac{L_e}{R} \cdot \omega}{\sqrt{\left(\dfrac{L_e}{R} \cdot \omega\right)^2 + 1}} \right) \\[4ex] \sin \varphi_2 = \dfrac{1}{m \cdot A} \left(\dfrac{\alpha \cdot V_0}{\sqrt{\left(\dfrac{L_e}{R} \cdot \omega\right)^2 + 1}} + c \cdot \omega \cdot z_0 \right) \end{cases} \tag{11.17}$$

From Eq. (11.16), it derives

$$z_0^2 = \frac{A^2}{\left(\omega^2 - \frac{k_1}{m} - \frac{3}{4}\cdot\frac{k_3}{m}\cdot z_0^2 - \frac{L_e\cdot B^2 l^2\cdot\omega^2}{\left(L_e^2\cdot\omega^2+R^2\right)\cdot m}\right)^2 + \left(\frac{B^2 l^2\cdot R\cdot\omega}{\left(L_e^2\cdot\omega^2+R^2\right)\cdot m} + \frac{c}{m}\cdot\omega\right)^2} \tag{11.18}$$

It is assumed that

$$\omega_N = \sqrt{\frac{m}{k_1}}\cdot\omega; \quad \xi = \frac{c}{2\cdot\sqrt{m\cdot k_1}}; \quad M^2 = \frac{k_3}{k_1}\cdot z_0^2; \quad ff^2 = \frac{m^2\cdot k_3}{k_1^3}\cdot A^2 = \frac{m^2\cdot k_3}{k_1^3}\cdot\omega^4\cdot y_0^2;$$

$$\sigma = \frac{B^2 l^2}{k_1}; \quad k_R = \sqrt{\frac{m}{k_1}}\cdot R; \quad \beta = \frac{k_3}{k_1}$$

$$\tag{11.19}$$

From Eq. (4.23), it gives

$$\begin{cases} R_N = \dfrac{R}{L_e\cdot\omega} \\[3mm] \alpha_N = \sqrt{\dfrac{\alpha^2\cdot L_e\cdot\omega}{c}} = \dfrac{B\cdot l}{R}\cdot\sqrt{\dfrac{L_e\cdot\omega}{c}} \end{cases} \tag{11.20}$$

Substituting Eqs. (11.19) and (11.20) into Eq. (11.18) gives

$$M^2 = \frac{ff^2}{\left(\omega_N^2 - 1 - \frac{3}{4}\cdot M^2 - \frac{2\cdot R_N^2\cdot\alpha_N^2\cdot\omega_N\cdot\xi}{\left(1+R_N^2\right)}\right)^2 + \left(\frac{2\cdot R_N^3\cdot\alpha_N^2\cdot\omega_N\cdot\xi}{\left(1+R_N^2\right)} + 2\cdot\omega_N\cdot\xi\right)^2} \tag{11.21}$$

where M is the dimensionless relative displacement response amplitude of the system. It is noted that Eq. (11.21) is cubic in M^2, thus there are three or one real root(s) for a given frequency.

To find the two loci of the jump points, Eq. (11.21) is written as

$$M^2\left\{\left(\alpha_N^2 - 1 - \frac{3}{4}\cdot M^2 - \frac{2\cdot R_N^2\cdot\alpha_N^2\cdot\omega_N\cdot\xi}{1+R_N^2}\right)^2 \right. $$
$$\left. + \left(\frac{2\cdot R_N^3\cdot\alpha_N^2\cdot\omega_N\cdot\xi}{1+R_N^2} + 2\cdot\omega_N\cdot\xi\right)^2\right\} = ff^2 \tag{11.22}$$

Differentiate the both sides of Eq. (11.22) with respect to frequency ω_N with ξ assumed as constants, and when $\frac{\partial M}{\partial\omega_N}\to\infty$, it gives

$$\frac{27}{8}\cdot M^4 - 6\cdot M^2\cdot\left(-\frac{2\cdot R_N^2\cdot\alpha_N^2\cdot\omega_N\cdot\xi}{1+R_N^2} + \omega_N^2 - 1\right) + 2\cdot\left(-\frac{2\cdot R_N^2\cdot\alpha_N^2\cdot\omega_N\cdot\xi}{1+R_N^2} + \omega_N^2 - 1\right)^2$$

$$+ 2\cdot\left(\frac{2\cdot R_N^3\cdot\alpha_N^2\cdot\omega_N\cdot\xi}{1+R_N^2} + 2\cdot\omega_N\cdot\xi\right)^2 = 0$$

$$\tag{11.23}$$

Two roots of Eq. (11.23) are given by

$$M^2 = 8 \cdot \left(-\frac{2 \cdot R_N^2 \cdot \alpha_N^2 \cdot \omega_N \cdot \xi}{1 + R_N^2} + \omega_N^2 - 1 \right)$$

$$\pm 4 \cdot \frac{\sqrt{\left(-\frac{2 \cdot R_N^2 \cdot \alpha_N^2 \cdot \omega_N \cdot \xi}{1+R_N^2} + \omega_N^2 - 1 \right)^2 - 3 \cdot \left(\frac{2 \cdot R_N^3 \cdot \alpha_N^2 \cdot \omega_N \cdot \xi}{1+R_N^2} + 2 \cdot \omega_N \cdot \xi \right)^2}}{9} \tag{11.24}$$

Eq. (11.24) describes the two loci of the jump points in the frequency–response curve. Therefore, for a given damping and given excitation amplitude, if there are three solutions for Eq. (11.21), Eq. (11.24) separates the loci of frequency–response curve into three sections. From Eqs. (11.14), (11.18)–(11.20), it gives

$$\frac{z_0}{y_0} = \frac{\omega_N^2}{\sqrt{\left[\left(\frac{2 \cdot R_N^2 \cdot \alpha_N^2 \cdot \omega_N \cdot \xi}{1+R_N^2} - \omega_N^2 \right)^2 + 1 + \frac{3}{4} \cdot M^2 \right]^2 + \left(\frac{2 \cdot R_N^3 \cdot \alpha_N^2 \cdot \omega_N \cdot \xi}{1+R_N^2} + 2 \cdot \omega_N \cdot \xi \right)^2}} \tag{11.25}$$

From Eqs. (11.13) and (11.25), it gives

$$\frac{V_0}{Bl \cdot \omega \cdot y_0} = \frac{1}{\sqrt{1 + R_N^2}}$$

$$\cdot \frac{\omega_N^2}{\sqrt{\left[\left(\frac{2 \cdot R_N^2 \cdot \alpha_N^2 \cdot \omega_N \cdot \xi}{1+R_N^2} - \omega_N^2 \right)^2 + 1 + \frac{3}{4} \cdot M^2 \right]^2 + \left(\frac{2 \cdot R_N^3 \cdot \alpha_N^2 \cdot \omega_N \cdot \xi}{1+R_N^2} + 2 \cdot \omega_N \cdot \xi \right)^2}} \tag{11.26}$$

and the converted resonant power is given by

$$\frac{P_h}{\left(\frac{m^2 \cdot \omega^4 \cdot y_0^2}{c} \right)} = \frac{4 \cdot \xi^2 \cdot R_N \cdot \alpha_N^2 \cdot \omega_N^2}{1 + R_N^2}$$

$$\cdot \frac{1}{\left[\left(\frac{2 \cdot R_N^2 \cdot \alpha_N^2 \cdot \omega_N \cdot \xi}{1+R_N^2} - \omega_N^2 \right)^2 + 1 + \frac{3}{4} \cdot M^2 \right]^2 + \left(\frac{2 \cdot R_N^3 \cdot \alpha_N^2 \cdot \omega_N \cdot \xi}{1+R_N^2} + 2 \cdot \omega_N \cdot \xi \right)^2} \tag{11.27}$$

Dimensionless converted resonant power for the nonlinear oscillator as shown in Eq. (11.27) is comparable to those for the linear electromagnetic and piezoelectric oscillators as shown in Eqs. (2.23) and (4.44). For the nonlinear oscillator, in addition to the dimensionless control variables of R_N and α_N, the mechanical damping ratio ξ and dimensionless relative displacement of the oscillator M are also control variables of the dimensionless converted resonant power which is different from that of the linear oscillator as illustrated in Eq. (2.23) and (4.44).

11.4 ANALYSIS OF A SINGLE DEGREE OF FREEDOM NONLINEAR OSCILLATOR IN A CYLINDRICAL TUBE GENERATOR USING THE PERTURBATION METHOD

The above cylindrical tube generator can also be studied using the perturbation method. Eqs. (11.5) and (11.6) can be combined into the following equation:

$$
\begin{cases}
m \cdot \ddot{z} + c \cdot \dot{z} + k_1 \cdot z + k_3 \cdot z^3 = -Bl \cdot \dfrac{V}{R} - m \cdot \ddot{y} \\[2mm]
V + \dfrac{L_e}{R} \cdot \dfrac{dV}{dt} - Bl \cdot \dot{z} = 0
\end{cases}
\tag{11.28}
$$

It is assumed that

$$
\omega_N = \frac{\omega}{\eta}, \quad \tau = \eta \cdot t, \quad \eta = \sqrt{\frac{k_1}{m}}
\tag{11.29}
$$

The base excitation is assumed as

$$
y = y_0 \cdot \sin(\omega \cdot t), \quad \ddot{y} = -y_0 \cdot \omega^2 \sin(\omega \cdot t) = -A \cdot \sin(\omega \cdot t)
\tag{11.30}
$$

Replacing variable t by τ in Eq. (11.28) gives

$$
\begin{cases}
\ddot{z} + \dfrac{c}{\eta \cdot m} \cdot \dot{z} + \dfrac{k_1}{\eta^2 \cdot m} \cdot z + \dfrac{k_3}{\eta^2 \cdot m} \cdot z^3 + \dfrac{Bl}{\eta^2 \cdot m \cdot R} \cdot V = \dfrac{A}{\eta^2} \cdot \sin(\omega_N \tau) \\[3mm]
\dfrac{dI}{d\tau} + \dfrac{R}{\eta \cdot L_e} \cdot I - \dfrac{Bl}{L_e} \cdot \dot{z} = 0
\end{cases}
\tag{11.31}
$$

It is assumed that

$$
z = l_m \cdot Z, \quad y = l_m \cdot Y, \quad I = i_m \cdot I_0,
\tag{11.32}
$$

where I is the current; i_m is a threshold or reference current; I_0 is a dimensionless current; l_m is a relevant physical dimension, such as the maximum displacement allowed by physical constraints; Z is a dimensionless relative displacement; Y is a dimensionless base (tube) input excitation displacement. It is further assumed that

$$
\rho = \frac{R}{\eta \cdot L_e}, \quad \theta = \frac{Bl \cdot l_m}{i_m \cdot L_e}, \quad \Gamma = \frac{A}{\eta^2}, \quad \mu = \frac{c}{m \cdot \eta} = 2 \cdot \xi, \quad \varepsilon = \frac{i_m^2 \cdot L_e}{m \cdot \eta^2 \cdot l_m^2},
$$
$$
\beta = \frac{k_3 \cdot l_m^2}{\omega_N^2 \cdot m}, \quad \theta \cdot \varepsilon = \frac{-i_m \cdot Bl}{m \cdot \eta^2 \cdot l_m}
\tag{11.33}
$$

where Γ is the dimensionless input base excitation displacement which is equivalent to ff in Eq. (11.21)

Substitution of Eqs. (11.32) and (11.33) into Eq. (11.31) gives

$$
\begin{cases}
\dot{I}_0 + I_0 \cdot \rho + \theta \cdot \dot{Z} = 0 \\[2mm]
\ddot{Z} + \mu \cdot \dot{Z} + Z + \beta \cdot Z^3 - \varepsilon \cdot \theta \cdot I_0 = \Gamma \cdot \sin(\omega_N \cdot \tau)
\end{cases}
\tag{11.34}
$$

where $\ddot{Z} = \frac{d^2 Z}{d\tau}$ and $\dot{Z} = \frac{dZ}{d\tau}$.

It is assumed that

$$\begin{cases} Z = a(\tau) \cdot \cos(\omega \cdot \tau) + b(\tau) \cdot \sin(\omega_N \cdot \tau) \\ I_0 = e(\tau) \cdot \cos(\omega_N \cdot \tau) + d(\tau) \cdot \sin(\omega_N \cdot \tau) \end{cases} \tag{11.35}$$

Then differentiating Z in Eq. (11.35) two times with respect to τ gives

$$\dot{Z} = (\dot{a} + b \cdot \omega_N) \cdot \cos(\omega_N \cdot \tau) + \left(\dot{b} - a \cdot \omega_N \right) \cdot \sin(\omega_N \cdot \tau) \tag{11.36}$$

$$\ddot{Z} = \left(\ddot{a} + 2 \cdot \dot{b} \cdot \omega_N - a \cdot \omega_N^2 \right) \cdot \cos(\omega_N \cdot \tau) + \left(\ddot{b} - 2 \cdot \dot{a} \cdot \omega_N - b \cdot \omega_N^2 \right) \cdot \sin(\omega_N \cdot \tau) \tag{11.37}$$

where the slowly varying assumption leads to $\ddot{a} = \ddot{b} = 0$.

Then Eq. (11.37) becomes

$$\ddot{Z} = \left(2 \cdot \dot{b} - a \cdot \omega_N \right) \cdot \omega \cdot \cos(\omega_N \cdot \tau) - (2 \cdot \dot{a} + b \cdot \omega_N) \cdot \eta \cdot \sin(\omega_N \cdot \tau) \tag{11.38}$$

From I_0 in Eq. (11.35), it gives

$$\dot{I}_0 = (\dot{e} + d \cdot \omega_N) \cdot \cos(\omega_N \cdot \tau) + \left(\dot{d} - e \cdot \omega_N \right) \cdot \sin(\omega_N \cdot \tau); \tag{11.39}$$

Substituting Eqs. (11.35), (11.36), and (11.39) into the first equation of Eq. (11.34) gives

$$\begin{aligned} &[\dot{e} + d \cdot \omega_N + c \cdot \rho + \theta \cdot (\dot{a} + b \cdot \omega_N)] \cdot \cos(\omega_N \cdot \tau) \\ &+ \left[\dot{d} - c \cdot \omega_N + d \cdot \rho + \theta \cdot \left(\dot{b} - a \cdot \omega_N \right) \right] \cdot \sin(\omega_N \cdot \tau) = 0 \end{aligned} \tag{11.40}$$

which leads to

$$\begin{cases} \dot{e} + \theta \cdot \dot{a} = -d \cdot \omega_N - e \cdot \rho - \theta \cdot b \cdot \omega_N \\ \dot{d} + \theta \cdot \dot{b} = e \cdot \omega_N - d \cdot \rho + \theta \cdot a \cdot \omega_N \end{cases} \tag{11.41}$$

In the same way, substituting Eqs. (11.35), (11.36), and (11.38) into the second equation of Eq. (11.34) gives

$$\begin{cases} 2 \cdot \dot{b} \cdot \omega_N + \mu \cdot \dot{a} = a \cdot \left(\omega_N^2 - 1 - \frac{3}{4} \cdot \beta \cdot r^2 \right) - \mu \cdot b \cdot \omega_N + \varepsilon \cdot \theta \cdot e \\ -2 \cdot \omega_N \cdot \dot{a} + \mu \cdot \dot{b} = b \cdot \left(\omega_N^2 - 1 - \frac{3}{4} \cdot \beta \cdot r^2 \right) + \mu \cdot a \cdot \omega_N + \varepsilon \cdot \theta \cdot d + \Gamma \end{cases} \tag{11.42}$$

where $r^2 = a^2 + b^2$. The steady-state amplitude of the periodic response is obtained from the fixed points of Eqs. (11.41) and (11.42) which require $\dot{a} = \dot{b} = \dot{e} = \dot{d} = 0$.

Rearranging Eq. (11.41) gives

$$\begin{cases} 0 = -\tilde{d} \cdot \omega_N - \tilde{e} \cdot \rho - \theta \cdot \tilde{b} \cdot \omega_N \\ 0 = \tilde{e} \cdot \omega_N - \tilde{d} \cdot \rho + \theta \cdot \tilde{a} \cdot \omega_N \end{cases} \tag{11.43}$$

From Eq. (11.43), it gives

$$\begin{cases} \tilde{e} = -\dfrac{\theta \cdot \omega_N}{\omega^2 + \rho^2} \cdot \left(\tilde{a} \cdot \omega_N + \rho \cdot \tilde{b} \right) \\[3mm] \tilde{d} = \dfrac{\theta \cdot \omega_N}{\omega^2 + \rho^2} \cdot \left(\rho \cdot \tilde{a} - \omega_N \cdot \tilde{b} \right) \end{cases} \tag{11.44}$$

Rearranging Eq. (11.42) gives

$$\begin{cases} 0 = \tilde{a} \cdot \left(\omega_N^2 - 1 - \dfrac{3}{4} \cdot \beta \cdot \tilde{r}^2 \right) - \mu \cdot \tilde{b} \cdot \omega_N + \varepsilon \cdot \theta \cdot \tilde{e} \\[3mm] 0 = \tilde{b} \cdot \left(\omega_N^2 - 1 - \dfrac{3}{4} \cdot \beta \cdot \tilde{r}^2 \right) + \mu \cdot \tilde{a} \cdot \omega_N + \varepsilon \cdot \theta \cdot \tilde{d} + \Gamma \end{cases} \tag{11.45}$$

where $\tilde{r}^2 = \tilde{a}^2 + \tilde{b}^2$.

Substituting Eq. (11.44) into Eq. (11.45) gives

$$\begin{cases} \tilde{a} \cdot \left[1 + \dfrac{3}{4} \cdot \beta \cdot \tilde{r}^2 + \left(\dfrac{\varepsilon \cdot \theta^2}{\omega_N^2 + \rho^2} - 1 \right) \cdot \omega_N^2 \right] \omega_N^2 + \tilde{b} \cdot \omega_N \cdot \left(\dfrac{\varepsilon \cdot \theta^2 \cdot \rho}{\omega_N^2 + \rho^2} + \mu \right) = 0 \\[4mm] \tilde{b} \cdot \left[1 + \dfrac{3}{4} \cdot \beta \cdot \tilde{r}^2 + \left(\dfrac{\varepsilon \cdot \theta^2}{\omega_N^2 + \rho^2} - 1 \right) \cdot \omega_N^2 \right] - \left[\dfrac{\varepsilon \cdot \theta^2 \cdot \rho}{\omega_N^2 + \rho^2} + \mu \right] \cdot \tilde{a} \cdot \omega_N = \Gamma \end{cases} \tag{11.46}$$

From Eq. (11.46), it gives

$$\tilde{r}^2 = \dfrac{\Gamma^2}{\left[1 + \frac{3}{4} \cdot \beta \cdot \tilde{r}^2 + \left(\frac{\varepsilon \cdot \theta^2}{\omega_N^2 + \rho^2} - 1 \right) \cdot \omega_N^2 \right]^2 + \left[\frac{\varepsilon \cdot \theta^2 \cdot \rho}{\omega_N^2 + \rho^2} + \mu \right]^2 \cdot \omega_N^2} \tag{11.47}$$

where $\tilde{r}$ is the steady-state dimensionless relative response displacement of the oscillator system, which is equivalent to M in Eq. (11.21). From Eqs. (11.14), (11.18)–(11.20), (11.29), Eq. (11.47) can be proved to be identical to Eq. (11.18).

To find the two loci of the jump points, Eq. (11.47) can be written as

$$\tilde{r}^2 \left\{ \left[1 + \dfrac{3}{4} \cdot \beta \cdot \tilde{r}^2 + \left(\dfrac{\varepsilon \cdot \theta^2}{\omega_N^2 + \rho^2} - 1 \right) \cdot \omega_N^2 \right]^2 + \left[\dfrac{\varepsilon \cdot \theta^2 \cdot \rho}{\omega_N^2 + \rho^2} + \mu \right]^2 \cdot \omega_N^2 \right\} - \Gamma^2 = 0 \tag{11.48}$$

Differentiate both sides of Eq. (11.48) with respect to frequency ω while μ, ε, ρ, θ, and Γ are assumed as constants, and when $\frac{\partial r}{\partial \omega_N} \to \infty$, it gives

$$\beta^2 \cdot \frac{27}{8} \cdot \tilde{r}^4 + 6 \cdot \beta \cdot \tilde{r}^2 \cdot \left[1 + \left(\frac{\varepsilon \cdot \theta^2}{\rho^2 + \omega_N^2} - 1\right) \cdot \omega_N^2\right] + 2 \cdot \left[\mu + \frac{\varepsilon \cdot \theta^2 \cdot \rho}{\rho^2 + \omega_N^2}\right]^2 \cdot \omega_N^2$$

$$+ 2 \cdot \left[1 + \left(\frac{\varepsilon \cdot \theta^2}{\rho^2 + \omega_N^2} - 1\right) \cdot \omega_N^2\right]^2 = 0$$

(11.49)

Two roots of Eq. (11.49) are given by

$$\tilde{r}^2 = \frac{-8 \cdot \left[1 + \left(\frac{\varepsilon \cdot \theta^2}{\rho^2 + \omega_N^2} - 1\right) \cdot \omega_N^2\right] \pm 4\sqrt{\left[1 + \left(\frac{\varepsilon \cdot \theta^2}{\rho^2 + \omega_N^2} - 1\right) \cdot \omega_N^2\right]^2 - 3 \cdot \left(\mu + \frac{\varepsilon \cdot \theta^2 \cdot \rho}{\rho^2 + \omega_N^2}\right)^2 \cdot \omega_N^2}}{9}$$

(11.50)

From Eqs. (11.14), (11.19), (11.20), (11.24), (11.33), Eq. (11.50) can be proved to be identical to Eq. (11.24). The results of the perturbation method have verified those of the harmonic balance method.

NOMENCLATURE

M	The mass of the oscillator
c	The short circuit damping coefficient
α	The equivalent force factor
z_0	The relative displacement amplitude of the oscillator with respect to the base
z	The relative displacement of the oscillator with respect to the base
$\dot{z}$	The relative velocity of the oscillator with respect to the base
$\ddot{z}$	The relative acceleration of the oscillator with respect to the base
x	The displacement of the oscillator with respect to the base
$\dot{x}$	The velocity of the oscillator with respect to the base
$\ddot{x}$	The acceleration of the oscillator with respect to the base
y_0	The excitation displacement amplitude of the base
y	The excitation displacement of the base
$\dot{y}$	The excitation velocity of the base
$\ddot{y}$	The excitation acceleration of the base
$U(z)$	The potential energy of the spring element
$\frac{dU(z)}{dz}$	The first differential of $U(z)$ with respect to the relative displacement z
$\frac{dI}{dt}$	The current change rate
$\frac{dV}{dt}$	The voltage change rate

k_1	The short circuit spring stiffness coefficient of the linear displacement term
k_3	The short circuit spring stiffness coefficient of nonlinear displacement term
k_{2n-1}	The short circuit spring stiffness coefficient of linear or nonlinear displacement term or the potential energy coefficient of the nonlinear oscillator
F_k	The elastic restoring force of the magnetic spring
F_c	The damping force of the magnetic field
F_e	The electromagnetic force
B	The magnetic field intensity constant
I	The current in the coils
l	The total length of the coil wire
N	The wiring number of each coil
D_0	The outer diameter of the tube
R	The external resistance
R_e	The internal resistance of the coil
L_e	Self inductance of the coil
V	The voltage across the two output terminals of the circuit
V_0	The amplitude of the voltage across the two output terminals of the circuit
μ_0	The permeability coefficient of the coil with air coil
μ_r	The permeability coefficient of the coil
h_c	The height of each coil
φ_1	The phase difference between the output voltage V and the relative displacement z
φ_2	The phase difference between the base (tube) input excitation displacement y and the relative displacement z
α_N	Dimensionless force factor
R_N	Dimensionless resistance
P_h	The harvested resonant power
i_m	The threshold or reference current
I_0	The dimensionless current
l_m	The relevant physical dimension such as the maximum displacement allowed by physical constraint
Z	The dimensionless relative displacement
$\dot{Z}$	The dimensionless relative velocity
$\ddot{Z}$	The dimensionless relative acceleration
Y	The dimensionless base (tube) input excitation displacement
H	The height of the ocean wave
T	The period of the ocean wave
A	The input excitation acceleration amplitude of the base (tube)
α	The force factor
M	The dimensionless relative displacement response amplitude of the system
ω_N	The dimensionless excitation frequency
ξ	The dimensionless damping ratio

ff	The dimensionless input excitation displacement amplitude of the base (tube) which is equivalent to Γ
$\sigma = B^2 l^2/k_1$	The ratio of the squared electromagnetic coupling strength over the spring stiffness coefficient
τ	The dimensionless time instance or variable
t	Time instance or variable
η	The radial natural frequency of the oscillator system
k_R	The ratio of the load resistance over the radial natural frequency of the oscillator system
β	The ratio of the linear and nonlinear term stiffness coefficients of the elastic restoring force for the magnetic spring
∞	Infinity
$\rho = R/(\eta \cdot L_c)$	The dimensionless resistance at the natural frequency of the oscillator system which is equivalent to R_N
$\theta = Bl \cdot l_m/(i_m \cdot L_c)$	The dimensionless force factor coefficient reflecting the relative velocity induced current
Γ	The dimensionless base input excitation displacement amplitude which is equivalent to ff
μ	The mechanical damping loss factor of the oscillator system
$\varepsilon = \left(i_m^2 \cdot L_c\right) \Big/ \left(m \cdot \eta^2 \cdot l_m^2\right)$	The dimensionless force factor coefficient reflecting the current induced electromagnetic force
a	The dimensionless relative displacement amplitude in one of the orthogonal axis directions
$\dot{a}$	The dimensionless relative velocity amplitude in one of the orthogonal axis directions
$\ddot{a}$	The dimensionless relative acceleration amplitude in one of the orthogonal axis directions
$\widetilde{a}$	The steady dimensionless relative displacement amplitude in one of the orthogonal axis directions
b	The dimensionless relative displacement amplitude in the other orthogonal axis direction
$\dot{b}$	The dimensionless relative velocity amplitude in the other orthogonal axis direction
$\ddot{b}$	The dimensionless relative acceleration amplitude in the other orthogonal axis direction
$\widetilde{b}$	The steady dimensionless relative displacement amplitude in the other orthogonal axis direction
e	The dimensionless current amplitude in one of the orthogonal axis directions
$\dot{e}$	The dimensionless current change rate amplitude in one of the orthogonal axis directions
$\widetilde{e}$	The steady dimensionless current amplitude in one of the orthogonal axis directions
d	The dimensionless current amplitude in the other orthogonal axis direction
$\dot{d}$	The dimensionless current change rate amplitude in the other orthogonal axis direction

$\tilde{d}$	The steady dimensionless current amplitude in the other orthogonal axis direction
r	The total dimensionless relative displacement amplitude
$\tilde{r}$	The total steady dimensionless relative displacement amplitude which is equivalent to M

Superscripts

- ·   The first differential
- ··   The second differential

Abbreviations

AWS   Archimedes Wave Swing
PA   Point absorber
OWC   Oscillating water columns
WD   Wave dragon
WEC   Wave energy converter
SDOF   Single degree of freedom

REFERENCES

[1] Cargo CJ, et al. Determination of optimal parameters for a hydraulic power take-off unit of a wave energy converter in regular waves. Proc Inst Mech Eng Part A J Power Energy 2011. http://dx.doi.org/10.1177/0957650911407818.

[2] http://www.nucleartourist.com/renewables/ocean_and_wave_energy.htm.

[3] Cruz J. Ocean wave energy - current status and future perspectives. Green energy and technology. Springer; 2008. ISSN1865-3529.

[4] Mei CC. Hydrodynamic principles of wave power extraction. Philos Trans R Soc Lond A Math Phys Eng Sci 2012;370(1959):208—34.

[5] Delauré Y, Lewis A. 3D hydrodynamic modelling of fixed oscillating water column wave power plant by boundary element methods. Ocean Eng 2003;30(3):309—30.

[6] Falcao A. Wave energy utilization: a review of the technologies. Renew Sustain Energy Rev 2010;14(2010):899—918.

[7] Güneya MS, Kaygusuzb K. Hydrokinetic energy conversion systems: a technology status review. Renew Sustain Energy Rev December 2010;14(9):2996—3004.

[8] Mueller MA. Electrical generators for direct drive wave energy converters. IEE Proc Gener Trans Distrib 2002;149(4):446—56.

[9] Beeby SP, O'Donnell T. Electromagnetic energy harvesting. US: Springer; 2009, ISBN 978-0-387-76463-4.

[10] Williams C, Yates RB. Analysis of a micro-electric generator for microsystems. Sensors Actuators A Phys 1996;52(1):8—11.

[11] Elliott SJ, Zilletti M. Scaling of electromagnetic transducers for shunt damping and energy harvesting. J Sound Vib 2014. http://dx.doi.org/10.1016/j.jsv.2013.11.036i.

[12] Foisal ARM, et al. Multi-frequency electromagnetic energy harvester using a magnetic spring cantilever. Sensors Actuators A Phys 2012;182:106—13.

[13] Munaz A, et al. A study of an electromagnetic energy harvester using multi-pole magnet. Sensors Actuators A Phys 2013;201:134—40.

[14] Cepnik C, Radler O, et al. Effective optimization of electromagnetic energy harvesters through direct computation of the electromagnetic coupling. Sensors Actuators A Phys 2011;167(2):416−21.

[15] Halbach K. Design of permanent multipole magnets with oriented rare earth cobalt material. Nucl Instr Methods 1980;169(1):1−10.

[16] Shahosseini I, Najafi K. Cylindrical Halbach magnet array for electromagnetic vibration energy harvesters. In: Mems 2015, Estoril, Portugal, 18−22 January, 2015, IEEE; 2015. ISBN:978-1-4799-7955-4/15.

[17] Zhu D, Beeby SP, Tudor MJ, et al. A planar electro-magnetic vibration energy harvester with a Halbach array. In: The 11th international workshop on micro and nanotechnology for power generation and energy conversion applications, November 15−18, 2011, Seoul, Republic of Korea (Power MEMS); 2011. p. 347−50.

[18] Cheung JT. Frictionless linear electrical generator for harvesting motion energy. 2004. Technical report-71222, (Contract number: MDA972-03-C-0025), sponsored by DARPA/Advanced Technology Office 3701, N. Fairfax Drive, Arlington, VA.

[19] Budal K, Falnes J. A resonant point absorber of ocean-wave power. Nature 1975;256: 478−9.

[20] Korde UA. Efficient primary energy conversion in irregular waves. Ocean Eng 1999; 26(7):625−51.

[21] Sabzehgar R, Moallem M. A review of ocean wave energy conversion systems. In: IEEE electrical power & energy conference; 2009.

[22] Tang X, Zuo L. Enhanced vibration energy harvesting using dual-mass systems. J Sound Vib 2011;330(21):5199−209.

[23] Gatti RR. Spatially varying multi degree of freedom electromagnetic energy harvesting (Ph.D. thesis). Australia: Curtin University; 2013.

[24] Harne RL, Wang KW. A review of the recent research on vibration energy harvesting via bistable systems. Smart Mater Struct 2013;22:023001.

[25] Beeby SP, Wang L, Zhu D, Weddell AS, Merrett G, Stark VB. A comparison of power output from linear and nonlinear kinetic energy harvesters using real vibration data. Smart Mater Struct 2013;22:075022.

[26] Owens BAM, Mann BP. Linear and nonlinear electromagnetic coupling models in vibration-based energy harvesting. J Sound Vib 2012;331(4):922−37.

[27] Mann B, Sims N. Energy harvesting from the nonlinear oscillations of magnetic levitation. J Sound Vib 2009;319(1):515−30.

[28] Dallago E, et al. Analytical model of a vibrating electromagnetic harvester considering nonlinear effects. Power Electron IEEE Trans 2010;25(8):1989−97.

[29] Barton DAW. Energy harvesting from vibrations with a nonlinear oscillator. J Vib Acoust Stress Reliab Des 2010;132(2):021009.

[30] Vocca H, Cottone F. Kinetic energy harvesting. In: Fagas G, Grammaitoni L, Paul D, Berini GA, editors. Ict - energy concepts towards zero - power information and communication technology; 2014. p. 25−48.

[31] Vocca H, Cottone F, Neri I, Gammaitoni L. A comparison between nonlinear cantilever and buckled beam for energy harvesting. Eur Phys J Spl Top 2013;222:1699−705.

[32] Leijon M, Danielsson O, Eriksson M, Thorburn K, Bernhoff H, Isberg J, et al. An electrical approach to wave energy conversion. Renew Energy 2006;31:1309−19.

Analysis of Multiple Degrees of Freedom Electromagnetic Vibration Energy Harvesters and Their Applications

12

CHAPTER OUTLINE

12.1 ANALYSIS OF A TWO DEGREES OF FREEDOM ELECTROMAGNETIC VIBRATION ENERGY HARVESTER OSCILLATOR SYSTEM

In the previous chapter, single degree of freedom (SDOF) electromagnetic vibration energy harvester (EMVEH) and its various design options were identified. However, the SDOF system was limited to only one resonant natural frequency. Ideally, the energy harvester should be able to resonate at every frequency present in the random excitations. By employing multiple degree of freedom systems, one can harvest vibration energy at multiple resonant frequencies and achieve the maximum power [2, 4–7, 9, 10–15]. The simplest form of a multiple degrees of freedom system (MDOF) is a two degree of freedom system (2DOF) with two oscillators. The study of the 2DOF is a prerequisite to design complex MDOF systems. Schematic of a typical 2DOF electromagnetic vibration harvester oscillator system is illustrated in Fig. 12.1 where the two coil oscillators consist of two cylindrical core irons wired by two coils. A tube is used to house the two coil oscillators. The two coil oscillators are connected to each other by a steel spring and connected by two steel springs onto

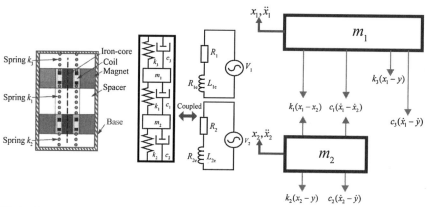

FIGURE 12.1

Schematic of a two degree of freedom electromagnetic vibration energy harvester oscillator system.

the two ends of the tube in the axial direction. Two ring magnets are evenly arranged inside the tube. The radial outer surfaces of the two ring magnets are fixed onto the inner surfaces of the concentric tube. The two coil oscillators are concentrically moving and passing through the radial inner surfaces of the two ring magnets inside the concentric tube. The two coils are respectively connected to two external resistances which form two harvesting circuits. The electromechanical system governing equations are given by:

$$
\begin{cases}
m_1 \cdot \ddot{x}_1 + c_3(\dot{x}_1 - \dot{y}) + k_3(x_1 - y) + k_1(x_1 - x_2) + c_1(\dot{x}_1 - \dot{x}_2) + B_1 \cdot l_1 \cdot \dfrac{V_1}{R_1} = 0 \\[2mm]
m_2 \cdot \ddot{x}_2 + c_2(\dot{x}_2 - \dot{y}) + k_2(x_2 - y) + k_1(x_2 - x_1) + c_2(\dot{x}_2 - \dot{x}_1) + B_2 \cdot l_2 \cdot \dfrac{V_2}{R_2} = 0 \\[2mm]
V_1 + R_{1e} \cdot \dfrac{V_1}{R_1} + \dfrac{L_{1e}}{R_1} \cdot \dot{V}_1 = B_1 \cdot l_1 \cdot (\dot{x}_1 - \dot{y}) \\[2mm]
V_2 + R_{2e} \cdot \dfrac{V_2}{R_2} + \dfrac{L_{2e}}{R_2} \cdot \dot{V}_2 = B_2 \cdot l_2 \cdot (\dot{x}_2 - \dot{y})
\end{cases}
$$

(12.1)

where m_1 and m_2 are the oscillator masses; k_1, k_2, and k_3 are the short circuit stiffness coefficients of three springs; c_1, c_2, and c_3 are the short circuit damping coefficients of the three dashpots; x_1 and x_2 are the displacement of the two oscillators; y is the base excitation displacement; V_1 and V_2 are the output voltages of the two harvesting circuits; B_1 and B_2 are the magnetic field constants or magnetic flux density constants near the radial inner surfaces of the two ring magnets; l_1 and l_2 are the length of the two coils; R_1 and R_2 are the external load resistances of the two harvesting circuits; R_{e1} and R_{e2} are the internal resistances of the two coils; L_{e1} and L_{e2} are the electric induction coefficients of the two coils.

Applying the Laplace transform onto Eq. (12.1) gives

$$
\begin{bmatrix}
m_1 \cdot s^2 + (c_1 + c_3) \cdot s + (k_1 + k_3) & -c_1 \cdot s - k_1 & -B_1 \cdot l_1 \cdot s & 0 \\
-c_1 \cdot s - k_1 & m_2 \cdot s^2 + (c_1 + c_2) \cdot s + (k_1 + k_2) & 0 & -B_2 \cdot l_2 \cdot s \\
-B_1 \cdot l_1 \cdot s & 0 & \left(1 + \dfrac{R_{1e}}{R_1} + \dfrac{L_{1e} \cdot s}{R_1}\right) & 0 \\
0 & -B_2 \cdot l_2 \cdot s & 0 & \left(1 + \dfrac{R_{2e}}{R_2} + \dfrac{L_{2e} \cdot s}{R_2}\right)
\end{bmatrix}
$$

$$
\cdot
\begin{Bmatrix}
\dfrac{X_1(s)}{Y(s)} \\[4pt]
\dfrac{X_2(s)}{Y(s)} \\[4pt]
\dfrac{V_1(s)}{Y(s)} \\[4pt]
\dfrac{V_2(s)}{Y(s)}
\end{Bmatrix}
=
\begin{Bmatrix}
c_3 \cdot s + k_3 \\
c_2 \cdot s + k_2 \\
-B_1 \cdot l_1 \cdot s \\
-B_2 \cdot l_2 \cdot s
\end{Bmatrix}
\tag{12.2}
$$

When $s = i\omega$, the above transform is the Fourier transform. The parameters of m_1, m_1, k_1, k_2, k_3, c_1, c_2, c_3, B_1, B_2, l_1, l_2, R_1, R_2, R_{e1}, R_{e2}, L_{e1}, L_{e2} in Eq. (12.2) can be identified from experiments. From Eq. (12.2), the ratios of X_1/Y, X_2/Y, V_1/Y, and V_2/Y can be calculated using the Matlab program codes similar to those in Figs. 1.8 and 4.4. The dimensionless output voltage, harvested power, and energy harvesting efficiency can be calculated from Eq. (4.23), and the similar procedures are illustrated in Section 4.7 of Chapter 4. Eq. (12.1) can be arranged as

$$
\begin{cases}
(\ddot{x}_1 - \ddot{y}) = -\dfrac{c_3}{m_1}(\dot{x}_1 - \dot{y}) - \dfrac{k_3}{m_1}(x_1 - y) - \dfrac{k_1}{m_1}(x_1 - y) + \dfrac{k_1}{m_1}(x_2 - y) - \dfrac{c_1}{m_1}(\dot{x}_1 - \dot{y}) + \dfrac{c_1}{m_1}(\dot{x}_2 - \dot{y}) \\[6pt]
\quad -\ddot{y} - \dfrac{B_1 \cdot l_1 \cdot V_1}{m_1 \cdot R_1} \\[8pt]
(\ddot{x}_2 - \ddot{y}) = -\dfrac{c_2}{m_2}(\dot{x}_2 - \dot{y}) - \dfrac{k_2}{m_2}(x_2 - y) - \dfrac{k_1}{m_2}(x_2 - y) + \dfrac{k_1}{m_2}(x_1 - y) - \dfrac{c_1}{m_2}(\dot{x}_2 - \dot{y}) + \dfrac{c_1}{m_2}(\dot{x}_1 - \dot{y}) \\[6pt]
\quad -\ddot{y} - \dfrac{B_2 \cdot l_2 \cdot V_2}{m_2 \cdot R_2} \\[8pt]
\dot{V}_1 = -\dfrac{R_{1e} + R_1}{L_{1e}} \cdot V_1 + \dfrac{B_1 \cdot l_1 \cdot R_1}{L_{1e}} \cdot (\dot{x}_1 - \dot{y}) \\[8pt]
\dot{V}_2 = -\dfrac{R_{2e} + R_2}{L_{2e}} \cdot V_2 + \dfrac{B_2 \cdot l_2 \cdot R_2}{L_{2e}} \cdot (\dot{x}_2 - \dot{y})
\end{cases}
\tag{12.3}
$$

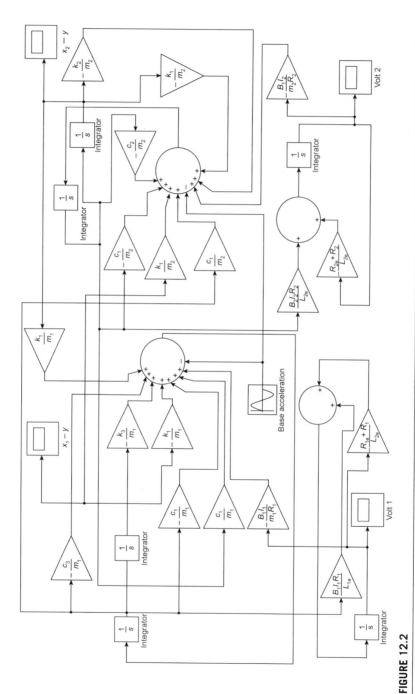

FIGURE 12.2

Simulation schematic for Eq. (12.3) for prediction of the oscillator relative displacement responses and output voltages from the base excitation acceleration using Matlab Simulink time domain integration method.

Eq. (12.3) can be solved using the time domain integration method as illustrated in Fig. 12.2 where the system parameters of m_1, m_1, k_1, k_2, k_3, c_1, c_2, c_3, B_1, B_2, l_1, l_2, R_1, R_2, R_{e1}, R_{e2}, L_{e1}, L_{e2} should be identified from experiments.

The output of the program code in Fig. 12.2 produces the sine wave time traces of the relative displacement and output voltage under a sine wave excitation input of one excitation frequency and fixed excitation amplitude. The frequency can be changed one by one in a predefined range, the sine wave amplitudes of the relative displacement and output voltage at each frequency can be read from the scope outputs of the program code. The results predicted by the time domain integration method should be able to verify those using the frequency response method as illustrated by Eq. (12.2) and the Matlab code similar to that in Fig. 4.4.

The performance of the 2DOF linear EMVEH can be analyzed and optimized using the methods illustrated in this chapter.

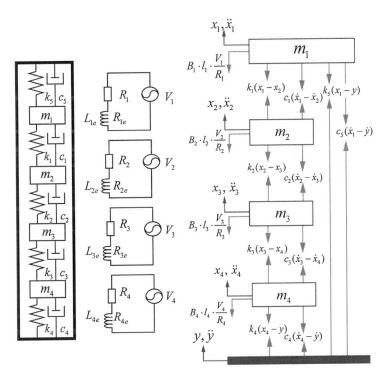

FIGURE 12.3

Schematic of a four degree of freedom electromagnetic vibration energy harvester oscillator system.

12.2 ANALYSIS OF A FOUR DEGREE OF FREEDOM ELECTROMAGNETIC VIBRATION ENERGY HARVESTER OSCILLATOR SYSTEM

Schematic of a typical four degree of freedom electromagnetic vibration harvester oscillator system is illustrated in Fig. 12.3 where four coil oscillators consist of four cylindrical core irons wired by four coils. A tube is used to house the four coil oscillators. The four coil oscillators are connected to each other by three steel springs and connected by two steel springs onto the two ends of the tube in the axial direction. Four ring magnets are evenly arranged inside the tube. The radial outer surfaces of the four ring magnets are fixed onto the inner surfaces of the concentric tube. The four coil oscillators are concentrically moving and passing through the radial inner surfaces of the four ring magnets inside the concentric tube. The four coils are respectively connected to four external resistances which form four harvesting circuits.

The electromechanical system governing equations are given by:

$$\begin{cases}
m_1 \cdot \ddot{x}_1 + c_5(\dot{x}_1 - \dot{y}) + k_5(x_1 - y) + k_1(x_1 - x_2) + c_1(\dot{x}_1 - \dot{x}_2) + B_1 \cdot l_1 \cdot \dfrac{V_1}{R_1} = 0 \\[2mm]
m_2 \cdot \ddot{x}_2 + c_2(\dot{x}_2 - \dot{x}_3) + k_2(x_2 - x_3) + k_1(x_2 - x_1) + c_1(\dot{x}_2 - \dot{x}_1) + B_2 \cdot l_2 \cdot \dfrac{V_2}{R_2} = 0 \\[2mm]
m_3 \cdot \ddot{x}_3 + c_3 \cdot (\dot{x}_3 - \dot{x}_4) + k_3 \cdot (x_3 - x_4) + k_2 \cdot (x_3 - x_2) + c_2(\dot{x}_3 - \dot{x}_2) + B_3 \cdot l_3 \cdot \dfrac{V_3}{R_3} = 0 \\[2mm]
m_4 \cdot \ddot{x}_4 + c_4 \cdot (\dot{x}_4 - \dot{y}) + k_4 \cdot (x_4 - y) + k_3 \cdot (x_4 - x_3) + c_3(\dot{x}_4 - \dot{x}_3) + B_4 \cdot l_4 \cdot \dfrac{V_4}{R_4} = 0 \\[2mm]
V_1 + R_{1e} \cdot \dfrac{V_1}{R_1} + \dfrac{L_{1e}}{R_1} \cdot \dot{V}_1 = B_1 \cdot l_1 \cdot (\dot{x}_1 - \dot{y}) \\[2mm]
V_2 + R_{2e} \cdot \dfrac{V_2}{R_2} + \dfrac{L_{2e}}{R_2} \cdot \dot{V}_2 = B_2 \cdot l_2 \cdot (\dot{x}_2 - \dot{y}) \\[2mm]
V_3 + R_{3e} \cdot \dfrac{V_3}{R_3} + \dfrac{L_{3e}}{R_3} \cdot \dot{V}_3 = B_3 \cdot l_3 \cdot (\dot{x}_3 - \dot{y}) \\[2mm]
V_4 + R_{4e} \cdot \dfrac{V_4}{R_4} + \dfrac{L_{4e}}{R_4} \cdot \dot{V}_4 = B_4 \cdot l_4 \cdot (\dot{x}_4 - \dot{y})
\end{cases}$$

$$(12.4)$$

Applying the Laplace transform onto Eq. (12.4) gives

$$
\begin{bmatrix}
a_{11}(s) & -c_1 \cdot s - k_1 & 0 & 0 & B_1 \cdot l_1 \cdot s & 0 & 0 & 0 \\
-c_1 \cdot s - k_1 & a_{22}(s) & -c_2 \cdot s - k_2 & 0 & 0 & B_2 \cdot l_2 \cdot s & 0 & 0 \\
0 & -c_2 \cdot s - k_2 & a_{33}(s) & -c_3 \cdot s - k_3 & 0 & 0 & B_3 \cdot l_3 \cdot s & 0 \\
0 & 0 & -c_3 \cdot s - k_3 & a_{44}(s) & 0 & 0 & 0 & B_4 \cdot l_4 \cdot s \\
B_1 \cdot l_1 \cdot s & 0 & 0 & 0 & a_{55}(s) & 0 & 0 & 0 \\
0 & B_2 \cdot l_2 \cdot s & 0 & 0 & 0 & a_{66}(s) & 0 & 0 \\
0 & 0 & B_3 \cdot l_3 \cdot s & 0 & 0 & 0 & a_{77}(s) & 0 \\
0 & 0 & 0 & B_4 \cdot l_4 \cdot s & 0 & 0 & 0 & a_{88}(s)
\end{bmatrix}
$$

$$
\begin{Bmatrix}
\dfrac{X_1(s)}{Y(s)} \\[2mm]
\dfrac{X_2(s)}{Y(s)} \\[2mm]
\dfrac{X_3(s)}{Y(s)} \\[2mm]
\dfrac{X_4(s)}{Y(s)} \\[2mm]
V_1(s) \\
V_2(s) \\
V_3(s) \\
V_4(s)
\end{Bmatrix}
=
\begin{Bmatrix}
c_5 \cdot s + k_5 \\
0 \\
0 \\
c_4 \cdot s + k_4 \\
-B_1 \cdot l_1 \cdot s \\
-B_2 \cdot l_2 \cdot s \\
-B_3 \cdot l_3 \cdot s \\
-B_4 \cdot l_4 \cdot s
\end{Bmatrix}
$$

$$(12.5)$$

where the diagonal elements are

$$
\begin{cases}
a_{11}(s) = m_1 \cdot s^2 + (c_1 + c_5) \cdot s + (k_1 + k_5) \\[1mm]
a_{22}(s) = m_2 \cdot s^2 + (c_1 + c_2) \cdot s + (k_1 + k_2) \\[1mm]
a_{33}(s) = m_3 \cdot s^2 + (c_3 + c_2) \cdot s + (k_3 + k_2) \\[1mm]
a_{44}(s) = m_4 \cdot s^2 + (c_3 + c_4) \cdot s + (k_3 + k_4) \\[1mm]
a_{55}(s) = 1 + \dfrac{R_{1e}}{R_1} + \dfrac{L_{1e} \cdot s}{R_1} \\[3mm]
a_{66}(s) = 1 + \dfrac{R_{2e}}{R_2} + \dfrac{L_{2e} \cdot s}{R_2} \\[3mm]
a_{77}(s) = 1 + \dfrac{R_{3e}}{R_3} + \dfrac{L_{3e} \cdot s}{R_3} \\[3mm]
a_{88}(s) = 1 + \dfrac{R_{4e}}{R_4} + \dfrac{L_{4e} \cdot s}{R_4}
\end{cases}
$$

$$(12.6)$$

When $s = i\omega$, the above transform is the Fourier transform. The system parameters in Eqs. (12.5) and (12.6) for the 4DOF oscillator system can be identified from experiments in the same way as those in Eq. (12.2) for the 2DOF oscillator system. From Eqs. (12.5) and (12.6), the ratios of X_1/Y, X_2/Y, X_3/Y, X_4/Y, V_1/Y, V_2/Y, V_3/Y, and V_4/Y can be calculated using the Matlab program codes similar to those in Figs. 1.8 and 4.4. The dimensionless output voltage, harvested power, and energy harvesting efficiency can be calculated from Eq. (4.23) and the similar procedures illustrated in Section 4.7 of Chapter 4. Eq. (12.4) can be arranged as

$$
\left\{
\begin{aligned}
(\ddot{x}_1 - \ddot{y}) &= -\frac{c_5}{m_1}(\dot{x}_1 - \dot{y}) - \frac{k_5}{m_1}(x_1 - y) - \frac{k_1}{m_1}(x_1 - y) + \frac{k_1}{m_1}(x_2 - y) \\
&\quad -\frac{c_1}{m_1}(\dot{x}_1 - \dot{y}) + \frac{c_1}{m_1}(\dot{x}_2 - \dot{y}) - \ddot{y} - \frac{B_1 \cdot l_1 \cdot V_1}{m_1 \cdot R_1} \\
(\ddot{x}_2 - \ddot{y}) &= -\frac{k_1}{m_2}(x_2 - y) + \frac{k_1}{m_2}(x_1 - y) - \frac{c_1}{m_2}(\dot{x}_2 - \dot{y}) + \frac{c_1}{m_2}(\dot{x}_1 - \dot{y}) \\
&\quad -\frac{k_2}{m_2}(x_2 - y) + \frac{k_2}{m_2}(x_3 - y) - \frac{c_2}{m_2}(\dot{x}_2 - \dot{y}) + \frac{c_2}{m_2}(\dot{x}_3 - \dot{y}) - \ddot{y} - \frac{B_2 \cdot l_2 \cdot V_2}{m_2 \cdot R_2} \\
(\ddot{x}_3 - \ddot{y}) &= -\frac{k_2}{m_3}(x_3 - y) + \frac{k_2}{m_3}(x_2 - y) - \frac{c_2}{m_3}(\dot{x}_3 - \dot{y}) + \frac{c_2}{m_3}(\dot{x}_2 - \dot{y}) \\
&\quad -\frac{k_3}{m_3}(x_3 - y) + \frac{k_3}{m_3}(x_4 - y) - \frac{c_3}{m_3}(\dot{x}_3 - \dot{y}) + \frac{c_3}{m_3}(\dot{x}_4 - \dot{y}) - \ddot{y} - \frac{B_3 \cdot l_3 \cdot V_3}{m_3 \cdot R_3} \\
(\ddot{x}_4 - \ddot{y}) &= -\frac{c_4}{m_4}(\dot{x}_4 - \dot{y}) - \frac{k_4}{m_4}(x_4 - y) - \frac{k_3}{m_4}(x_4 - y) + \frac{k_3}{m_4}(x_3 - y) \\
&\quad -\frac{c_3}{m_4}(\dot{x}_4 - \dot{y}) + \frac{c_3}{m_4}(\dot{x}_3 - \dot{y}) - \ddot{y} - \frac{B_4 \cdot l_4 \cdot V_4}{m_4 \cdot R_4} \\
\dot{V}_1 &= -\frac{R_{1e} + R_1}{L_{1e}} \cdot V_1 + \frac{B_1 \cdot l_1 \cdot R_1}{L_{1e}} \cdot (\dot{x}_1 - \dot{y}) \\
\dot{V}_2 &= -\frac{R_{2e} + R_2}{L_{2e}} \cdot V_2 + \frac{B_2 \cdot l_2 \cdot R_2}{L_{2e}} \cdot (\dot{x}_2 - \dot{y}) \\
\dot{V}_3 &= -\frac{R_{3e} + R_3}{L_{3e}} \cdot V_3 + \frac{B_3 \cdot l_3 \cdot R_3}{L_{3e}} \cdot (\dot{x}_3 - \dot{y}) \\
\dot{V}_4 &= -\frac{R_{4e} + R_4}{L_{4e}} \cdot V_4 + \frac{B_4 \cdot l_4 \cdot R_4}{L_{4e}} \cdot (\dot{x}_4 - \dot{y})
\end{aligned}
\right.
\tag{12.7}
$$

Eq. (12.7) can be solved using the time domain integration method from the program code similar to Fig. 12.2 where the system parameters should be identified from experiments. The output of the program code produces sine wave time traces of the relative displacement and output voltage for a sine wave excitation input of one excitation frequency and amplitude. The results predicted by the time domain integration method should be able to verify those using the frequency response method as illustrated by Eqs. (12.5) and (12.6) and the Matlab code similar to that in Fig. 4.4.

12.3 ANALYSIS OF AN *N* DEGREE OF FREEDOM ELECTROMAGNETIC VIBRATION ENERGY HARVESTER OSCILLATOR SYSTEM

Schematic of a typical N degree of freedom electromagnetic vibration harvester oscillator system is illustrated in Fig. 12.4 where N coil masses are connected to each other by $N-1$ steel springs and connected to the two ends of a tube which houses the N oscillators. N ring magnets are evenly arranged and fixed onto the inner surface of the tube. The N coil oscillators are concentrically moving and passing through the radial inner surfaces of the N ring magnets in the axial direction. The N coils are respectively connected to N external resistances which form N harvesting circuits. Following similar procedures as above, the ratios or frequency response functions of the relative displacements can be calculated by:

$$
\begin{bmatrix}
a_{11}(s) & -c_1 \cdot s - k_1 & 0 & \cdots & 0 & B_1 \cdot l_1 \cdot s & 0 & \cdots & \cdots & 0 \\
-c_1 \cdot s - k_1 & a_{22}(s) & -c_2 \cdot s - k_2 & 0 & \cdots & 0 & B_2 \cdot l_2 \cdot s & \cdots & \cdots & 0 \\
0 & -c_2 \cdot s - k_2 & \vdots & -c_{N-2} \cdot s - k_{N-2} & 0 & \cdots & \cdots & \cdots & \cdots & 0 \\
0 & \cdots & -c_{N-2} \cdot s - k_{N-2} & a_{N-1N-1}(s) & -c_{N-1} \cdot s - k_{N-1} & 0 & 0 & 0 & B_{N-1} \cdot l_{N-1} \cdot s & 0 \\
0 & \vdots & \cdots & -c_{N-1} \cdot s - k_{N-1} & a_{NN}(s) & \vdots & \vdots & 0 & 0 & B_N \cdot l_N \cdot s \\
B_1 \cdot l_1 \cdot s & 0 & \cdots & 0 & 0 & \vdots & \vdots & \vdots & \vdots & 0 \\
0 & B_2 \cdot l_2 \cdot s & \cdots & \vdots & \vdots & \vdots & \vdots & \vdots & \vdots & \vdots \\
0 & \vdots & \cdots & \cdots & \vdots & \vdots & 0 & \vdots & \vdots & \vdots \\
\vdots & \cdots & \cdots & \cdots & 0 & \cdots & \cdots & 0 & 0 & 0 \\
0 & 0 & \cdots & \cdots & B_N \cdot l_N \cdot s & 0 & \cdots & \cdots & 0 & 0
\end{bmatrix}
$$

$$
\begin{Bmatrix}
\dfrac{X_1(s)}{Y(s)} \\
\dfrac{X_2(s)}{Y(s)} \\
\vdots \\
\vdots \\
\dfrac{X_N(s)}{Y(s)} \\
V_1(s) \\
V_2(s) \\
\vdots \\
V_N(s)
\end{Bmatrix}
=
\begin{Bmatrix}
c_{N+1} \cdot s + k_{N+1} \\
0 \\
0 \\
\vdots \\
0 \\
c_N \cdot s + k_N \\
-B_1 \cdot l_1 \cdot s \\
-B_2 \cdot l_2 \cdot s \\
\vdots \\
-B_N \cdot l_N \cdot s
\end{Bmatrix}
$$

$$(12.8)$$

From Eq. (12.8), the ratios or frequency response functions of the relative displacements and output voltages can be calculated and plotted using the frequency response method and the program code similar to that in Fig. 4.4.

Following similar procedures as above, the converted integration equation is given by Eq. (12.9). Eq. (12.9) can be solved using the time domain integration method and the Matlab Simulink code program code similar to that in Fig. 12.2 where the system parameters should be identified from experiments.

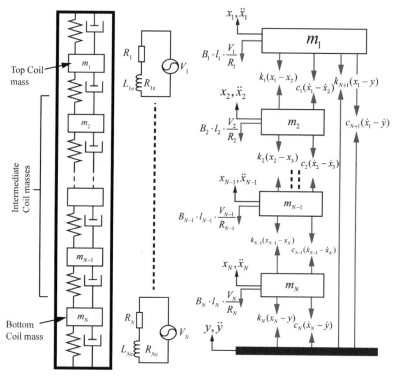

FIGURE 12.4

Schematic of an *N* degree of freedom electromagnetic vibration energy harvester oscillator system.

$$
\begin{cases}
(\ddot{x}_1 - \ddot{y}) = -\dfrac{c_{N+1}}{m_1}(\dot{x}_1 - \dot{y}) - \dfrac{k_{N+1}}{m_1}(x_1 - y) - \dfrac{k_1}{m_1}(x_1 - y) + \dfrac{k_1}{m_1}(x_2 - y) - \dfrac{c_1}{m_1}(\dot{x}_1 - \dot{y}) \\[2mm]
\qquad\quad + \dfrac{c_1}{m_1}(\dot{x}_2 - \dot{y}) - \ddot{y} - \dfrac{B_1 \cdot l_1 \cdot V_1}{m_1 \cdot R_1} \\[3mm]
(\ddot{x}_2 - \ddot{y}) = \dfrac{k_1}{m_2}(x_2 - y) + \dfrac{k_1}{m_2}(x_1 - y) - \dfrac{c_1}{m_2}(\dot{x}_2 - \dot{y}) + \dfrac{c_1}{m_2}(\dot{x}_1 - \dot{y}) - \dfrac{k_2}{m_2}(x_2 - y) + \dfrac{k_2}{m_2}(x_3 - y) \\[2mm]
\qquad\quad - \dfrac{c_2}{m_2}(\dot{x}_2 - \dot{y}) + \dfrac{c_2}{m_2}(\dot{x}_3 - \dot{y}) - \ddot{y} - \dfrac{B_2 \cdot l_2 \cdot V_2}{m_2 \cdot R_2} \\[2mm]
\dots \\[2mm]
(\ddot{x}_N - \ddot{y}) = -\dfrac{c_N}{m_N}(\dot{x}_4 - \dot{y}) - \dfrac{k_N}{m_N}(x_N - y) - \dfrac{k_{N-1}}{m_N}(x_N - y) + \dfrac{k_{N-1}}{m_N}(x_{N-1} - y) - \dfrac{c_{N-1}}{m_N}(\dot{x}_N - \dot{y}) \\[2mm]
\qquad\quad + \dfrac{c_{N-1}}{m_N}(\dot{x}_{N-1} - \dot{y}) - \ddot{y} - \dfrac{B_N \cdot l_N \cdot V_N}{m_N \cdot R_N} \\[3mm]
\dot{V}_1 = -\dfrac{R_{1e} + R_1}{L_{1e}} \cdot V_1 + \dfrac{B_1 \cdot l_1 \cdot R_1}{L_{1e}} \cdot (\dot{x}_1 - \dot{y}) \\[3mm]
\dot{V}_2 = -\dfrac{R_{2e} + R_2}{L_{2e}} \cdot V_2 + \dfrac{B_2 \cdot l_2 \cdot R_2}{L_{2e}} \cdot (\dot{x}_2 - \dot{y}) \\[2mm]
\dots \\[2mm]
\dot{V}_N = -\dfrac{R_{Ne} + R_N}{L_{Ne}} \cdot V_N + \dfrac{B_N \cdot l_N \cdot R_N}{L_{4e}} \cdot (\dot{x}_N - \dot{y})
\end{cases}
$$

$$(12.9)$$

12.4 FIELDS OF APPLICATION

The MDOF EMVEH systems can be applied in the real world to harvest not only vibration energy but also any kinetic energy that is from periodic motion. The sources of vibration energy that can be harvested using the MDOF EMVEH systems can be classified as (1) low power sources in the order of microwatts to milliwatts, (2) renewable energy sources such as wind power and wave power in the order of kW to MW.

Low power sources will be used to power only low-power microelectronics. Therefore, at microscale and mesoscale level, the MDOF EMVEH systems can be designed to harvest energy for microelectronic devices such as sensors, smart sensors, or wireless sensors since their power requirement is in the order of microwatts to milliwatts. These sensors will normally be subject to ambient vibrations and thus can be either industrial sensors for condition monitoring applications, automotive sensors connected to CAN, or bridge sensors for structural health monitoring. Multi-frequency harvesters were developed and characterized by using three-dimensional (3D) excitation at different frequencies [4,5]. The 3D dynamic behavior and performance analysis of the device show that the first vibration mode of 1285 Hz is an out-of-plane motion, while the second and third modes of 1470 and 1550 Hz, respectively, are in-plane at angles of 60 degrees (240 degrees) and 150 degrees (330 degrees) to the horizontal (x-) axis. For an excitation acceleration of 1 g, the maximum power densities achieved are 0.444, 0.242, and 0.125 μW/cm^3 at the three different vibration modes. Nevertheless, the flux density change rate is not the most in this situation and the arrangement of magnets and coils is important [3].

At larger scale, the MDOF EMVEHs can be applied to convert renewable energy from any periodic motion to useful electric energy. Typically, these energy sources are in the form of fluid motion like wind, ocean wave energy, etc., the MDOF EMVEH systems can be used provided suitable mechanical to mechanical conversion mechanisms are available.

Energy harvesting from vehicle suspensions is a typical example for the application of the 2DOF EMVEHs. There are two types of regenerative shock absorbers which are still under the research and development stage. They are linear electromagnetic generator as shown in Fig. 12.5 and rotational generator with the rack pinion mechanism as shown in Fig. 12.6 or the ball–screw mechanism as shown in Fig. 12.7.

The harvested power depends on the magnetic field intensity or magnetic flux density, flux change rate around the coils, and the length of coils. Design of the linear regenerative shock absorber should be focused on the stacking patterns of magnets and coils where alternating magnet polarity or Halbach arrangement, multiple coils, and coil phases should be used to maximize the magnetic field intensity or magnetic flux density and its change rate around the coils. The iron core of the coils also contributes largely to the flux density. Movable iron cores contribute to the change rate of the flux density. The switch-controlled energy harvesting interface circuits such as the synchronized magnetic flux extraction circuit with DC–DC inverter should be used for improving energy harvesting efficiency.

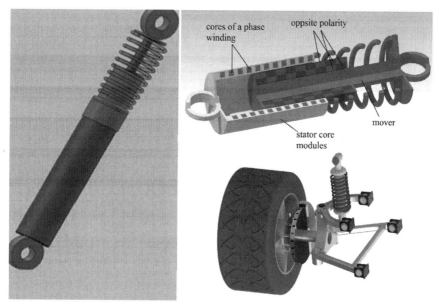

FIGURE 12.5

Regenerative shock absorber built with a linear generator with longitudinal or transverse magnets and highly magnetic conductive casing.

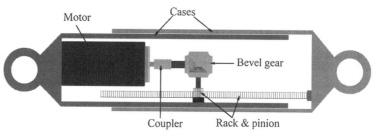

FIGURE 12.6

Regenerative shock absorber built with the rack and pinion mechanism.

The harvested power depends on the relative velocity of the magnet and coils. The speed of the linear generator is limited by its stroke. The rack and pinion or ball—screw mechanism is used to convert the bidirectional linear motion of the shock absorbers into a unidirectional high speed rotation of a rotational generator through a special mechanism. The special mechanism substantially increases the relative speed between the magnets and coils.

When the vehicle travels on the road, the road roughness, accelerations, decelerations, and unevenness will excite the undesired vibration. Traditionally, oil shock absorbers are used in parallel with the suspension springs to ensure the ride comfort,

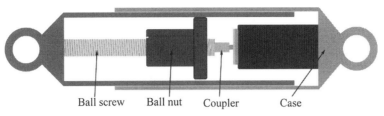

Ball screw   Ball nut   Coupler   Case

FIGURE 12.7

Regenerative shock absorber built with the ball—screw mechanism.

road handling, and safety by dissipating the undesired vibration energy into waste heat. Regenerative shock absorbers have been proposed to be energy harvesting transducers to convert the kinetic energy of the undesired vibration into useful electricity and to reduce the vibration. According to Ref. [8], for a middle-sized passenger car (1.5 tons) with four shock absorbers, average powers of 100, 400, and 1600 W are available for harvesting while driving 97 km/h (60 mile/h) on Class B (good), Class C (average), and Class D (poor) roads, respectively. It is assumed that average roughness of the road surface of Class C (average) is 0.2 mm. The car is assumed to be installed with four 15″ wheels of outer diameter 350 mm, the averaged up-down speed of the four shock absorbers is calculated to be 31 mm/s. If the energy harvesting efficiency is assumed to be 50%, the harvested power is then calculated to be 226 W. The energy potential for trucks, railcars, and off-road vehicles is in the order of 1—10 kW. The harvested power of 226 W means 2.4—9% increase of fuel efficiency for the vehicles. This estimation is consistent with the literature in the thermoelectric waste heat recovery, for example, 390 W means 4% fuel efficiency for a BMW [1].

Another promising alternative energy technology is to harvest vibration energy from civil structures. Large vibration exists in these civil structures due to the dynamic loadings of wind, earthquake, traffic, and human motions. These civil structures include tall buildings, communication towers, and long-span bridges which are very susceptible to the dynamic loadings. Large vibration amplitudes can damage the structures or the secondary components or cause discomfort to its human occupants. The wind-induced vibration could be huge. For example, some tall buildings could be installed with the 100 kW—1.5 MW wind turbines, which give a sense of wind forces and power acting on the tall buildings. The tuned mass dampers are currently used to dissipate and control the vibration energy of modern buildings and bridges. The viscous dissipative tuned mass dampers can also be converted into regenerative mass dampers or replaced by energy transducers such as the MDOF EMVEHs to simultaneously harvest vibration energy and control vibration from tall buildings.

Civil structures can be built with structure frame with rods and ball joints. The rod element can be designed to be a sandwiched steel tube where the inner tube wall should be made of nonmagnetic materials. Inside the sandwiched steel tube,

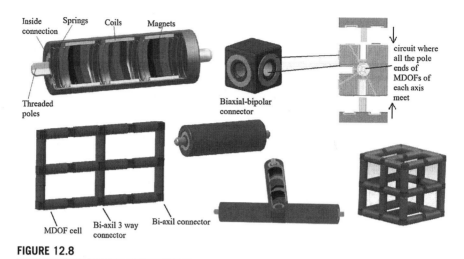

FIGURE 12.8

Structure frame built with the multiple degrees of freedom electromagnetic vibration energy harvester cells.

multiple cylindrical magnets are connected by springs. Multiple coils are wired outside the tube around the magnets. The coils are shielded by aluminum foil to avoid the flux leakage of the air gap between the magnets and the steel tube. The joints can be designed as trial axial bipolar connectors which connect the outputs of the coils from one rod tube to another rod tube and finally to the energy harvesting interface circuits. The springs can also be replaced by the magnetic springs from the magnetic fields between two magnets of opposite-placed polarities. The two impinging magnets should be installed at the end caps of the rod tube. The polarities of the two impinging magnets should be opposite to those of the facing magnets nearby. To reduce the sliding friction, a ferrofluid of nanosized magnet powders and lubricating oil should be filled into the rod tube to form ferrofluid bearings near the edges of the magnets. To increase the flux density change rate at each of the coils, one magnet should be replaced by two magnets of opposite polarities. The two magnets are separated by a thin carbon steel spacer. The structure frame built with the rod tube MDOF EMVEHs is shown in Fig. 12.8.

When the train moves on the track, the track will deflect vertically. Railcars can weigh from 30 ton (empty) to 140 ton (loaded) each. A typical freight train exerts a load around 900−1360 kg on the track surface and induces 3−10 mm track deflection. The frequency at which the track deflects depends on the distance between the two bogies of the train. Usually, the freight trains have different cart lengths, so the distance between bogies is not uniform. The frequency is around 0.6−1.8 Hz calculated by assuming that the train is moving at a speed of 40−120 km/h. Assuming that a four-wheel railcar has 100 ton weight and 24 m length, the four wheels will pass over in 1.36 s at a train speed of 64 km/h. The average power potential is calculated to be about 2 kW under 6-mm track deflection. If 5% of the support force is provided to a harvester at the trackside, the harvestable energy from up and down track motion will be 200 W. The actual harvested power available depends on the speed of the train, the weight of the train, the type of railway track, ballast, road foundation, and so on.

NOMENCLATURE

y, $y(t)$	The base excitation displacement
Y, $Y(s)$	The Laplace transform of $y(t)$
t	The time instance or variable
s	The Laplace variable
i	The square root of -1

For the Two Degrees of Freedom Harvester Oscillator System

m_1	The first oscillator mass
m_2	The second oscillator mass
c_1	The short circuit damping coefficient between m_1 and m_2
c_2	The short circuit damping coefficient between m_2 and the base
c_3	The short circuit damping coefficient between m_3 and the top
k_1	The short circuit stiffness coefficient between m_1 and m_2
k_2	The short circuit stiffness coefficient between m_2 and the base
k_3	The short circuit stiffness coefficient between m_3 and the top
R_1	The external load resistance of the first harvesting circuit
R_2	The external load resistance of the second harvesting circuit
L_{e1}	The electric induction coefficient of the first coil
L_{e2}	The electric induction coefficient of the second coil
R_{1e}	The internal resistance of the first coil
R_{2e}	The internal resistance of the second coil
V_1, $V_1(t)$	The output voltage of the first harvesting circuit
V_2, $V_2(t)$	The output voltage of the second harvesting circuit
$\overline{V}_1$, $\overline{V}_1(s)$	The Laplace transform of V_1, $V_1(t)$
$\overline{V}_2$, $\overline{V}_2(s)$	The Laplace transform of V_2, $V_2(t)$
B_1	The magnetic field constant or magnetic flux density of the first ring magnet
B_2	The magnetic field constant or magnetic flux density of the second ring magnet
l_1	The length of the first coil
l_2	The length of the second coil
x_1, $x_1(t)$	The displacement of the first oscillator mass
x_2, $x_2(t)$	The displacement of the second oscillator mass
X_1, $X_1(s)$	The Laplace transform of x_1, $x_1(t)$
X_2, $X_2(s)$	The Laplace transform of x_2, $x_2(t)$

For the Four Degrees of Freedom Harvester Oscillator System

m_1	The first oscillator mass
m_2	The second oscillator mass
m_3	The third oscillator mass
m_4	The fourth oscillator mass

c_1	The short circuit damping coefficient between m_1 and m_2
c_2	The short circuit damping coefficient between m_2 and m_3
c_3	The short circuit damping coefficient between m_3 and m_4
c_4	The short circuit damping coefficient between m_4 and the base
c_5	The short circuit damping coefficient between m_1 and the top
k_1	The short circuit stiffness coefficient between m_1 and m_2
k_2	The short circuit stiffness coefficient between m_2 and m_3
k_3	The short circuit stiffness coefficient between m_3 and m_4
k_4	The short circuit stiffness coefficient between m_4 and the base
k_5	The short circuit stiffness coefficient between m_1 and the top
$V_1, V_1(t)$	The output voltage of the first harvesting circuit
$V_2, V_2(t)$	The output voltage of the second harvesting circuit
$V_3, V_3(t)$	The output voltage of the third harvesting circuit
$V_4, V_4(t)$	The output voltage of the fourth harvesting circuit
$\overline{V}_1, \overline{V}_1(s)$	The Laplace transform of $V_1, V_1(t)$
$\overline{V}_2, \overline{V}_2(s)$	The Laplace transform of $V_2, V_2(t)$
$\overline{V}_3, \overline{V}_3(s)$	The Laplace transform of $V_3, V_3(t)$
$\overline{V}_4, \overline{V}_4(s)$	The Laplace transform of $V_4, V_4(t)$
B_1	The magnetic field constant or magnetic flux density of the first ring magnet
B_2	The magnetic field constant or magnetic flux density of the second ring magnet
B_3	The magnetic field constant or magnetic flux density of the third ring magnet
B_4	The magnetic field constant or magnetic flux density of the fourth ring magnet
R_1	The external load resistance of the first harvesting circuit
R_2	The external load resistance of the second harvesting circuit
R_3	The external load resistance of the third harvesting circuit
R_4	The external load resistance of the fourth harvesting circuit
R_{1e}	The internal resistance of the first coil
R_{2e}	The internal resistance of the second coil
R_{3e}	The internal resistance of the third coil
R_{4e}	The internal resistance of the fourth coil
l_1	The length of the first coil
l_2	The length of the second coil
l_3	The length of the third coil
l_4	The length of the fourth coil
L_{e1}	The electric induction coefficient of the first coil
L_{e2}	The electric induction coefficient of the second coil
L_{e3}	The electric induction coefficient of the third coil
L_{e4}	The electric induction coefficient of the fourth coil
$x_1, x_1(t)$	The displacement of the first oscillator mass
$x_2, x_2(t)$	The displacement of the second oscillator mass
$x_3, x_3(t)$	The displacement of the third oscillator mass
$x_4, x_4(t)$	The displacement of the fourth oscillator mass
$X_1, X_1(s)$	The Laplace transform of $x_1, x_1(t)$
$X_2, X_2(s)$	The Laplace transform of $x_2, x_2(t)$
$X_3, X_3(s)$	The Laplace transform of $x_3, x_3(t)$
$X_4, X_4(s)$	The Laplace transform of $x_4, x_4(t)$
$a_{11}(s) =$	$m_1 \cdot s^2 + (c_1 + c_5) \cdot s + (k_1 + k_5)$
$a_{22}(s) =$	$m_2 \cdot s^2 + (c_1 + c_2) \cdot s + (k_1 + k_2)$

$$a_{33}(s) = \quad m_3 \cdot s^2 + (c_3 + c_2) \cdot s + (k_3 + k_2)$$
$$a_{44}(s) = \quad m_4 \cdot s^2 + (c_3 + c_4) \cdot s + (k_3 + k_4)$$
$$a_{55}(s) = \quad 1 + \frac{R_{1e}}{R_1} + \frac{L_{1e} \cdot s}{R_1}$$
$$a_{66}(s) = \quad 1 + \frac{R_{2e}}{R_2} + \frac{L_{2e} \cdot s}{R_2}$$
$$a_{77}(s) = \quad 1 + \frac{R_{3e}}{R_3} + \frac{L_{3e} \cdot s}{R_3}$$
$$a_{88}(s) = \quad 1 + \frac{R_{4e}}{R_4} + \frac{L_{4e} \cdot s}{R_4}$$

For the *N* Degrees of Freedom Harvester Oscillator System

m_{N-1}	The $N-1$th oscillator mass
m_N	The Nth oscillator mass
c_N	The short circuit damping coefficient between m_N and the base
c_{N+1}	The short circuit damping coefficient between m_1 and the top
k_N	The short circuit stiffness coefficient between m_N and the base
k_{N+1}	The short circuit stiffness coefficient between m_1 and the top
$V_{N-1}, V_{N-1}(t)$	The output voltage of the $N-1$th harvesting circuit
$V_N, V_N(t)$	The output voltage of the Nth harvesting circuit
$\overline{V}_{N-1}, \overline{V}_{N-1}(s)$	The Laplace transform of $V_{N-1}, V_{N-1}(t)$
$\overline{V}_N, \overline{V}_N(s)$	The Laplace transform of $V_N, V_N(t)$
B_{N-1}	The magnetic field constant or magnetic flux density of the $N-1$th ring magnet
B_N	The magnetic field constant or magnetic flux density of the Nth ring magnet
R_{N-1}	The external load resistance of the $N-1$th harvesting circuit
R_N	The external load resistance of the Nth harvesting circuit
R_{N-1e}	The internal resistance of the third coil
R_{Ne}	The internal resistance of the fourth coil
l_{N-1}	The length of the $N-1$th coil
l_N	The length of the Nth coil
L_{eN-1}	The electric induction coefficient of the $N-1$th coil
L_{eN}	The electric induction coefficient of the Nth coil
$x_{N-1}, x_{N-1}(t)$	The displacement of the $N-1$th oscillator mass
$x_N, x_N(t)$	The displacement of the Nth oscillator mass
$X_{N-1}, X_{N-1}(s)$	The Laplace transform of $x_{N-1}, x_{N-1}(t)$
$X_N, X_N(s)$	The Laplace transform of $x_N, x_N(t)$

$$a_{(N-1)(N-1)}(s) = \quad m_{N-1} \cdot s^2 + (c_{N-1} + c_{N-2}) \cdot s + (k_{N-1} + k_{N-2})$$
$$a_{NN}(s) = \quad m_N \cdot s^2 + (c_{N-1} + c_N) \cdot s + (k_{N-1} + k_N)$$
$$a_{(2N-2)(2N-2)}(s) = 1 + \frac{R_{N-1e}}{R_{N-1}} + \frac{L_{N-1e} \cdot s}{R_{N-1}}$$
$$a_{(2N)(2N)}(s) = \quad 1 + \frac{R_{Ne}}{R_N} + \frac{L_{Ne} \cdot s}{R_N}$$

Superscripts

- The first differential
- .. The second differential

Abbreviations

SDOF	Single degree of freedom
MDOF	Multiple degrees of freedom
DOF	Degree of freedom

REFERENCES

[1] Fairbanks J. Vehicular thermoelectrics: a new green technology. In: Thermoelectric applications workshop, Coronado, CA, 3—8 January; 2011.

[2] Gatti RR. Spatially Varying multi degree of freedom electromagnetic energy harvesting [Ph.D. thesis]. Australia: Curtin University; 2013.

[3] Kulkarni S, et al. Design, fabrication and test of integrated micro-scale vibration-based electromagnetic generator. Sens Actuators A 2008;145:336—42.

[4] Liu H, et al. Feasibility study of a 3D vibration-driven electromagnetic MEMS energy harvester with multiple vibration modes. J Micromech Microeng 2012;22(12):125020.

[5] Liu H, et al. A multi-frequency vibration-based MEMS electromagnetic energy harvesting device. Sens Actuators A 2013;204:37—43.

[6] Yang B, et al. Electromagnetic energy harvesting from vibrations of multiple frequencies. J Micromech Microeng 2009;19(3):035001.

[7] Zhang PS. Design of electromagnetic shock absorbers for energy harvesting from vehicle suspensions [Master degree thesis]. USA: Stony Brook University; 2010.

[8] Zhou K. Robust and optional control. Upper Saddle River (NJ): Prentice Hall; 1996.

[9] Zuo L, Tang XD. Large-scale vibration energy harvesting. J Intell Mater Syst Struct 2013;24(11):1405—30.

[10] Wang X, Liang XY, Wei HQ. A study of electromagnetic vibration energy harvesters with different interface circuits. Mech Syst Signal Process 2015;58—59(2015):376—98.

[11] Bawahab M, Xiao H, Wang X. A study of linear regenerative electromagnetic shock absorber system. SAE 2015-01-0045. 2015.

[12] Wang X, Liang XY, Hao ZY, Du HP, Zhang N, Qian M. Comparison of electromagnetic and piezoelectric vibration energy harvesters with different interface circuits. Mech Syst Signal Process 2016;72—73.

[13] Wang X. Coupling loss factor of linear vibration energy harvesting systems in a framework of statistical energy analysis. J Sound Vib 2016;362(2016):125—41.

[14] Wang X, Liang XY, Shu GQ, Watkins S. Coupling analysis of linear vibration energy harvesting systems. Mech Syst Signal Process March 2016;70—71:428—44.

[15] Wang X, John S, Watkins S, Yu XH, Xiao H, Liang XY, et al. Similarity and duality of electromagnetic and piezoelectric vibration energy harvesters. Mech Syst Signal Process 2015;52—53(2015):672—84.

Index

W

Printed in the United States
By Bookmasters